Insights from research and practice

A handbook for adult literacy, numeracy and ESOL practitioners

Edited by

Margaret Herrington

Visiting Professor of Education,
University of Wolverhampton

Special Lecturer in Continuing Education,
University of Nottingham

and

Alex Kendall

Associate Dean, Education,
University of Wolverhampton

Published by the National Institute of Adult Continuing Education
(England and Wales)
21 De Montfort Street, Leicester LE1 7GE
Company registration no: 2603322
Charity registration no: 1002775

First published 2005
Reprinted 2006
Articles in Part 2 have been reproduced from original RaPAL Bulletins/Journals.

NIACE has a broad remit to promote lifelong learning opportunities for adults. NIACE works to develop increased participation in education and training, particularly for those who do not have easy access because of barriers of class, gender, age, race, language and culture, learning difficulties and disabilities, or insufficient financial resources.

For a full catalogue of NIACE's publications, please visit

www.niace.org.uk/publications

Cataloguing in Publications Data
A CIP record for this title is available from the British Library

ISBN 10: 1 86201 202 4 (pbk)
ISBN 10: 1 86201 244 X (hbk)
ISBN 13: 978 1 86201 202 8 (pbk)
ISBN 13: 978 1 86201 244 8 (hbk)

Cover design by Creative
Designed and typeset by Avon DataSet, Bidford-on-Avon, Warwickshire
Printed and bound in Great Britain by Antony Rowe Limited, Chippenham

Contents

Acknowledgements

In a work of this kind there are always many acknowledgements to make. We would like to start by acknowledging the work of all past and present members of the Research and Practice in Adult Literacy (RaPAL) network and, in particular, of the authors represented and re-presented in this volume.

Particular thanks also go to the groups of members who have devoted considerable time and energy to editing, producing and despatching the RaPAL bulletins over 20 years. Leading names include David Barton, Julia Clarke, Fiona Frank, Mary Hamilton, Gaye Houghton, Roz Ivanič, Jane Mace, Wendy Moss, Fiona Ormerod, Carol Taylor and Mary Wolfe, but thanks go to the many collectives (with too many names to list here) that have sustained the work over many years. Without their work, the writers' insights could not have been disseminated.

We would both like to thank professional colleagues for their support. Alex would like to thank her colleagues in the PCE subject group at Wolverhampton University for being supportive, critical friends. Margaret thanks her former colleagues at the University of Nottingham Study Support Centre, and in particular Barbara Taylor and Mark Dale, who positively welcomed her investigative stance. She also wants to thank the dyslexic students in adult literacy classes and in higher education, who taught her to remain ever-curious about different ways of interacting with texts.

Major thanks go to the editorial and production staff at NIACE and, in particular, to Virman Man and David Shaw, without whose enthusiasm and editorial skill the book would not have emerged in its present form.

The compilation of the book has also involved extensive scanning and proof reading of the original articles and we would like to thank Nancy and Paul Herrington for their sustained work on this. The articles have been reproduced from the original RaPAL bulletins/

journals as faithfully as possible. Author biographical details have been trimmed as necessary, but not updated.

There are also personal acknowledgements. We would like to acknowledge the personal support and inspiration from our families. Margaret would like to thank Paul, and her children and partners for their humorous forbearance. Alex would like to thank Julian, Lydia, Barrie, Val and Jan for affording her time out and away to devote to this project.

Needless to say, the responsibility for the selection and analysis of the papers is ours alone, as is that for any omissions or errors.

Contributing authors

Marie-Helene Adrien
David Archer
John Arnett
David Barton
Richard Barwell
Sue Bergin
Arvind Bhatt
Stephen Black
Blackfriars Literacy Scheme
Stephen Brookfield
Geraldine Castleton
Doreen Chappell
Julia Clarke
Julian Clissold
Fie van Dijk
Celia Drummond
Sue Erlewyn-Lajeunesse
Ann Finlay
Zoe Fowler
Jane Freeland
Peter Goode
Michael Gray
Annette Green
Mary Hamilton
Barbara Hately-Broad
Tricia Hegarty
Margaret Herrington
Gaye Houghton
Roz Ivanič
Catherine Jamieson
Margaret Jessop
Kathryn Jones
Joy Joseph
Tom Joseph
John Karlik
Ann Kelly
Jill Kibble
Linda Kirkham
Gunther Kress
Crissie Laugeson
Gillian Lawrence
Theresa Lillis
Monica Lucero
Charles Lusthaus
Jane Mace
Catherine Macrae
Juliet Merrifield
Jean Milloy
Ellen Morgan
Wendy Moss
Alan Murdoch
Rebecca O'Rourke
Susie Parr
Josie Pearse
Rob Peutrell
Jane Pinner
Kathy Pitt
Melanie Ramsay
Guilherme Rios
Chris Rishworth
Denise Roach
Jill Ross
Rosemary Rouse
Foufou Savitsky
Jean Searle
Jan Sellers
Iffat Shahnaz
John Simpson
Sandra Southee
Brian Street

Helen Sunderland
Sean Taylor
Susan Taylor
Jan Thompson
Adele Tinman
Alison Tomlin
Kate Tomlinson
Dave Tout
Val Watkinson
Tanya Whitty
Rosie Wickert
Chris Wild
Anita Wilson
Mary Wolfe

Foreword

I am very pleased to write this foreword on behalf of the RaPAL Management Group for this collection drawn from the Research and Practice in Adult Literacy bulletins and journals. Throughout my own practice I have often looked towards the RaPAL journals for inspiration, information and sometimes for affirmation!

Many of you opening this collection may be new to the world of adult literacy. You may be considering your own starting point and preparing to work with adults who will seek your guidance and support. This collection will be useful to you to generate ideas for working with adult learners and may give you food for thought about carrying out your own research.

In writing this foreword I revisited the very first RaPAL Bulletin[1] which opened with an article entitled 'Why research and practice…?'. The article explained that 'Research is to help us explore and improve what we do, become more thoughtful and effective teachers and learners. It is a way of recording and sharing knowledge and experience'. The article also highlighted the need for participative research that would 'break down the roles of the "researcher" and the "researched".' It asked us to consider the process of research as well as the end product. These aims are as laudable now as they were then and this collection tracks these processes over the past 25 years. It presents evidence of the range of research carried out by adult literacy practitioners and therefore is an historical record of the professional approach adopted by them as a means to make provision more effective for learners.

This publication will also be good news in the further and higher education sectors, where I believe it will be useful as a key text for a range of adult literacy qualifications. As you use it, please remember that this is one aspect of RaPAL and I would urge you to meet us at our annual conference and visit us on our website at www.literacy.lancaster.ac.uk/rapal/sites.htm. Most of all, however, I would urge you to join us and become involved. An organisation that doesn't involve, doesn't evolve.

Lots of people have contributed throughout the years to RaPAL and many of them will not be named within this publication. I would like to take the opportunity of thanking them and the editors of this publication for bringing together a very valuable resource.

Fiona Macdonald
RaPAL Chair, 2005

[1] Research and Practice in Adult Literacy *Bulletin* No. 1, Summer 1986.

Introduction

Purpose and rationale

This book is intended for educators in adult literacy, numeracy and ESOL. It draws together insights from British and overseas researchers, practitioners and students who have published their ideas in the journal of the Research and Practice in Adult Literacy (RaPAL) network between 1986 and 2004.

We wanted to compile this book because many new practitioners are entering this field following the government's *Skills for Life* initiative in England and Wales (2004). New challenges have been set in terms of a core curriculum, standards, national tests, targets, graduate teacher training standards and qualifications, materials development, and research and development programmes, and we wanted to distil for practitioners the central philosophical, political and pedagogical insights that can be seen in the work of RaPAL over a 20-year period. We wanted to link past experience explicitly to the present policy context.

Why is this so important? There are three main reasons. First, literacy, numeracy and ESOL are highly charged in political, economic and cultural terms, and very different models of literacy and numeracy are constructed and employed in research, policy and practice contexts. For example, the current policy context in the countries of the UK is informed by a global economic discourse (Farrell, 2001). Improving the literacy levels of workers is seen as an essential aspect of economic advancement. Literacy as a commodity becomes central to a political agenda that links literacy with economic productivity (Gee, 2000; Sanguinetti, 2001). Certainly the UK government is keen to assert a correlation between an individual's literacy level and the level of income they might expect to command

(Pember, 2001).[1] Thus literacy and economic 'enlightenment' seem cast in a symbiotic relationship, a straightforward logic within which one's skills portfolio – of which literacy skills are an important aspect – is directly and unproblematically linked to earning capacity. The comprehensiveness of the current *Skills for Life* initiative can be seen as a managerialist response to this.

We believe that practitioners, as professional educators, must be able to contextualise – theoretically and politically – any new policies they are asked to implement, both for themselves and for the sake of their students. They must be able to draw on past experience if they are to be aware of their own position within the new policy context and within the varied teaching and learning situations they experience. The RaPAL network provides many examples of practitioners attempting to make sense of competing ideas about literacy and attempting to construct appropriate professional responses in the light of these.

Hence, the second reason for this compilation concerns the pedagogical/andragogical legacy from the recent past. Many new practitioners say, 'We know now what you want us to teach, but how do we actually teach this curriculum?'. It is important for practitioners to be able to draw on the wealth of ideas and reflections of colleagues within the history of practice, and this volume provides them with an opportunity for doing so.

Third, researchers and practitioners in this field have effectively developed a new body of knowledge about adult literacy. Some of this huge outpouring of investigative energy is reflected in the RaPAL network. Members – researchers, students, tutors, managers – have asked important questions about the nature of adult learning, about language and literacy, about the nature of research itself and about how practitioners and students could be involved in this. It is important for new staff to draw on this experience, reviewing the processes and outcomes, if they are to see the value of, and feel confident about, building research into the infrastructure of their own practice.

What is RaPAL?

This network of adult literacy researchers, practitioners and students emerged from conferences in Lancaster and London, in 1984 and 1985 respectively. By this time, adult literacy educators in some areas had started to engage critically with the literacy policy challenges set in motion by the adult literacy campaign. They were unravelling concepts of literacy, 'shock horror' claims about illiteracy and economic, personal and civic decline, and simplistic measures of literacy standards. They were asking rigorous questions about what the policy challenge actually amounted to and how those in the field should engage with it. In problematising the issue of 'fighting illiteracy', researchers and practitioners became engaged with fundamental ideas about who determines what literacy and literateness shall mean in any situation, about who determines the kinds of levels and standards applied in many contexts, and about the literacy educator's function in relation to these concepts.

RaPAL sought to contribute to these developments, in an inclusive and intellectually coherent way, by creating a mechanism – a network – that improved communication between practitioners and researchers (with an annual conference and a thrice-yearly journal) and, more profoundly, by encouraging 'a much broader view of what counts as research' (*RaPAL Bulletin* No. 1, 1986). It identified some of the key intellectual processes involved in research as accessible to all, and invited all members, including practitioners and learners, to engage in these.

For example, in the first edition of the journal:

> *Research does not have to be remote, written in mystifying language and published in obscure (expensive) books and journals that few people ever get to see. Research can be closer to the everyday practice of what we all do as we learn. It involves asking questions, trying to answer them, asking other people, recording what they say, developing ideas, changing them, writing and sharing ideas in many different ways. It's looking at learning in a reflective, critical way.*

> *Research is to help us explore and improve what we do . . . It is a way of recording and sharing knowledge and experience.* (No. 1, 1986, p. 1)

RaPAL also recognised that much researching activity in practice settings was often not acknowledged as such:

> *A great deal of reflective and evaluative work already goes on in adult basic education, often informally, under the name of outreach or report writing. This work needs to be made visible and these activities developed into procedures that can be generalised, compared and used by others. A good evaluative description can generate research and this process is as important as research that sets out with hypotheses and aims to prove something.* (No. 1, 1986, p. 1)

RaPAL continued to develop its distinctive profile in terms of research and practice and, over a 20-year period, members have valued this highly in terms of their own professional development. Being connected with others who were thinking deeply about this work has been experienced as supportive, liberating and inspiring (comments at annual RaPAL conferences 1996–2004). The network has provided spaces in which members risked saying what they did in practice, argued with others about the relative merits of particular stances, challenged policies from radical perspectives and developed new research ideas. In the absence of a professional association for adult literacy workers, it has provided a mechanism for sustaining independent criticality in the face of what some would assert is myth-driven literacy policy. It has also communicated a strong sense of practice as an important research site for practitioners and learners as well as for researchers (Ivanič, *RaPAL Bulletin* No. 2, 1986, p. 1).

The significance of this network in terms of its impact on the field as a whole has yet to be fully assessed. Suffice it to say, for now, that RaPAL members became leading writers, researchers, policy analysts, managers, staff developers and tutors in this field.

The RaPAL journals reflect something of the multiple narratives about research in/and practice that have emerged in the network. This volume provides a selection of the journal articles, including some that are currently out of print.

Who are we?

We are both members of RaPAL: one very long-standing and one relatively new.

Margaret has been a member since 1988 and served successively as membership secretary, chair and journal editor (1996–2004). She worked as a literacy educator/manager in adult basic education (ABE) in Leicestershire following the first adult literacy campaign in the mid-1970s, and more recently in higher education at the universities of Leicester and Nottingham. She is currently an educational consultant and Visiting Professor of Education at the University of Wolverhampton. Alex has served as secretary of RaPAL since 2001 and has worked in the post-16 field since 1993. She started as an ESOL teacher and taught English to students in post-16 education institutions. She joined the University of Wolverhampton in 2000 and is now Associate Dean, Education.

We bring very different kinds of experience to the task. Margaret brings insights from literacy, numeracy and ESOL practice and her ongoing attempts to draw theory into and from practice. She has engaged in different types of research activity within numerous practice contexts, such as adult education and community literacy settings, a probation day centre, further education colleges, open and distance learning settings and libraries, and is interested in the philosophy of practice as well as research. She has also focused in recent years on trying to make sense of theory and practice about dyslexia as it has emerged across many disciplines. Alex, with a recent background as an English teacher, comes to the task with up-to-date insights from critical theory and a willingness to use these in relation to theory and practice in the field. In particular, she brings theoretical ideas about 'otherness' to this field in which the adult learner is often still characterised as 'other' in a deficit sense.

Currently, we are both engaged in delivering the new teacher training courses in literacy and ESOL and feel very strongly that the experience of RaPAL members has something to say to new practitioners. Although the *Skills for Life* framework is new, the structural and political issues around literacy education persist.

The process of compiling the book

We know that readers will approach this book from many directions, and we first want to clarify some of our initial assumptions about how new knowledge is created in this field, about literacy itself and about the professional development of practitioners.

Initial assumptions

We place our thinking within a postmodern, critical thinking tradition, which views knowledge as always contingent in relation to how and by whom it is generated. We therefore see ideas and 'stories' about literacy as governed and shaped at different times, in different ways and by different agents. Hence current policy in England and Wales which considers literacy as:

> *the ability to read, write and speak in English and to use mathematics at a level necessary to function at work and in society in general* (DfEE, A Fresh Start, 1999)

must be seen in relation to the current particular, political and social context.

Our own view of literacy encompasses notions of skill and functionality but also acknowledges a deeper reality. At different times and in different ways both of us had experienced an incredulity about the *meta-narrative* (Barry, 2002, referring to Lyotard's essay 'Answering the question: What is postmodernism?', 1982) of literacy which appeared to conceptualise it as primarily a set of context-free skills. We had both valued the different narratives emerging about literacy and literacies which had emphasised the situated, social practice of literacy and the issues of power and identity arising therein (New Literacy Studies *et al.*: see Barton, Hamilton and Ivanič, 2000). We therefore use a multi-narrative framework in relation to literacy and assume that any knowledge generation about it must involve gathering data from and by students, tutors and researchers.

However, we are not uncritical of postmodern positions here. They, too, represent particular perspectives (Houghton, 2004) and we recognise that such complexities can be seen as problematic in the real world of teaching individual students. If, for example, someone wants to be able to write a note to school about their child, there is work that can go on in producing writing which seems to have no bearing whatsoever on making knowledge about literacy. We would argue though that, since writing is always about social relations, the writing produced will affect how the writer is viewed by the teacher; and this in turn is informed by the values people hold about writing and ability, educatedness and place in society. In teaching an adult how to construct such a letter for the child's school, you can teach someone how to construct a very basic communication but not necessarily alert them to how they will appear to the teacher concerned. You can teach them how to write in ways that will send very different signals to the reader about their confidence with standard English, because this is a way of teaching the writing of the powerful, or you can teach them in ways that alert them to all the literacy and power relationships involved, giving curriculum outcomes that combine literacy competence and literacy awareness. Knowledge about this form of literacy practice is thus constructed at a number of possible levels, and literacy staff and students may decide to engage with one or more of these.

A further set of assumptions also requires clarification. This book focuses on professional development in its broadest sense, the rigorous analysis of which has been rather patchy in this field. Marginal working conditions for many years and, more profoundly, the fact that this work involved some challenge to the ways of thinking about literacy that dominated the school system, may both have contributed to the infrequent articulation of what professional development should embrace.

Yet several generations of staff in this field did engage in a multi-layered mix of staff development processes, including:

1. a 'bottom-up', 'learning by doing' approach;
2. sharing expertise within training events and within regular newsletters (ALBSU, for example);
3. non-graduate training programmes;
4. learning through attending and delivering volunteer training;

5. undertaking local and special development projects (ALBSU);
6. establishing regional networks;
7. attending national conferences;
8. contributing to national and international networks such as RaPAL.

This mixed mode approach had its advantages in terms of avoiding undue prescriptiveness and requiring an explicit learning from the learners. For staff working on the margins with little help from existing knowledge about how to do this work effectively, it was a pragmatic solution which, though rarely articulated in full, often produced inspiring results.

One of the important gaps, however, was the infrequency of any explicit reference to the theoretical hinterland that practitioners required in building their distinctive professional role. Individuals, of course, read their preferred theorists (Freire, for example). However, RaPAL members in universities such as Goldsmiths, Lancaster, Leicester and, later, Sheffield (Hamilton, *RaPAL Journal*, No. 30, 1996) started to create diploma and further degree opportunities which encouraged course participants to engage in critical reflection on literacy theory, policy and practice. The RaPAL network encouraged colleagues to look at their professional roles and contexts, and to be aware of the role of both teacher and change agent in relation to literacy ideology. Above all, it encouraged research and practice processes within a kind of 'open collective model' of professional development (Huberman, 1995).

Our assumptions, therefore, at the time of writing in 2004 are that:

- the new graduate training programmes for England and Wales, currently being developed within the *Skills For Life* initiative, provide an important opportunity to encourage the explicit building and dissemination of theory in general and of theories of practice in particular (Mace, 1988). New staff in the field are entitled to know how this work is to be described in ways that reflect the depth of practice, and should be encouraged to construct models of investigative practice with students;
- we need to retain a sense of the multi-layered dynamic of

> professional development and the place of research in this activity. The new training courses must be seen as part of longitudinal processes which involve interaction in practice, sharing, arguing, building and analysing. Reflexivity is relevant here as educators self-consciously consider their own positions within institutions and cultures and as they engineer their own ways forward, over time, using research and practice.

This volume will serve to support both developments.

Making the selection

We both felt that as a 'collection' the RaPAL journals since 1986 presented a complex, multi-voiced narrative in which students, practitioners, academics and researchers worked alongside each other, often in dialogue, to make sense of the experience of ABE. Themes occurred and re-occurred, re-framing and challenging the contemporality of current debate. For the newcomer to the field this collection thus offered something of a genealogy of ABE, and the beginnings of a response to questions about where we are now. By bringing together diverse and sometimes dissenting voices, it provides a refreshing and important model of possibility, of how curriculum reform and development might evolve and of how practice-based research can proceed.

Hence, our initial idea was simply to reprint all the RaPAL papers. But this would have required many volumes and would have appeared unwieldy to new staff. We therefore had to consider how to select in order to re-present the work, and recognised early on that this was no easy task.

Our first activity was to read the whole collection and select articles that we particularly wished to see in such a volume as this. So far, so good. But applying our first criterion produced a vast number of articles, as we rediscovered their importance and recognised that so many of them could still be used in teacher education. We at once recognised the particular importance of including those articles that were position papers about concepts of literacy, but this did not remove any from our list. As a way of temporarily avoiding the wielding of the knife, we organised the material thematically and focused on the areas that the writers had covered (see p. xxii). We found it fruitful to read the articles again in these sections and to identify their particular relevance to current policy and research questions. We could not avoid the final cut to make a more manageable tome, and ultimately this was only made easier by an accommodation between us about what we were individually trying to achieve.

We had openly acknowledged our personal ideas about the book to each other (we had acknowledged that this was an exercise in analytical reflexivity). Alex wanted to ensure that her initial recognition of the relevance of RaPAL writing to current policy was represented. She felt that as policy and practice in adult literacy, numeracy and ESOL (ALNE) were currently being re-negotiated, the newcomer to basic skills teaching was presented with a dominant rhetoric that constructed the present as a new 'golden era', an optimistic future distinct from an amateurish, disorganised, under-developed past. She had been surprised to see that the current issues such as curriculum and assessment were not new and in fact had recurred over a lengthy period. She wanted to challenge the tendency in policy circles to obscure the past work with its traditions of research and practice and network dialogue. Alex wanted to give readers a flavour of these.

Margaret shared this view but also wanted to reflect and speak from within her own long experience within RaPAL. She was aware of the diverse audience within the network who in some sense were being represented in this volume. She also wanted to acknowledge the exceptional roles that key members had taken within the network; to attempt to describe the particular dynamic that facilitated the active encouragement of new writers alongside academics with international reputations; and to analyse the central concerns of the network over time and the relationship between this writing and the development of policy and practice in general. In a nutshell, Margaret wanted the selection also to illustrate a *history* of the network; and thus leaving out many important articles was much more problematic for her.

An accommodation was reached by attempting a brief analysis of the whole collection and then selecting articles that literally did give a flavour of the themes covered and that connected with current policy priorities. We acknowledged firmly that this was not the place for the 'compleat history' of RaPAL and that not all the leading figures could be represented.

We also wanted to start to explore the issue of longitudinal professional development among RaPAL members. The collection shows the continuing contributions of many members, over many years. These were more than individual examples of reflective practice and appeared to reflect deepening development processes in terms of knowledge-making and professional skills and sensitivities over time. We wanted the selection to give some hint of these developments. Given the importance of longitudinal development, we also decided to include Margaret's newly constructed personal case study about research in practice in relation to dyslexia.

The organisation of the book

The book amounts to an open invitation to practitioners to develop their strengths as investigative practitioners within the research in practice tradition and to generate new knowledge during a process of professional development.

In Part 1 we discuss what RaPAL has meant by research in/and practice. There are two important caveats here. First, the journal collection does not represent the RaPAL experience in its entirety. Many leading members who have exerted great influence are barely represented within the journals because they have chosen to contribute in other vital ways. Many members have not actually written at all, even though they have been regular readers over many years. Some have written but have not been included because our criteria did not allow us to prioritise their particular pieces.

Second, the collection itself can be approached in a number of ways. It can be viewed as an accessible lively mix, an unpredictable coming together of questions, challenges and findings, most of which have an immediate resonance in the present policy context. However, it can also be viewed more analytically in terms of, for example, its relationships with the changing policy and practice contexts in which the writers were operating. It could be analysed in terms of the actual exercise of power within the collective and in terms of the significance of intellectual leadership. It can also be viewed in terms of absence as well as presence: the absence of following through the implications of the more radical literacy and power insights, and the lack at times of challenge and contest.

Our primary purpose here is not to undertake a wide-ranging analysis of the collection but rather to draw the range of research and practice activity from the data, and to locate this within the broader debates about research and practice. We propose that the RaPAL approach amounts to a distinctive model, and practitioners are invited to consider their own research and practice patterns in the light of this.

Part 2 provides a selection of papers from the collection. These have been organised thematically as follows:

General themes

1. Perspectives
2. Policy
3. Practitioner roles and identity
4. Literacy, numeracy and ESOL learners
5. Managing provision
6. Curriculum content and process
7. Assessment and accreditation

Specific issues, contexts and themes

1. Numeracy
2. ESOL
3. Dyslexia
4. Disability
5. International literacy
6. Family literacy
7. Workplace
8. Prisons
9. Literacy and gender

However, in selecting these themes and the articles that illustrated them we were highly conscious of what and whom we were excluding. We both wanted, for example, to include an individual section on the history of literacy but there were insufficient papers in the collection. This should not be interpreted as RaPAL having little interest in this subject; indeed there is considerable evidence of teachers discussing the history of literacy conventions and practices with their students, and major researchers within RaPAL have produced other publications which focused heavily in this area.

We also knew that we could not include everyone or even every piece of great merit. We took the practical decision not to abridge the pieces in order to include more items, and they appear in this volume in their original wording.[2] Ultimately we wanted to show not only the individual pieces but also the interactions between writers, and thus some responses were included at the expense of single pieces. Similarly, if there was a major dearth of articles on a theme, any such article would have a higher priority than an equally good article on a well-trodden theme. Where appropriate we have alerted readers to further references on each theme within the collection.

A further key decision was not to attempt to contextualise each article, in terms either of the circumstances in which it had been written or of the particular writer's subsequent work. We simply provided the date of publication and anticipated that readers would pursue further details if necessary.

In selecting these themes, we could identify a set of principles that resonate with the literacy as social practice position identified by Barton, Hamilton and Ivanič (2000, p. 8). These principles include the following:

- The ways in which literacy is conceptualised have a direct effect on how research and practice is developed and enacted.
- The historical context shapes such conceptualisations.
- Adult learners have unique knowledge about their own learning which teachers in the field must access in order to teach with most effect. The students are the teachers as well as the learners in this sense; and the teachers, too, become learners.

- There is a power dynamic at play at all levels within this field: in the cultural construction and valuation of certain forms of literacy, in the policy initiatives, in the classroom, and in research practices themselves. Literacy educators have to recognise this if they are to make sensible decisions about sharing and exercising power and thence to achieve the necessary mutuality with students.
- A professional stance demands that we position disability and specific learning difficulties as central to mainstream practice.
- International thinking and experience consistently inform our practice in the UK.

Part 3, which concludes the book, invites practitioners to move beyond individual examples and to link these processes together within a longitudinal staff development profile. It therefore provides an individual research and practice case study by one of the editors, who has worked in both basic and higher education. In this part, Margaret draws out the key episodes and 'moments of development' in generating her expertise about dyslexia over a 20-year period.

How to use the book

The book is designed for ease of access, with each section connected to the next but also capable of standing alone. Some readers will want to dip into the selection for particular articles, whereas others will value the attempt to explore research and practice in general. Anyone who is particularly interested in dyslexia may find the longitudinal case study a useful account against which to consider their own development.

Notes

1. As head of the Adult Basic Skills Strategy Unit at the Department for Education and Skills, Susan Pember has claimed that an individual might

expect their salary to increase as their basic skills improved – in particular, she suggested in 2001 that an increase of £50,000 over 20 years was a realistic expectation for those improving their numeracy skills (RaPAL conference, University of Nottingham, 2002).

2. The location details of the authors have been left as they were at the time of original publication, with only minor editing.

References

Barry, P. (2002) *Beginning Theory: An Introduction to Literary and Cultural Theory*. Manchester University Press.

Barton, D., Hamilton, M. and Ivanič, R. (eds) (2000) *Situated Literacies: Reading and Writing in Context*. Routledge, p. 8.

Department for Education and Employment (DfEE) (1999) *A Fresh Start: Improving Literacy and Numeracy. The Report of the Working Party Chaired by Sir Claus Moser*. DfEE.

Farrell, L. (2001) 'The New Word Order: Workplace Education and the Textual Practice of Economic Globalisation', *Pedagogy, Culture and Society*, Vol. 9 No. 2, pp. 57–75.

Gee, J. P. (2000), 'New people in new worlds: Networks, the new capitalism and schools', in B. Cope and M. Kalanzis (eds), *Multi-Literacies*. Routledge.

Hamilton, M. (1996) 'Higher education and professional development opportunities for ABE in the UK', *RaPAL Bulletin*, No. 30, pp. 27–31.

Houghton, G. (2004) Private communications about about her Ph.D. research on theoretical research narratives.

Huberman, M. (1995) 'Professional careers and professional development: Some intersections', in T.R. Guskey and M. Huberman (eds), *Professional Development in Education: New Paradigms and Practices*. Teachers College.

Ivanič, R. (1986) *RaPAL Bulletin* No. 2.

Mace, J. (1988), *Talking about Literacy: Principles and Practice of Adult Literacy Education*. Routledge.

Pember, S. (2001) Keynote address given to the RaPAL conference, University of Nottingham, June 2002.

RaPAL (1986) *RaPAL Bulletin* No. 1, 1986.

Sanguinetti, G. (2001), 'Pedagogy, Performativity and Power: Teachers

Engaging with Competency Based Literacy Education'. Paper presented at the International Literacy Conference, Literacy and Language in Global And Local settings: New Directions for Research and Teaching. University of Cape Town, Nov 13–17.

Part 1

Building knowledge and enhancing expertise through research in/and practice

Introduction

This book is an invitation to practitioners to embrace research within their practice as a means of developing their critical, investigative, professional[1] stances. In this chapter, we elicit relevant insights from the whole RaPAL collection.[2] In particular, we shall draw out both the main theoretical, policy and practice issues with which RaPAL members have concerned themselves and the RaPAL perspectives on the actual processes of research and practice. With regard to the latter, we shall examine:

1. the activities involved in research and who can participate;
2. the reasons why members engaged in research: motivations and drivers;
3. the forms of enquiry they used;
4. what 'practice' in this field appears to mean;
5. the relationships between research and practice;
6. the dialogues between learners, teachers and researchers;
7. longitudinal research and practice processes.

Finally, we shall attempt to locate the RaPAL experience within the wider conventions about research and practice.

RaPAL themes and why they are still relevant

Across the whole collection it is possible to see that RaPAL has:

- problematised the whole issue of literacy/illiteracy/maths/language crises and problems, and so consistently challenged simplistic theoretical narratives about language, literacy and adult learners and any policies based upon these. RaPAL stands for criticality in relation to policy, and many

members can be seen critically reviewing new developments, speaking out when they see major failings and constructing alternative proposals;
- worked on reconceptualising literacy, literacies and numeracy and so produced new perspectives about how literacy/literacies/numeracies work, the power and identity issues involved, the theoretical bases of policies, and both general and specific theories about literacies;
- problematised the making of knowledge in this field by raising questions about the role of researchers, practitioners and students and experimenting with new possibilities;
- focused on issues of power and identity within and around learner and tutor positions within practice in many adult learning domains and contexts, for example the family, the community, social and leisure settings, prisons, libraries and the media. Hence RaPAL encourages a critical analysis of practice: What does student-centred practice really mean? In what sense is open learning open? What is the tutor doing during language experience processes? How does participation actually work in participatory practice?;
- recognised the broader literacy and numeracy curricula which students seek as producing a wider range of learning outcomes than is often recognised, thus allowing a wider range of criteria for success. The example of producing an ongoing critical analysis of process and power in writing, leading to an emphasis on student publishing, led to a deeper awareness of what is actually learned by writing and publishing;
- engaged with, and drawn heavily from, the work of international theorists, policy-makers, practitioners and students at all levels. Knowledge and process outcomes can be seen within the collection, and the sheer joy of these connections is alive in the present lives of RaPAL members.

All of these are relevant for new practitioners in the middle of this decade.

The model of questioning has a wide significance. The new literacy, numeracy and ESOL policies in England and Wales are based on particular concepts of language, literacy and numeracy and should be critiqued accordingly. The RaPAL model of encouraging members to build new theories and new practices is as important as ever, and the creation of what appears as a highly prescriptive set of curriculum structures need not preclude such work. As a new generation of practitioners enters this field, working in many contexts, RaPAL offers an accessible treasure trove of examples of effective practice. The invitation to make new knowledge is as urgent as ever, and for adult learners to be seen to be engaged in this is vital. Finally, international connections continue to offer practitioners the opportunity to pursue their own work in relation to ideas and practices elsewhere.

Research and practice processes

What does the RaPAL collection say about research . . . what it involves and who can participate?

We have already noted briefly the ways in which RaPAL members have attempted to demystify research processes. The collection illustrates the RaPAL challenges to some traditional research stories by claiming that:

- the everyday processes of raising questions, seeking answers, recording, analysing, reflecting, arguing, discussing and disagreeing are central to research activity;
- they are not the exclusive preserve of paid researchers and are open to anyone involved in literacy, numeracy and ESOL education;
- practitioners and learners are knowledge-makers too. RaPAL creates a forum for knowledge-makers, organises training in research methods for students (for example, Doing Research Weekend, *RaPAL Bulletin* No. 5) and provides opportunities for collaborative work:

  Students and tutors observing, analysing experimenting,

T

> *evaluating and reporting back is research-through-practice of exactly the type RaPAL wants to promote. (Ivanič, No. 2, 1986, p. 1)*

- there are many ways of doing research we can all engage in, creating new possibilities:

  > *It is also important to remember that research method is not fixed. Adult basic education needs to explore new methods, to be involved in developing these methods and to press for their increased acceptability. In particular we need to explore ways of participative research that break down the roles of researcher and researched.* (RaPAL editors, No. 1, 1986, p. 1)

  It is important to keep asking questions and checking assumptions about what is actually going on during research (Lee, No. 28/29, 1995/6; Macrae, No. 35, 1998).

- some research processes are already going on within practice but are not fully acknowledged:

  > *A great deal of reflective and evaluative work already goes on in adult basic education, often informally, under the name of 'outreach' or 'report writing'. This work needs to be made more visible and these activities developed into procedures that can be generalised, compared and used by others. A good evaluative description can generate research and this process is as important as research that sets out with hypotheses and aims to prove something.* (RaPAL editors/Ivanič, No. 1, 1986, p. 1)

  Jill Barnett argues that the source materials for research are already available within practice – for example, reports, notes, materials, discussions – and that there are some relatively straightforward ways of building it in (Barnett, No. 6, 1988, pp. 2–4).

For many members this amounted to a direct challenge to some traditional research practices with their strict divisions of labour between researchers and practitioners, with 'approved' research methodologies in which learners were the researched rather than researchers. Here practitioners and students were being invited to share the challenge and show what could be done. Although these positions looked clear enough at the time (Hamilton, No. 9, 1989), at least one founding member expressed the kind of uncertainty that practitioners often feel as they embark on research:

> *Jane Lawrence talked about some of the questions which came up for her when doing a research project: what is the research about? how was the research carried out? how many people did you interview? who paid for it? what did you find out? I was asked all these questions during and after I was doing a research project in 1980. The questions were unnerving and, except for the fourth one, difficult to answer simply. I was, at the time, an adult literacy tutor organiser with a couple of years' experience. Although I had done research before, it never quite fitted the model most people have of research and this project didn't fit that model either.* (Baynham, No. 1, 1986, pp. 3–4)

Why members got involved with research

Busy practitioners might well wonder why they should get involved in research at all, especially when it is not part of their paid responsibility. In the RaPAL collection we find some possible answers to this question. We can discern a range of overlapping proactive and reactive, personal and instrumental, individual and institutional research drivers. These can be seen as a series of stages and/or levels, from first questions through to undertaking further degrees or projects within employment contracts.

But this view does not adequately convey the sheer range of processes involved in generating research questions or in selecting which ones to pursue. The collection suggests a number of catalysts and, most importantly, shows how everyday reactions within practice can fuel investigations. These are just as likely in 2005 as they have been over the last 20 years.

Starting with reactions . . .

Unusual comments, descriptions. Margaret's own experience of dyslexic students describing their experience of reading and writing in ways that jolted her as a practitioner. She wondered what comments like 'reading is like creating a path through the undergrowth . . . and I have to keep treading through it each day to be able to keep reading. If I stop the path gets overgrown' (Herrington, 2001) actually meant for her practice. How different were these experiences to those of others? Was this new knowledge? What to make of such experiences? What to do about them?

Persisting niggles. Wanting to relieve the discomfort of confusion and unanswered issues. Some members give an impression of writing to resolve or assuage persistent niggles, questions and issues that would not go away and had to be addressed. The 'inner voice of disquiet' about the widely held adult education narrative is evident in Stephen Brookfield's article about the myths of adult education. His critique walks us through all the elements of the myth and, as he debunks them, he gives a strong impression of having to get these things said and clear them up for himself (Brookfield, No. 10, 1989).

Unconvincing accounts of theory and practice. Tom Joseph (a former student and ex-prisoner) did not find Anita Wilson's account of informal literacies in prison entirely convincing (Barton, Hamilton and Ivanič, 2000). Tom reviewed his own experience of literacy in prison (learning to read, discovering poetry, working on an Access course . . .) in relation to Anita's theory about informal literacies and wrote a rejoinder that contributed some of his personal evidence. He felt that he had to contribute from the inside his account, which focused on the power of traditional formal literacy for his personal redemption and as a route to enhancing his power outside prison. Anita, a leading researcher in this area, engaged with the critical suggestions and responded to Tom.

The driver here was about representation. Tom did not feel that his experience was being represented within Anita's model and, as he wanted to encourage other prisoners to try and escape the hell of recidivism, he felt his story was an important one for people to hear. The interchange showed how very different views could be expressed, heard and challenged within the network and, more profoundly, how students could exercise power in challenging significant research findings from prominent researchers (Joseph, No. 43, 2000/01).

Questions that are hanging in the air, waiting for their time, become too pressing. In 1998, The Blackfriars Literacy Scheme wrote:

> *. . . we decided it was time to look at what we actually meant by 'student control' in our literacy scheme. Workers had talked about it for years. Volunteers were trained in how to achieve it. Students were supposed to be taking it. But what did the term . . . actually mean? Were we all talking about the same concept? Were some of us meaning 'student participation' rather than control?* (No. 6, 1988, p. 4)

These are examples of being perhaps prodded into research.

More proactive drivers are also evident

Curiosity, wanting to know and often actively wanting to learn from others. One of the earliest pieces of research reported in the journal was Mary Hamilton's research on the literacy difficulties of those involved in the National Child Development Study. This was important quantitative and qualitative work at a time when headlines still focused on the 'millions unable to read' and showed that adults with literacy problems were not all unemployed, without qualifications or experiencing difficulties in everyday life (RaPAL editors, No. 3, 1987).

A particular strand of 'finding out' involved learning from overseas colleagues and students. Members attended conferences and workshops and travelled to explore policy and practice in Europe, the USA, Canada and Australia, and sometimes groups of students travelled to other countries to find out about their provision. Fiona Frank, in her groundbreaking work on workplace basic skills, sought evidence from other countries (Frank, No. 17, 1992).

Needing or wanting to make sense of one's own personal or professional experience. Members often showed a desire to think an issue out. Gaye Houghton, for example, wanted to get to the bottom of what it actually meant to be changing one's writing to fit new conventions (Houghton, No. 28/29, 1995/6).

Wanting to improve practice and provision. Many members got involved in research projects that were designed to improve provision. For example, the recruitment of adults with literacy difficulties in rural areas was the subject of a research project in County Offally in the Republic of Ireland (Derbyshire and Hensey, No. 8, 1989).

General issues of accessibility, quality and funding were examined. When, for example, changes in funding arrangements occurred, Pablo Foster wrote a piece to help ABE staff respond effectively to the new arrangements under the Education Reform Act. Though ostensibly a piece of explanation, it was clearly fired by a desire to ensure that staff did not experience the decimation of their schemes. He was trying to save provision (Foster, No. 8, 1989).

Wanting to make connections. Wanting to fill in gaps. An example of this was asking about the extent to which the ABE tutor role involved an explicit guidance and counselling function. Given the general problem of tutors' roles being complex but unarticulated, this exploration with literacy organisers filled in some useful information about how they viewed guidance within their roles (Herrington and Clayfield, No. 18, 1992).

Wanting to critique policy, especially when it appeared to disadvantage students. Over many years, a steady flow of articles did critique aspects of policy. A recent example focused on methods of multiple choice literacy assessment and challenged the new methods in terms of traditional assessment criteria of validity and reliability (Heath, No. 50, 2003).

Wanting to counter injustice? An example of students using research to support community action was that written by Raphael McDuffus *et al.* The authors described and analysed their work on fighting funding cuts and building closure (McDuffus *et al.*, No. 32, 1997).

Wanting to think more deeply about language, literacy and numeracy within the culture, start new discourses and develop new theory. A very important development in this respect was the ethnography of literacy work which started in Lancaster during 1988. Barton and Padmore wanted to 'find out about people's everyday uses of literacy and to build up a notion of "community literacy" which is distinct from school views of literacy or work views of literacy' (Barton and Padmore, No. 6, 1988).

Sometimes these drivers overlapped with each other and also with more instrumental purposes of career advancement. The value of drawing them out is to see how close the drivers feel to everyday working life, linking together feelings with political stances and intellectual exploration.

Forms of enquiry used

The collection shows how RaPAL members used many different forms of enquiry. The central processes of asking and answering questions are shown within different forms, and we have identified a selection below in order to illustrate how new practitioners can easily find their way into and through investigative research/practice. The sheer variety of methods should encourage practitioners not to see all teacher research as action research within the classroom, and the distinctive range of student involvement serves as a consistent reminder of the RaPAL priority of opening up research processes to adult learners.

Asking short, individual questions in the journal

RaPAL argues that asking questions is a fundamental part of research. Hence it acknowledged that this could happen in different forms and at many levels. It encouraged practitioners and learner members to send to the journal questions that were emerging from their experience in practice. Four examples are quoted here.

Learners managing provision

> *When learners run things, including the budget, how long does/should a management group last? What kind of training is necessary for managing? What kinds of problem have arisen as a result of professionals and learners having different definitions of what's to be done?* (No. 1, 1986, p. 11)

Serial or holistic thinking.

> *Pask says people either think one way or the other, and this affects learning. It could start to explain some of the barriers to learning we have seen if only we could understand what he says. Does anyone know of a translation?* (No. 1, 1986, p. 11)

Group versus individual tuition.

> *We polled our students about which they preferred. All those who usually worked in groups (language students, writing group students) preferred working in groups; those who usually worked individually preferred that. We had to brief students about each mode before they could decide which they preferred for what. Has anyone else tried this? What happened?* (No. 1, 1986, p. 11)

What are the links between literacy and numeracy?

> (No. 5, 1988, p. 19)

These questions attempted to link theory and practice, consider barriers to learning, explore student thinking styles and consider the roles of students within the service alongside professional staff. They also specifically asked for résumés of research on particular topics (No. 5, 1988) and revealed an interest in numeracy right from the start of the network (No. 4, 1987).

In addition to individuals posing questions, RaPAL editors also sent out general questions. It could be a simple matter of information gathering, as when Mary Hamilton sought information about any HE courses being developed for ABE staff (No. 30, 1996), but was more likely to occur at times of policy change. For example, at the time of the Education Reform Bill in 1988 members were asked, 'How would you be reformed?'. It was not just a survey question but an invitation to engage critically with policy change.

Students, too, asked research questions. In her evaluative comments on the Doing Research Training Weekend, Wendy Moss commented, 'The weekend showed very clearly that students know which questions to ask. Tutors and researchers in universities don't have this knowledge' (No. 9, 1989, p. 11).

These questions hint at the underbelly of questions in and from practice which practitioners and learners wanted to know more about.

Writing answers to individual questions asked by others is also investigative

Researchers, practitioners and students were moved to write in response to individual questions:

> ***Group versus individual tuition.***
> *As an adult literacy student . . . who has attended both individual and group sessions, I feel most strongly that students' choice is essential. Any scheme which is not offering both individual and group work, in my opinion, is not taking the students' needs into account.*
>
> *My reason for saying this is this. We have been through a school system that did not work for us and we usually see ourselves as having been the failure, not the other way round. Despite this we have decided that we want to try again. We have decided to trust new teachers. This has taken an enormous amount of courage but this new found confidence and enthusiasm can be smashed by anyone, especially those in an authoritative position. To be met by a lack of choice at an early stage can be extremely damaging.*
>
> *I think choice is the key issue for all students throughout their learning. Choice should be available at every stage and in every scheme.*
>
> *Personally I preferred individual tuition for reading and once I had gained confidence at reading, I now enjoy group work for writing.*

Serialist/holist thinking. Mary Hamilton replies . . .
There have been many attempts to group people according to their learning style based on the idea that people approach learning in different ways, and each person has a characteristic learning style. People have been described, for example, as being field dependent or independent, surface or deep processors, convergers or divergers.

Gordon Pask invented the terms 'serialist' and 'holist' as a result of his observations of people trying to solve problems. He laid a problem on the table, watched what a group of students did and asked them why they were doing it. He found that people fell into two clear groups: the serialists (or step by step learners) and the holists (or global learners). The serialists concentrated on the immediate details and individual parts of the overall problem, building up a picture step by step as they went along. Holists on the other hand focus on the overall goal that they are working toward and search in a more complicated way for the clues they need.

Each style has its problems: serialists sometimes cannot see the wood for the trees, while holists risk making wild over-generalisations.

Although Pask has some evidence that the holists are better learners than the serialists because they are more flexible, both styles can be effective. What causes problems is being forced to learn in a style that is not your preferred one.

Two drawbacks to Pask's ideas. First, the fact that everyone who looks at student learning comes up with a different pair of learning styles makes me think, why just two styles? Why not three or four or 104? Second, perhaps people use different learning styles for different tasks, varying their approach according to what they think is needed. How would the problems Pask used compare with the kinds of learning tasks facing an ABE student? We need to watch and ask a lot more students at work before we could use these labels with any certainty.

Despite these uncertainties, Pask's work draws attention to the fact that people may approach a given task in very different ways and, which may be obvious, we can learn about how our students learn by noticing what they do. (No. 2, 1986, p. 15)

These dialogic responses are thoughtful and evaluative, drawing in general ideas about learning and about the learner's direct role in evaluating literacy provision.

Another version of answering questions was the testimony of Stella Fitzpatrick who wrote an extensive letter about the effects of funding cuts on her practice (No. 15, 1991). No one had posed the question directly but it was an important question in the field at the time and her evidence was important.

Questions within 'conversational' enquiry

Sometimes the asking of questions or the writing of commentaries turned into 'conversations' but there were also direct investigative conversations. Wendy Moss, for example, interviewed Chrissie Maher about the Plain English Campaign and the literacy and power issues at the heart of such work (Moss, No. 4, 1987).

Questions and answers within analytical commentary

This form of enquiry is extensively represented in the collection as members drew on their existing empirical and theoretical knowledge and created analytical and synthesising responses.

1. **About literacy/literacies.** An example of this was a number of pieces about spelling. Issues around the standardisation of spelling and its implications for students were explored, as were the recurring ideas about simplifying the whole system (Upward, No. 5, 1988).
2. **About relating theory to practice.** Mary Norton provides an excellent exploration of how she related ideas from *The Give and Take of Writing* by Jane Mace to her own practice as a tutor (Norton, No. 50, 2003).
3. **About methods of doing participatory research.** Harold Rosen briefly analyses the role of narrative within his participatory research project course (Rosen, No. 2, 1986) and three students

evaluated workshops they had attended on 'Student Involvement in Literacy Research' (Golightly, Nicola and Stone, No. 5, 1988).

4. **About types of provision.** Students and staff raised and answered questions about open learning (Andrews, Fisher and O'Mahoney, No. 18, 1992).
5. **About policies and policy development.** Barbara Ratcliffe explored learning outcomes of ABE students and noted the loss of student control over their curriculum:

   > *The power of ABE students has diminished over the last decade in terms of how much control they have over what they learn and how it is taught. During the 1980s student needs were seen as paramount in designing the course, in terms of location, accessibility and suitability of material . . . The 1990s is seeing a change in this. The government is now the main influence mediating its policy through local colleges and funding council.* (No. 23, 1994, p. 19)

   And at a time when current policies were being forged, post-Moser, Jay Derrick analysed the work on developing the new national standards for basic skills (Derrick, No. 41, 2000).
6. **About books.** Individual practitioners provide analytical reviews of a vast array of books over this period but Ellayne Fowler gives us an example of students and tutors working together on book reviews (Fowler, No. 48, 2002).
7. **About social and cultural change.** Many writers tackle the bigger issues of literacy in society.

Occasionally these analyses emerged as ground-clearing exercises about particular issues, for example identifying different approaches within emerging areas of work (such as family literacy) or clearing ways through particular issues and processes (Ivanič and Richards, No. 3, 1987; Ivanič and Simpson, No. 6, 1988).

Seeking, gathering and analysing 'new' data

Some articles included analysis but also focused on gathering new data. This included life history and autobiographical material (Beevers, No. 7, 1988) and documentary research and analysis (Courtman, No. 49, 2002/03) as well as a range of surveys and interviews designed to gather evidence from others. Mandy McMahon, for example, gathered data from adult education management teams about what they did (McMahon, No. 6, 1988).

Gary Roberts and Jane Prowse reported on the fun of research as they worked with family literacy groups to gather data about literacy events represented in soap operas and then compared these with literacy events in people's homes:

> *It became apparent from the figures collated that as the co-researchers progressed from soap watching to look at events in their own homes and community lives, they became more aware of the many literacies they practised and how these are inter-related.* (Roberts and Prowse, No. 38, 1999, p. 27)

These drivers and methods of doing research in practice, evident in the collection, reflect something of the life and variety of this work as the desire to question aspects of practice, to communicate and to change the world found expression.

Where does this fit with RaPAL's view of practice?

If the collection shows a picture of multiple research stories and knowledge being generated on many levels, what does the collection suggest about practice?

For RaPAL, practice has never been simply about generating a set of methods for teaching reading and writing. The work of practice has always involved a sensibility and criticality about literacy/literacies in the culture, policy discourses, and the problematic position of the learners in relation to these. It has always addressed the question of what is the appropriate position of the teacher in these situations. For example, are tutors representatives of 'literate' society and gatekeepers for highly valued forms of literacy, and should they be representing the literacy problems of others (Baynham, No. 1, 1986)? It has also consistently explored the ways in which literacy learners could exercise power within the curricula. In teaching adult literacy learners, practitioners had to make sense of why people continued to fail to read and write and why only some sought help with this, where to place themselves in relation to this, and what to do about it on a day-to-day basis.

However, in the early years this complexity was not always articulated in print or even among themselves. Eric Appleby describes his reaction to having to explain the basis of UK practice to US colleagues:

> *. . . in rushing around getting on with the job we may have neglected to keep track of exactly what we are doing and why. By contrast our American hosts demonstrated an awe-inspiring ability to discuss and evaluate their work in theoretical terms. The process of trying to explain the basis of our work to those from such a different background . . . forced me . . . to go a long way to some of the most fundamental principles which underlie good practice.* (Appleby, No. 7, 1998, p. 13)

The RaPAL collection as a whole shows members articulating their positions, and a core of ideas emerged as central to practice:

- The centrality of the student role in creating curriculum (Clarke, No. 16, 1991). Occasional examples of students setting up their own provision (Pecketwell College, No. 3, 1987) and students being seen as authors rather than just receivers of text from others, and as researchers, reflective evaluators and teachers (Mace, Moss and Pinner, No. 7, 1988; Chappell, No. 50, 2003).
- A broad but negotiated curriculum which included the reframing of past failure with literacy (Warrington, No. 13, 1990), empowerment and renewal across a range of personal, family, social and economic criteria and sometimes issues of social action and social justice (Duffus, Sharp and Nolan, No. 32, 1997).
- A willingness to unravel, analyse and evaluate pedagogic/ andragogic processes in a range of adult learning contexts.
- An understanding about critical literacy practices and critical

language awareness, including emancipatory practice (Merrifield, No. 38, 1999).
- Challenges to new, narrowly based policies about practice, especially when they work against the students (RaPAL Working Group, No. 37, 1999).
- Networking with colleagues and students at home and abroad.

The articles show how research-laden practice is, even when it is not acknowledged as such. At the heart of practice there are always the ongoing research questions about why someone cannot do what they wish to do and how best to facilitate their learning. Practitioners grapple with these on a day-to-day basis, generating and working with hypotheses and evaluating outcomes. The work of researchers, in clarifying ways of thinking about literacy and about the impact of literacy concepts on the everyday lives and prospects of learners, directly informs the work of practitioners in reframing with learners their own past 'failure' and in developing new methods. Similarly, learners pose questions that researchers cannot yet answer. And so there is an ongoing set of questions to be explored within practice. Figure 1.1 reflects some of the RaPAL questions about who is doing the researching in practice. In general, we are interested in exploring the top two quadrants in the diagram, that is, in research which involves the practitioners and learners as researchers themselves.

This is highly relevant to current practice. Some practitioners can feel that they are simply being asked to apply a system to learners and that there is no space for deeper processes. However, the systematic recording of data about practice required under the *Skills*

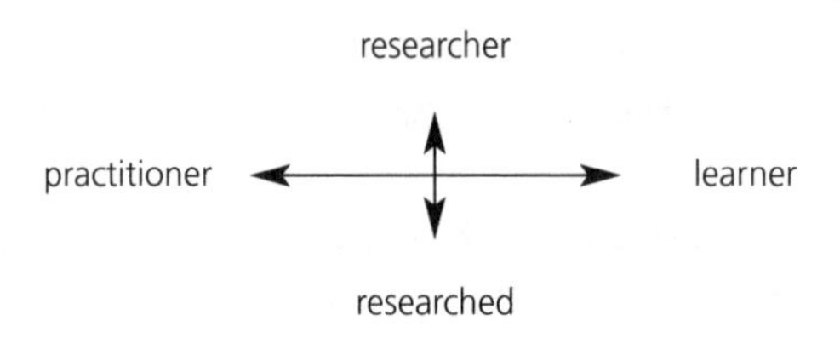

Figure 1.1 Locating learners and practitioners in relation to research

for Life initiative offers a mechanism for reflective work, and policy-makers themselves are continuing to question and evaluate the initiative. It is not fixed in stone, and RaPAL members have demonstrated the importance of engaging critically with a whole range of policy initiatives.

Research and practice links

The RaPAL perspective invites us to construct a range of possible connections between research and practice. We need not restrict ourselves to thinking of these as two separate domains, nor only of a top-down relationship between them – that is, using research findings in practice settings. We are shown that practice settings themselves are research sites in which practitioners and learners can engage in critical thinking, gather data, theorise, report, challenge and debate. And knowledge is created in the sense that implicit knowledge can be made visible and new knowledge which could not emerge in any other way can be generated.

The collection also suggests that research narratives – 'stories' about who is allowed to do what, why, and with what kind of knowledge outcomes – contain implicit and explicit relationships with practice, and that practice narratives too are research-laden. This means that practitioners should not see their work as separate and essentially different from that of researchers.

The overlaps and commonalities between research and practice are suggested in Figure 1.2. It can be helpful to see research and practice as moving, overlapping sets with a close alignment between investigative practice and research in practice, and involving activities that may be difficult to classify or that may be both research and practice. The diagram also suggests that common processes may run through both.

This overlap is illustrated by Kallenbach and Viens (2002), who claim that for practice to be called research it must be systematic, intentioned and documented. However, when they are describing research they include the kind of processes that are so common in investigative practice:

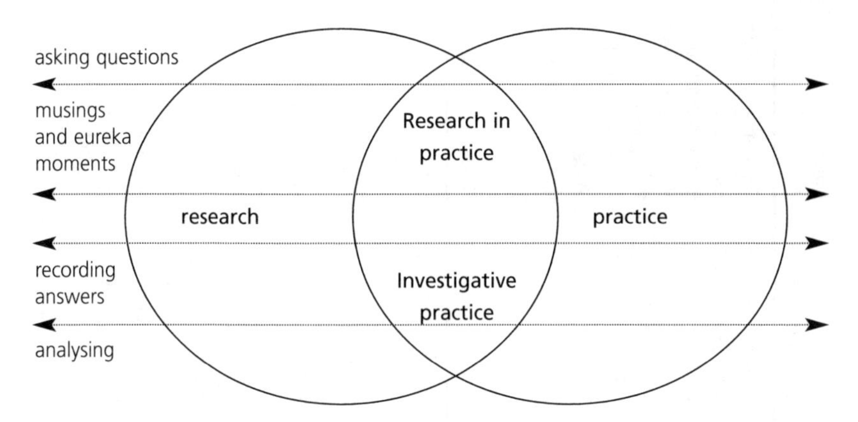

Figure 1.2 Overlap and commonalities between research and practice

> *Teacher research is constructivist in that practitioners learn by making information personally relevant, developing and trying new approaches, reflecting on the process and the results and questioning their own and the theory's assumptions.*

Hence the usual determiners of whether something can be called research themselves become problematic. Common criteria such as:

- intention;
- systematic data collection;
- systematic rethinking of the known;
- abstraction;
- must be written down;
- must be open to critical public scrutiny

often raise more questions than they answer. Surely the intention to find answers is intrinsic to good practice too? What counts as data? Notes, finished articles, photos, measurements? Does all research involve collecting data? Are oral forms not included? Doesn't reflective practice involve levels of abstraction? Would the critical scrutiny criteria rule out the BSA archive material on local development projects, for example, or reports on Pathfinder projects? And in relation to time, when does research actually materialise? Is it demonstrated by private notes or does it only count when the work is complete and scrutinised? Even more importantly, these increasingly

Table 1.1 A selection of research (and practice) narratives

Who	What	Why	How	Knowledge outcome	Impact/ dissemination	Comment
I Researchers: individuals/ institutions	Own questions/ funders' questions. About theory, policy and practice.	Employment requirement. Curiosity.	Range of possible methodologies. Sometimes discipline specific. Quantitative. Qualitative. Longitudinal. Experimental.	Epistemologically valid ... always contests about this.	Dissemination to other researchers primarily. Practitioners can learn from researchers.	Separation of R&P roles. Research prior to and above practice. Knowledge generated outside these paradigms not accepted as such. Fictional linearities. Hidden processes.
II Practitioners	Own questions about self, learners, literacy therory, policy and practice [responding to problems in practice].	Own professional development.	By themselves. By themselves *in situ* of practice. With learners in practice. Autobiographical. Action research cycles. Reflective practice cycles. Discipline of noticing leading to research.	Epistemologically valid if under the aegis of researchers (as above). Creation of knowledge that is not always acknowledged as such: thought of as too context bound and not generalisable. Theory generated rather than tested.	Often kept within institutions. Owned by practitioners and fed into ongoing practice. Sometimes disseminated to other practitioners and students but may stop there. Not always published in academic journals.	Better fit between research outcomes and development of practice. Better fit between research subjects and real-life problems? May lead to bids for research funding to go beyond own situations.

III Practitioners	Job requirements.	Employment requirement.	By themselves *in situ*, in practice. With learners in practice. With other colleagues in teams.	Production of reports for the employer or the development agency.	May stop there and never see the light of day. May be widely circulated. Not seen as the same level of knowledge in academic journals.	R&D interface. May lead to more research and development.
IV Practitioners and Learners	Generate questions and try to solve together in practice.	To answer learner and practitioner questions about learning.	Informal narratives. Surveys? Life histories?	Creates knowledge but who is listening?	Internal documents. Network articles.	Learners more involved in setting the research agenda. Practice develops directly and generatively.
V Learners	Own questions.	Curiosity/ evangelism.	Informal methods: co-operatives, autobiographical accounts.	Creates knowledge. May challenge researchers and practitioners.	Speaks directly to other learners ... and to teachers.	Learners set their own agenda.

overlap with practitioner activity as tutors are required to systematically record in written form all stages of student progress and also to reflect and evaluate on, and in, action.

Given these interesting juxtapositions of practices and roles, it can be helpful to think of research narratives as actually *research and practice* narratives – a series of 'stories' set in time and place about who is allowed to do what, with whom, why, with what knowledge outcomes and with what kind of implicit or explicit relationships with practice (Herrington, 2004).

Table 1.1 sketches out a selection of such stories, some more familiar than others, indicating continuing hierarchies in knowledge-making. The stories are problematic in many ways. At times they may be seen as mixing up unhelpfully job positions and research motivation and roles. But they are not meant to encapsulate the different living stories within RaPAL. Indeed the network exemplifies the ways in which these have been problematised, challenged and altered. They do, however, serve to remind readers that not all forms of research are actually open to everyone within this field and may help to explain that, even when practitioners and learners accept a multi-narrative approach to research and knowledge generation, some experience continuing insecurity about the nature of the knowledge they are producing. Darville's (2003) request for parity of esteem is as important for practitioners to hear as it is for researchers.

Individual stories such as these can be found among RaPAL members (Herrington, 2003), but the network encourages parity of esteem by creating a space in which these stories can interact with and shape each other. For example, the 'knowledge' space which is the *RaPAL Journal* can be seen as open to insights and reflections that do not always fit with prevailing views and hierarchies. It reflects the messiness of both research and practice processes and invites first-level descriptions and questions alongside more profound theorising.

Longitudinal research and practice patterns

We can see in the journal how RaPAL contributors were not restricted to particular research stories over time but created different longitudinal patterns of research in/and practice. To some extent these were governed by different professional roles and opportunities, but the variations, especially among practitioners, were marked.

Some practitioners undertook research connected with further degrees or nationally funded development projects. These included one-off pieces of research which then informed practice or produced insights on theories of practice – stepping aside and looking beyond in some way. Others appeared to build research activity into their practice in an organic way, creating a research-within-practice gestalt.

Some worked on particular themes over time, whilst for others coherence rested in their political stance or in seeing research processes as integral to their professional stance as practitioners. Some had literally produced one piece of writing while others wrote on different matters over a lengthy period.

These patterns were clearly connected with the practitioners' lives and values in different ways, and a deeper conceptualisation of longitudinal development is required here. For now we can say that the patterns could not simply be characterised as sequential cycles of reflective action and practice. They included the personal feelings, qualities and stances, the professional positions, the urgency of resolving practical problems and the need to react to political contexts.

We can say less about the longitudinal patterns of the literacy and numeracy students involved here. While Pecketwell College at Hebden Bridge, the organisation established by adult literacy students for adult literacy students, was still continuing to undertake research in 2003 (Gardener and Glynn, No. 48, 2002), many individuals and groups of students who did research do not emerge with long-term profiles. This does not necessarily mean that they did not have them, simply that they are not recorded in the collection.

About processes within the network . . .

Much of the writing can be seen as involving internal and external dialogue. The writers are articulating their ideas both to themselves and to others. Individuals try to make sense of their ideas and situations through writing, and for some there is an impression of catharsis – the release of feelings and beliefs – as arguments are constructed and testimonies shaped. Others, at times, appear to focus more heavily on communicating ideas, principles and evidence to others within the network and beyond.

However, there are far more dialogic connections than this conveys, and here we note three examples. First, there are examples of loose, asynchronous interchanges and occasionally a sequential debate. In an early bulletin, for example, Ken Levine wrote a powerful piece about the relationships between the adult literacy curriculum and widespread changes in literacy practices (Levine, No. 2). He argued strongly that, given the new complexities, it would be important to establish a Community Information Service 'to promote a common recognition that providing information for the public and literacy work are two sides of the same coin' (page 5). In short, he felt it was important to focus on the production of official information and to connect literacy schemes with this. John Booker and Lorraine Brook both subsequently challenged Levine's conclusions by providing new information about what was actually happening in their organisations (No. 3, 1987). Other readers could observe this dialogue, making new connections with their own ideas at each stage.

There is also evidence of dialogue in the production of pieces of writing: students writing with each other, with tutors and with researchers. Sometimes the actual nuts and bolts of the dialogue are laid open to the reader. Practitioners, too, frequently write together and so we see reflections of previous dialogues often emerging as single voices.

Further, in some editions editors have constructed a written dialogue by seeking responses from readers and arranging these alongside the original article: tightening time around the response rather than waiting for a sequence to develop. This was often done to expose competing views within the network. It was a way of pulling hidden dialogues into print.

In terms of the power relations discussed earlier, the RaPAL collection provides examples of dialogue between leading thinkers and writers in the field and newcomers at all levels. This has sent an important signal about parity of esteem.

Where does this fit with research and practice in general?

So far we have mentioned the particular stories about research and practice evident within the RaPAL collection. The final question for us was about how and where this experience fits with general views about and the relationships between research and practice. Where in the burgeoning panoply of action research, teacher research, emancipatory research and practice, reflective practice and critical pedagogy can we place the RaPAL narrative (Brown and Jones, 2001; Carr and Kemmis, 1986; Clarke *et al.*, 2003), and does it represent something distinctive?

Though it can be seen as embracing all of the above, and other more recent theories of practice such as Mason's (2002) *Researching Your Own Practice: The Discipline of Noticing*, it seems to us to be distinctive. It is not a new model of individual research and practice, though it offers new stories about how this work can be done and in particular reveals something more about motivational factors. So far as we can see, it seems rather to be a distinctive collective model. RaPAL is a virtual, knowledge-making space or domain, with material presence in the journal and in the annual conference. It is a space infused with a range of values (democratic, radical, humanist) and inhabited in an imaginative sense by a range of literacy, numeracy and ESOL investigators, working with a whole range of research narratives. In this sense it is not so different to some other professional networks. However, the distinctiveness (and bravery) lie in:

- the membership of the space (literacy students, leading paid researchers, part-time tutors and senior managers all inhabit it);
- the explicit challenge to traditional power relationships between the different groups in the space and the new relationships forged;
- the openness to each other's work;
- the movements between roles: for example, students as researchers and co-researchers;
- a refusal to control the literacy outcomes too tightly, with the result that the network is always open to new and unpredictable directions;

- the nature of the communications – the culture of encouraging and assisting all members to express their ideas.

The distinctive values, processes, fluidity, dynamic and openness to the unknown characterise a model for research and practice that could potentially have far wider applications. The selection of papers in Part 2 gives a flavour of this.

Postscript

At a time when the UK government has recognised the importance of research in this field, the NRDC (National Research and Development Centre) is positively encouraging practitioner researchers. Regional and local developments are complementing this work. At the time of writing in late 2004, a group of practitioner researchers at Wolverhampton University (West Midlands LSC) are engaged in exploring the impact of the core curriculum on their practice. This aspect of the RaPAL agenda has thus become centre stage and the experience of members particularly relevant to present day practitioners.

Notes

1. We acknowledge that professionalism is problematic for many in this field. It can be viewed as a way of sustaining powerful elites and hence seems particularly inappropriate in a field of emancipatory practice. We consider that, though it is problematic, the interests of learners are best served by having educators who are more fully alerted to exploring issues of power in the curriculum. Forging a professionalism that encourages this is part of our rationale for compiling this work.
2. The details of the complete collection can be viewed on the RaPAL website (www.literacy.lancaster.ac.uk/rapal). All references to RaPAL journals in this text will give the author(s) and number of the issue.

References

Barton, D., Hamilton, M. and Ivanič, R. (eds) (2000) *Situated Literacies: Reading and Writing in Context*. Routledge, p. 8.

Brown, T. and Jones, L. (2001) *Action Research and Postmodernism: Congruence and Critique*. Open University Press.

Carr, W. and Kemmis, S. (1986) *Becoming Critical. Education, Knowledge and Action Research.* Cited at: www.infed.org/research/b-actres.htm

Clarke, J., Edwards, R., Harrison, R. and Reeve, F. (2003) *(Re)presenting Research in Lifelong Learning*. Online at: http://education.cant.ac.uk/canterburypapers/Edwards.htm

Darville, R. (2003) 'Making connections', *Literacies*, vol. 1, pp. 3–5. Online at: www.literacyjournal.ca

Herrington, M. (2003) Research with the RaPAL Network, No. 1, 'Difficult to Reach' Research, NRDC, March.

Herrington, M. (2004) 'Literacies? Wonderful, great, please carry on! Joining in the conversation about research and practice', *Literacies*, vol. 2, online at: www.literacyjournal.ca

Kallenbach, S. and Viens, J. (2002) *Adult Multiple Intelligences (AMI) Study: Teacher Research Methods*. Paper given at The Edmonton Gathering. Details available on the AMI website: http://pzweb.harvard.edu/ami

Mason, J. (2002) *Researching Your Own Practice: The Discipline of Noticing*. Routledge Falmer.

Part 2

Relevant examples from the RaPAL collection

Contents

A General themes

1. Perspectives

2. Policy

3. Practitioner roles and identity

B Specific issues, contexts and themes

1. Numeracy

2. ESOL

3. Dyslexia

A1 Perspectives

Further RaPAL references about perspectives

Street, B.V. 'Adult literacy in the United Kingdom. A history of research and practice', *RaPAL Bulletin* No. 33, 1997.

Situating literacies

David Barton

RaPAL Bulletin No. 43, Winter 2000

David Barton is Professor of Linguistics at the University of Lancaster. He is a founder member of RaPAL.

If we think of an example of someone reading or writing, there is a wide range of things they might be doing. The person might be filling in a form, writing a letter, checking a timetable, following a recipe, or doing any one of a host of everyday activities. In each of these cases, the person will be reading and writing for a purpose, with an aim, with the point of achieving something. It is also true that in each of these examples the reading and writing will be carried out in a particular place at a particular point in time. People don't just randomly fill in forms or write letters. They fill in a particular form at a particular time and place.

For example, I have just filled in a register of electors' form. It came through the post last week and has lain on the stairs for several days. I sat on the stairs and filled it in, on a Sunday morning, partly as a way of tidying the house. These forms are delivered every year in early September and are used to make up the register of electors for the following year. On the front of the form in large letters it says *Don't lose your right to vote*. So that's another reason I fill it in, in order to participate as a citizen. In smaller letters on the back it says *You must by law, fill in this form*. There is a lot which could be said about this form: the mixture of legal and everyday language; the strangeness of form filling as an example of writing (how it is both collaborative and imposed); the range of different purposes it serves; and the changes in recent years, with a web-site being mentioned on this year's form.

However, the central point I want to make is how my filling in this form is situated in a particular time and place; I did this in Lancaster, England in September 2000. In a different place or at a different time it would have been a different activity.

Activities such as form filling may seem very specific, but when we examine them in detail, we find a wide range of practices. To move on to the example of letter writing, this can mean very different things in different situations. It can be a very, private, personal activity between two people, such as sending a love letter, or it can be very public activity, such as when members of a campaigning group decide communally to write a letter to a politician. If we examine particular examples of people writing letters, we will find the ways in which the activities are socially situated. So, for example, teenage boys are not usually thought of as avid letter writers, but in the Young Offenders Institution in Lancaster, where there are delinquent teenagers who are often stereotyped as having no interest or ability in literacy, one finds several lads who keep up regular and extensive correspondence with friends and relatives. The time and the place can make a great deal of difference.

Looking closely at specific situations can be very revealing and can help us understand more about the significance of reading and writing in people's lives, the problems people experience and how they deal with them, and how literacy demands are changing. I will give some examples of research I have been involved in. First, in order to understand more about everyday reading and writing Mary Hamilton and I spent several years studying Lancaster as a case study. We focused on one particular neighbourhood and followed a group of people for several months, interviewing them, observing them and participating in their lives in various ways. In this research (Barton and Hamilton 1998), we found a wide range of everyday activities where people read and write. Literacy plays a significant role in everyday communication, record-keeping, leisure activities and social participation. People use reading and writing to make sense of their lives and to deal with problems and difficulties they encounter. We noted the social patterning of activities, such as the complex ways in which participation in many literacy activities is gendered; we were also struck by the individual uses of reading and writing, the idiosyncratic activities, where each person seemed to have their own uses. People are also constantly coming across new uses of reading and writing throughout their lives, as they change and as society changes; learning is an integral part of using literacy in everyday life.

More broadly, the book entitled *Situated Literacies* (Barton, Hamilton and Ivanič 2000) contains a dozen studies of specific situations. To give three very different examples: one study covers the reading and writing of prisoners and how their practices are shaped by their specific situation; another investigates the literacy practices of bilingual Welsh farmers and how reading and writing links very local activities of cows in fields with global activities of European Community record keeping; thirdly, another study examines how members of a Catholic community use reading and writing to assert and maintain their identity. Another book (Barton and Hall 2000) explores the range of activities in letter writing. Articles cover what letter writing means in different cultures and social situations and how it is different in different historical periods. Chapters cover the different ways in which the activity of letter writing has been brought into school and adult classrooms. Successful and unsuccessful experience is described.

The focus on situating literacy is part of a broader approach to reading and writing which is becoming known as Literacy Studies. This is an approach which starts from specific situations, identifies particular literacy events and examines people's literacy practices. It studies the role of specific social institutions in supporting particular literacies and examines the historical development of literacy in different domains.

This approach of starting from the specific and starting from the everyday provides a coherent perspective on education. It can be used to address educational issues at any level and it can be quite illuminating when applied to adult literacy work. To illustrate this, it may be worth sketching out some similarities and differences between a Literacy Studies approach and current governmental approaches to literacy, embodied for instance in the Moser Report. As a starting-point it is worth identifying the considerable overlap in the different approaches. Both emphasise that literacy relates to real life needs, that these needs are changing, and that reading and writing are increasingly intertwined with other media. The government approaches focus on what people have in common and look for common needs; a situated approach looks in more detail at people's actual uses of reading and writing; it places more emphasis on individual needs and differences between people. It draws attention to how individual skills are located in social practices and it takes account of the different roles people take up in relation to literacy and the different identities associated with reading and writing.

A crucial question for any approach to learning is how people move from individual instances to more general activities, how filling in one form relates to filling in other forms; the dominant government approach might discuss this in terms of learning a set of skills, including transferable skills. A situated approach emphasises that people learn in specific situations and that one of the most powerful ways in which people generalise from specific situations is by reflecting on their practices, by talking about them, discussing them, identifying them and naming them. The situated approach moves the focus from individual skills to the understanding of more general principles. Learning then becomes much more than having a set of skills. Understanding has a central role and the learner does the transferring as part of an active process.

The current government approach emphasises learning in classrooms, with a strong curriculum framework and forms of assessment. A situated approach emphasises the importance of informal learning, how people are *learning constantly* in their everyday lives and it looks for ways of supporting this everyday learning. It understands that a task changes in different situations. Crucially for education, when a task is moved from everyday life into becoming a classroom activity or part of a test, it is transformed, or recontextualised, into a different activity. Letter-writing, form filling or timetable reading become different activities when carried out in classrooms or when embedded in tests. Often the most successful classroom activities are those which, like everyday activities, are real and purposeful. People learn by participating in situated practices.

Underlying the situated approach is a belief in the need for research into people's literacy practices. This is an approach which is always asking questions about what is going on in a particular situation, what it means for the participants and how it is changing. This reflectiveness is something which teachers and learners (as well as politicians and planners) need to be involved in. Certainty about practices is misplaced, especially given constant changes in contemporary practices. An emphasis on the situated nature of literacy draws attention to the wide range of activities people are involved in and the many things they use reading and writing for, whether or not people have a problem with reading and writing.

References

Barton, D. and Hall, N. (2000) [eds] *Letter writing as a social practice*. London: John Benjamins.

Barton, D. and Hamilton, M. (1998) *Local Literacies*. London: Routledge.

Barton, D., Hamilton, M and Ivanič, R. (2000) [eds] *Situated Literacies*. London: Routledge.

An interview with Paulo Freire

Sean Taylor

RaPAL Bulletin No. 34, Autumn 1997

Sean Taylor is a writer and teacher currently living in East London. He has written a number of books for adults and children and has a chapter on teaching writing in *Literacy, Language and Community Publishing*, edited by Jane Mace (Multilingual Matters, 1995).

In 1994 I gave up working in literacy teaching and writing development in east London and went with my wife to live in São Paulo in Brazil. Once I spoke fluent Portuguese, I decided to try to find Paulo Freire.

It turned out he lived in quiet street three or four miles from us. I asked to interview him, saying his thoughts might be some inspiration for people involved in literacy in the UK. We met. Realising I knew little of his work other than *Pedagogy of the Oppressed*, he sent me away with six new books and rearranged the interview three weeks later. *"At seventy-five I haven't got time to tell people where I was born"*, he said, with a twinkle in his eyes. What follows is a shortened version of the conversation we had the second time we met, on 19th November 1996.

How do you define an illiterate person?
Well, first of all I think I need to draw attention to something obvious, which is that only *lettered* cultures, which have mastered and which use graphic language, know illiteracy. That's to say that *unlettered* cultures, full of voices and spoken words, do not have illiteracy. An illiterate person is only a man or a woman who participates in a culture which knows and uses letters, but does not have command

over those letters. I think this is what illiteracy is. It is ignorance of graphic language within a culture which already uses graphic language. And look: a culture which knows graphic language and coexists with people who have not mastered graphic language is a form of violence. So illiteracy is a restriction and a violence to the right of people to participate in culture, which those who know and use the written word must not carry out.

Why is Brazil so full of this violence?
I don't think it's just Brazil. I think that this violence, this transgression of ethics, is a phenomenon proven to exist throughout the world of men and women. As well as the existence of a great number of illiterate Brasilians I would highlight for you the existence of illiterate people in the First World. Canada, today, has a lot of illiterate people. So does the United States.

So does England.
Yes, and this is *exactly* the same violence that you find here in Brazil from the point of view of having people with a limited mastery of the written word. But I would like to back up this theory by showing that this violence against illiterate people is part of a violence against the *being* of a person and one of the horrific options we have for transgressing human ethics. Could you have a violent act worse than the killing of a black person because they are black? Could you have a violence more offensive than scientists saying that black people are genetically inferior? Could you have a violence worse than the ethnic restrictions which Europe, which your country, lives and exercises? My conclusion – my sad conclusion – is that violence accompanies human lives. Violence is an option we have as human beings, but our fight as human beings should be the fight to overcome our potential for transgressions of ethics. In your maturity which you will give in the next century, and in fact in the next millennium, you will have to fight for the preserving of ethics, but not this ethic which is championed by the world – the ethic of the market. The market is an ethic that is narrow and perverse. It is an ethic that runs counter to human nature, the desire to be, and the beauty of life. Our fight has got to be *against* the ethic of the market and *for* the ethic of the human being,

You hear a lot about ways in which medicine has been revolutionised by information globally available on the internet. Do you see potentials in these new technologies to help education on a world scale?

Definitely. I think that technology, in itself, is neither bad nor good. It all depends on the political use that you put technology to. So I am very clear that pedagogues and intellectuals with a sense of their historical position cannot afford to turn their backs on advanced technology. For me the question is how to put this advanced technology to the service of extending people's understanding, their respect and their creativity. You can't deny, for example, what can be done with a computer in schools. When I was Education Secretary for the state of São Paulo, I introduced computing in the state school network of this city and saw the results. The importance of television, the importance of fax, the importance of photocopying, of the internet in the field of education – there is no doubt about it. They are things which can shorten the distances between people, establish and enable greater intimacy between human beings, and the exchange of ideas, experiences and points of view about what is beautiful in this world and how to make it even more beautiful. But the point is, that all this has to lie on two pillars which cannot be separated: ethics and politics.

Technology on its own has no power to take decisions. It is the ethical and political orientation of those who use the technology that will decide whether we have a humane globalisation or we have a globalisation which destroys us.

TECHNOLOGY ON ITS OWN HAS NO POWER TO TAKE DECISIONS. IT IS THE ETHICAL AND POLITICAL ORIENTATION OF THOSE WHO USE THE TECHNOLOGY THAT WILL DECIDE WHETHER WE HAVE A HUMANE GLOBALISATION OR WE HAVE A GLOBALISATION WHICH DESTROYS US.

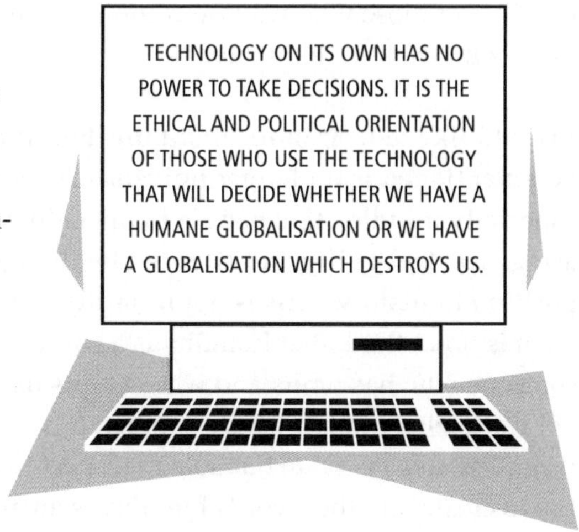

Are there ways that people working in education in the north of the planet can help the development of literacy and good educational practice in the south?
Look, I think it's the south of the world which should be helping the north of the world. I am against the supremacy of either the north or the south. For me, one of the serious problems we will continue to face in the next millennium is the relationship between the north and the south. Marcio Campos, at the University of Campinas in Brazil, has carried out a fascinating analysis of the verb *nortear*. This is a verb in Portuguese meaning to *give direction* and also meaning *to point someone towards the north!* This is profoundly ideological, because no one says *sulear* or *point towards the south*. In my opinion the question you have raised is not to do with preserving the *northerning* nor is it for the south to have to start *southerning*. What I think is that north and south should respect one another and should take up their historical roles as subjects, not as objects. For this to happen it is necessary for the north to be humble. The north has to invent a humbleness for itself, and discard its characteristic arrogance. Europe, for example, thinks that it started history. The north must discover that the south also exists, and is active and powerful. The south also creates. The south also produces. The south also thinks. It is not only the north that thinks. So there is no doubt to my mind, that the deeper the dialogue between the north and the south, the better for them both. But there has to exist in the north and in the south the humility of those who want to compliment each other rather than those who want to dominate.

I would like to hear more about this humility. I think the arrogance you mentioned is no longer universal. There are many people, especially middle-class people in my culture who, in place of arrogance, feel guilty and silenced by their privilege and their position in history. This is not humility, is it?
No it is not. I think that humility is a conscious position held by someone who has limits and who accepts their limits and who does not feel guilty when faced by another. It is the position of someone who feels able to create but able to respect the fact that the other also has to create. In other words humility is an attitude which makes me

know my limits and also my possibilities and which means I do not intend to impose the results of my dreams on others, nor put myself before others in a position of inferiority.

One of your great achievements is the way that you have combined theory and practice, and achieved a balance between them. I think a lot of people in my culture, who wish to see humane progress, either get engrossed in the thinking or in the direct action. Do you have any advice for people who are struggling to balance political theory and political action?
Look, I think it is impossible to separate theoretical reflection from practical action. In other words, in all practice there is always an intrinsic theory built in. Sometimes the person who acts, in fact, has no idea of the existence of the theory within their practice. What is necessary is that those who act are capable of attempting moments of reflection on their practice and on what lies behind their practice. It is necessary that those who theorise and reflect, see the theoretical thought that is present within their action. In the end it is necessary to avoid a rupture, a separation, a dichotomy between theoretical thought and concrete practice.

One of the problems that educational practitioners in Britain always seem to come up against is the way that the enthusiasm and positive energy of projects in their early stages frequently falls. There are internal fights and disillusionment. Do you have any advice on how to sustain positive energy in this work?
I don't think there are formulas for this. But I think it is necessary for those who have given themselves to a project to develop their persistence: the virtue of persisting in action even when the action is not bringing forth the results that were hoped for. If my action does not bring forth the results that were expected, instead of giving up I need to evaluate. I need to evaluate my actions and find out where I am going wrong. Is it or is it not my error? Is it or is it not the error of others? Is it possible or not to re-orientate the actions? All this is part of the worldwide process of evaluating practice. This, again, requires a humbleness among those who practice, and it is necessary that those carrying out the action have no fear of considering themselves

wrong. If you do something and have a fear of making an error and you do make an error, you want to hide your error and therefore are unworthy of your action. So I think all practice needs to develop, in its own practitioners and subjects, specific qualities without which the practice may fail. The quality of clarity. The capacity to analyse what you do. The capacity to compare the results being achieved and the results being dreamed. The ability to be realistic about how long things take. The ability to perceive deviations away from the desired path. All these are creative, fruitful qualities which develop understanding. This is why good practitioners have to be good thinkers. They have to produce new understanding as they act it out, and they have to be able to evaluate the efficacy of their practice.

Should this reflective evaluation be collective or personal?
I think evaluation should be collective. In other words those acting should evaluate their practice. But it can also be an individual experiment, so long as the individual's conclusions are brought for analysis to the group.

The National programme that you and others designed for Brazil in the 1960s was never implemented. Why?
A few elements of the programme were put into practice but then the whole project was simply aborted by the Military Coup. We were sent to prison for our programme, so we had no time to evaluate it!

If you could re-create the programme today, what form would it take?
Today, thirty years later, history has moved on and the emphasis would be different. We would have new things to do, to say; new ways of operating, new politics of action. From the point of view of the understanding of the role of the educational practitioner, the role of the literacy teacher, the role of the students learning to read and write, from the point of view of respect for the language and syntax of the student, from the point of view of understanding literacy learning as an act of creation, and in terms of ensuring that the literacy student is also the subject, it would all be the same.

I think that in the next millennium these things will remain constant in a progressive, democratic and humanist perspective. I say now, with more force than I said thirty or forty years ago, that the literacy student has to be seen as a creative subject of his or her process of learning to read and write, and not as a patient under the orientation of the educational practitioner. For this reason I emphasise a critical understanding of people's language. It's necessary to respect the syntax of the literacy student, which is also a syntax of his or her social class. It is from the universe of the student's thought and language that the process of literacy learning, the process of mastering the written representation of your language, should begin.

Paulo Freire 1921–1997

Born Sept 19th in Recife, Brazil; educator and political philosopher of education. During the early 1960s Paulo Freire developed his radical educational policy in Brazil and was involved in planning a national strategy for literacy.

Putting literacies on the political agenda

Brian Street
RaPAL Bulletin No. 13, Autumn 1990

Brian Street is from Sussex University and wrote this article for International Literacy Year. It also appears in Open Letter, the Australian Journal for Adult Literacy Research and Practice

1990 is International Literacy Year: According to the Task Force set up to coordinate activities around the world, the main objective is to create public awareness and "develop an atmosphere of positive attitudes towards the problem of illiteracy as a cultural problem and the need to tackle and combat it". Worthy sentiments but do they reveal serious flaws in the way that literacy is treated in public discussion? Do problems in the construction of literacy and "illiteracy" themselves lie at the root of many of the "problems" ILY is supposed to address? In their coverage of literacy issues, whether in the Third World or in the UK, both politicians and the press have a few simple stories to tell that deflect attention from the complexity and real political difficulties these issues raise. Attention is frequently restricted to scare stories on the numbers of "illiterates" both in the Third World and within "advanced" societies; patronising assumptions about what it means to have difficulties with reading and writing in contemporary society; and the raising of false hopes about what the acquisition of literacy means for job prospects, social mobility and personal achievement.

Campaigners as well as agencies and governments still make great play of figures that show, say "25% of the UK to be illiterate", or 25 million people in, the US; a reflex in assessing the degree of "development" in Third World countries remains their literacy "rate"; and UN statements highlight the increasing absolute numbers of "illiterates" in the world, and call meaninglessly for the "eradication of illiteracy by the year 2000". The figures are, of course, counters in a political game over resources: if campaigners can inflate the figures then the public will be shocked and funds will be forthcoming from embarrassed governments, or Aid Agencies can be persuaded to resource a literacy campaign.

Blaming the victim: literacy and employment

The reality is more complex, is harder to face politically, and requires qualitative rather than quantitative analysis. Recent studies have shown, for instance, that when it comes to job acquisition the level of literacy is less important than issues of class, gender and ethnicity: lack of literacy is more likely to be a symptom of poverty and deprivation than a cause. Researchers also point out that the literacy tests which firms develop for prospective employees may have nothing to do with the literacy skills required on the job: their function is to screen out certain social groups and types, not to determine whether the level of literacy skill matches that of the tasks required. Some employers, for instance, have the somewhat mythical belief that employees who have learnt literacy are less likely to be antagonistic to new technology, computers etc., and use literacy tests as a screen for these supposed attitudinal qualities. Whilst some individuals find that attendance on literacy programmes does lead to jobs they would not have got otherwise, the number of jobs in a country does not necessarily increase with literacy rates, so in many cases other people are simply being pushed out – those with literacy difficulties may be leap-frogging each other for scarce jobs. Governments have a tendency to blame the victims at a time of high unemployment and "illiteracy" is one convenient way of shifting debate away from the lack of jobs and onto people's own supposed lack of fitness for work. But many jobs require minimal literacy or a different kind of literacy skill than taught at school, and employers can sometimes teach these on the job fairly easily: lack of literacy skills is less often a real barrier to employment than the public accounts suggest.

Scribes and gatekeepers

Lack of literacy skills may also be less of a handicap in daily life than is often represented. The media likes to tell heroic stories of the "management" of illiteracy, how "illiterates" get around the city, or bypass written exercises like form filling or reading labels etc. The situation, however, need not be represented as though people are suffering from some disease or handicap. A researcher in the US, for instance, has shown how communities develop networks of exchange and interdependence in which literacy is just one skill amongst many being bartered: a mechanic without literacy skills may exchange skills in car maintenance for a neighbour's ability to fill in a form; a businessman may speak all of his letters into a tape for a friend to write up, much as medieval monarchs used scribes. In this situation the acquisition of literacy skills is not a first order priority at the individual level, so long as it is available at the community level. Many immigrant groups have found themselves in similar situations, with "gatekeepers" learning specific literacy skills relevant to their particular situation and often mediating with agencies of the host community. Amongst the Hmong of Philadelphia, for instance, sensitive literacy teachers have abandoned traditional, exam-oriented literacy teaching in favour of helping particular women develop sufficient commercial literacy to enable them to market their weaving and create an independent economic base.

Literacies, not literacy

Such examples have led researchers and practitioners to talk of "literacies" rather than of a single, monolithic "literacy". It is not only meaningless intellectually to talk of "the illiterate", it is also socially and culturally damaging. In many cases it has been found that people who have come forward to literacy programmes because they think of themselves as "illiterate" have considerable literacy skill but may be needing help in a specific area. This could be treated as no different from any potential students applying to educational institutions, whether adult education or university postgraduate work. When applying to a literacy programme, adults who have defined themselves as "illiterate" are often asked to read some written material so that their level and needs can be assessed: in one study in the U.S. they were asked to read texts produced by adult literacy students on the programme rather than "top-down" published material and many found, to their own surprise, that they could read them fairly easily.

Some, indeed, continued to read whole student magazines and publications ignoring the "test" aspect of the situation. Familiarity with content and context affected what were thought of as context-free, neutral skills in literacy decoding. At an adult education institute in the UK, students worried that they could not learn literacy properly because they "could not speak properly", that is their dialect or pronunciation differed from Standard English. The stigma of "illiteracy" is a greater burden than the actual literacy problems evident in such cases.

In many developing countries this stigma is still in the process of being constructed. People who have been accustomed to managing their daily lives, intellectual and emotional as well as practical and economic, through oral means have not required the elaborate definitions and distinctions associated with literacy and illiteracy in the west. In fact there are very few cultures today in which there is not some knowledge of literacy: children, for instance, learn to interpret logos on commercial goods and adverts, or to "read" television with its often sophisticated mix of script, pictures and oral language. Islamic societies have long been used to forms of reading and writing associated with, religious texts and with scholarly and commercial activities, whilst in other contexts people have developed their own "indigenous" writing system, used perhaps for specific purposes such as letter writing, sermons, love notes etc. Literacy campaigns, however, have generally ignored these local literacies and assumed that the recipients are "illiterate", beginning from scratch.

> *Even Paulo Freire, the most influential radical literacy campaigner, has tended to believe that people without western-type literacy are unable to "read the world"; his crusade to raise consciousness through literacy campaigns has been a leading challenge to dominant, authoritarian campaigns run by governments to do precisely the opposite, but it often rests on similar assumptions about the ignorance and lack of self awareness or critical consciousness of "non-literate" people.*

Taking hold of literacy

Recent research on the "reception" of western literacies in various parts of the Third World has attempted to shift the focus from the "impact" of literacy to the ways in which people "take hold" of a particular literacy; this stresses the active rather than passive character of the recipients. Frequently, it has been recognised, people absorb literacy practices into their own oral conventions, rather than simply mimic what has been brought. Where, for instance, there were rules in oral communication about not thrusting oneself forward, not offending others, but still getting your own way through subtle self-effacing uses of language, then the introduction of writing often leads to similar conventions being used for letters, political documents for love notes Similarly, linguistic and political self awareness is often expressed through subtle forms of speech making and oratory in which participants express the difference between the surface message and inner meaning in various coded ways. Again this often passes over into the communicative repertoire introduced by the literacy campaign. To judge by the kinds of texts and writing introduced by some literacy campaigns, it might appear that the recipients have a better sense of this than the campaigners themselves: far from being passive and backward illiterates grateful for the enlightenment brought by western literacy, indigenous peoples have their own literacies, their own language skills and conventions and their own ways of making sense of the new literacies being purveyed by the agencies, the missionaries and the national governments.

Rejecting the great divide

It is evidence such as this which has led both researchers and those involved on the ground in teaching on literacy programmes to revise the basic assumptions on which much literacy work has been conducted. The main shift in Literacy Studies that needs to be addressed now, in the public debates being stimulated by International Literacy Year, has been the rejection of "Great Divide" theory. According to this theory there is a "great divide" between "illiterates" and literates. For individuals this is taken to mean that ways of thinking, cognitive abilities, facility in logic, abstraction and higher order mental operations are all integrally related to the achievement of literacy: the corollary is that "illiterates" are presumed to lack all of these qualities, to be able to think less abstractly, to be more embedded, less critical, less able to reflect upon, for instance, the nature of the language they use or the sources of their political oppression. It appears obvious, then, that "illiterates" should be made literate in order to give them all of these characteristics and to "free" them from the oppression and "ignorance" associated with their lack of literacy skills.

At the social level great divide theory assumes that there is a difference of kind as well as degree between societies with mass literacy and those with only minority or elite literacy. For economic take off, it is claimed, a "threshold" of literacy is necessary for social progress: developing countries must therefore be brought up to this level (sometimes cited as 40% literacy in a population) in order for them to participate in the benefits of modernisation, progress, industrialisation and participation in the world economic order. Similarly, it is assumed that social groups that lack literacy but live within a country that is majority literate will be disadvantaged or "backward" and that their "illiteracy" is a major cause of this: give them literacy and they will achieve social mobility, economic and political equality and participation in the social order.

These ideas appear so obvious to common sense that it seems both foolhardy and perverse to challenge them. Yet work in the field of literacy studies during the 1980s has forced many researchers and practitioners in literacy programmes to revise their basic assumptions and to develop more complex theories of literacy that reject "great divide" thinking. These new theories have important implications for policy and practice in this field, but do not always get adequate attention in public discussions.

Literacy, schooling and language

Researchers investigating the cognitive consequences of literacy, for instance, have come to recognise that what is often attributed to literacy per se is more often a consequence of the social conditions in which literacy is taught. Literacy needs to be distinguished from education. Some kinds of education, though by no means all, may well inculcate critical self-awareness and facility with abstract concepts, but this is not so much to do with inherent characteristics of literacy as with the character of the programme. There is massive counter-evidence of literacy learning in a variety of contexts, in which abstraction, logic and critical thought play little part; one cannot therefore theorise that literacy in itself will lead to these things. Conversely, anthropological evidence demonstrates that self-reflection and critical thought are to be found in supposedly non-literate societies. One crucial variable, for instance, in child development is what psychologists refer to as "meta-linguistic awareness", the degree of self- consciousness about language. Many educationalists and developers have assumed that the acquisition of literacy is a key factor in developing this awareness. Anthropologists and sociolinguists, however, have questioned this and suggested that the focus on literacy as though it were a neutral, "autonomous" variable deflects attention away from more complex social variables. One such variable, for instance, may be the number of languages spoken in a region. Where people are in contact with or themselves speak a variety of different languages, they are likely to have developed a language for talking about language, to be aware of the character of different kinds of speech (and writing) and of the subtlety of meanings in different contexts. Play on figures of speech, skill in rhetoric as well as ability to develop and appreciate different genres are all features of so-called oral societies.

Anthropologists who have worked in these contexts soon find that "great divide" thinking is no help in understanding the complexity and subtleties of issues such as mesa-linguistic awareness, which nevertheless dominate much contemporary educational practice and thinking. Differences in individual cognitive skills, then, are more likely to stem from such differences in social and cultural experience than from the presence or absence of literacy.

Literacy and social context

Literacy itself, moreover, varies with social context. It is difficult to lay down a single objective criterion for a skill that is nevertheless widely represented as the key to individual and social progress. In medieval England, for instance, ability to read Latin earned the label "literatus"; in later periods, the major test was ability to read a prayer; many academic studies until recently used the ability to sign a marriage register as an index of both individual literacy and for assessing literacy rates in the whole society; in many developing countries the returns to international agencies for the nation's literacy rate rest on numbers of children completing the first three, or five, grades of schooling. In the UK too the goal posts have been shifting and what was considered adequate literacy at the turn of the century would merit the label "illiterate" by great divide standards today. Recognition of the variable and contested criteria used to define literacy and illiteracy is probably more widespread now and one of the aims of International Literacy Year will be to develop popular sensitivity to the contextual nature of literacy skills. But the implications of this for broad pronouncements on literacy rates, proportions of "illiterates" in a population etc., have not always been recognized.

Everyone has literacy difficulties

Misconceptions about their own "illiteracy", for instance, continue to debilitate many adults in situations where the "stigma" derives from a mistaken association of literacy difficulties with ignorance, mental backwardness and social incapacity. When it is discovered that many who are labelled in this way, whether by themselves or others, have in fact minor difficulties with spelling, decoding, sentence or paragraph structure (or simply non-standard pronunciation!), it seems remarkable that so much cultural and emotional weight could have been placed on them. One reason has been the prevalence of great divide theory: if these difficulties are associated with the category "illiteracy" and that category is associated with lack of cognitive functions, backwardness etc., then the stigma is inevitable. If, on the other hand, they are located in a theoretical framework that assumes there to be a variety of literacies in different contexts, no one line between literate and illiterate, and a range of cognitive and social skills associated with orality and literacy equally, then the agenda shifts and the stigma becomes meaningless. Everyone in society has some literacy difficulties in some contexts: (the classic "middle class" difficulty is with income tax forms but in this context the emotional charge is drawn off through humour and jokes rather than reinforced through categorical labelling and stigma).

Questioning the framework

Current theory, then, tells us that literacy in itself does not promote cognitive advance, social mobility or progress: literacy practices are specific to the political and ideological context and their consequences vary situationally. This does not lead us to abandon attempts to spread and develop the uses and meanings of literacy: it does force us to question whether the current framework in which such activities are conducted is the most fruitful. The political task is then a more complex one: to develop strategies for literacy programmes that address the complex variety of literacy needs evident in contemporary society. This requires the policy makers and the public discourses on literacy to take greater account of people's present skills and own perceptions; to reject the dominant belief in uni-directional progress towards western models of language use and of literacy; and to focus upon the ideological and context specific character of different literacies. International Literacy Year should be used to help open up this debate and establish clearer. Concepts and frameworks on which practice can be based, not to reiterate worn clichés and patronising stories about "illiteracy".

References

Fingeret, A. (1983) "Social network: a new perspective on independence and illiterate adults", in Adult Education Quarterly, 33(3), 133–146.

Graff, H. (1987) The Legacies of Literacy: continuities and contradictions in western culture and society, Indiana UP: Bloomington, Indiana.

Levine, K. (1986) The Social Context of Literacy, London, Routledge & Kegan Paul.

Scribner, S. and Cole, M. (1980) The Psychology of Literacy. Harvard University Press.

Street, B. (1985) Literacy in Theory and Practice, Cambridge University Press.

The futures of literacy

Gunther Kress
RaPAL Bulletin No. 42, Summer 2000

Gunther Kress is a Professor of Education at the University of London Institute of Education

Some preliminary remarks

In this paper I will attempt to deal with what I regard as perhaps the central issues relating to literacy at the moment and into the next two decades. My approach is to look at what I call, using somewhat non-technical terminology, the "stuff" of literacy. By this I mean the forms, the texts, the shapes, the very materiality of literacy – for instance, the fact that speech uses sound, the physical stuff of air being perturbed, and that writing uses light and graphic means, surfaces, and tools for inscribing surfaces. This very materiality has consequences, in my view, for what cultures may or may not easily do with the stuff. That is intended to set it off from what is currently the dominant approach, which is to look at what it is that people do with the "stuff". I think that *what people do with the stuff shapes the stuff* but I also think that the very materiality of the stuff, *what it is like as material,* what any one culture has done with it, how a culture has shaped it, has deep effects on what people may do with it.

The position I take is therefore somewhat uncomfortably between those who see the forms and structures of language-as-writing as simply *there,* as given, and demanding that they be used as they are (in fact seeing no real possibility of human social intervention), and those who see that nearly anything may in fact be done with the forms, the materials of 'literacy'. I think there is a constant tension between two forces. On the one hand there is the shaped stuff that our culture(s) have provided for us – itself the product of the socially located work of people in innumerable acts of communication always in specific settings, with their demands, their structures, their participants, their relations of power, and their histories. On the other hand there is the new, and for me always transformative action of individuals using the shaped stuff, but reshaping it – I will say "transforming it" for the rest of this article – in the light of their interests in the moment of interaction.

If I had to characterise my own interests in this, I would say that I want to have a theoretical position in which 1) the always transformative actions of socially formed humans can be recognized and understood; and 2) which sees at the same time the factors and the forces which act as constraints on that transformative action. Because my background was in linguistics and is now in semiotics ("the science of the life of signs in society" as Ferdinand de Saussure described it at the beginning of the last century), I focus on the constraints that arise out of the shape of the stuff with which we make our representations, our messages. If my background had been a sociological one, I would no doubt focus on the constraints which arise as an effect of social factors: the structurings of power and of its effects; the effects of cultural environments; of age as a social and as a biological category; of sex and gender; of class, still; of profession; of region; of ethnicity; and of many others. I am entirely interested in these issues, and I think that it is these factors which have the fundamental shaping effect; but I am also interested in the sedimentation – the layering effect of the social meaning – making work of those who reshaped the resources – which we have for representing ourselves to ourselves and to others; the matter that, broadly speaking, we call "literacy".

A terminological issue

We are at the moment at one of these points in human history where a real revolution is taking place. It is not the sort of revolution that freed the Americas from British colonial rule, or led to the formation of the French Republic, or to the establishment of the Soviet Union in the last century. It is a revolution of a profounder kind, which is reshaping the basic arrangements of social and cultural life in a more thorough-going way; and this revolution is changing communication, its possibilities and forms. Among other changes, it is affecting the role of writing as the hitherto central means of communication in many domains of public life. As I shall show later, image is taking the place of writing in many cases, or is jostling into an equal position with writing in equally many places, and texts. It is that which raises the issue of terminology: shall we call all forms of public, durable and storable communication 'literacy' – not to speak of the innumerable extensions of the metaphor of literacy to nearly any area of human practice (as in emotional, cultural, IT, sexual, literacy) – or shall we retain 'literacy' as a label for what I shall call "lettered representation"?

My preference is for "literacy as the term for representation through letters". I won't argue the case here, but will do so as and when it arises in this paper. For quite different reasons I will also avoid the pluralising – a very common contemporary practice – of the term to "literacies", because I think that it has consequences in practice and action which I would like to avoid.

In this paper I will focus on six issues:

1. the contexts of thinking about literacy;
2. practical and theoretical issues around (the stuff of) literacy;
3. the emergence of image and the shift from page to screen;
4. frames and genres: from text to connectivity;
5. multimodal communication and the design of texts;
6. educating for instabilities: literacy curricula for unknowable futures.

1 The contexts of thinking about literacy

"Literacy" always relates to and is a product of the economic, social, technological and cultural conditions of the society in which it is used. It is essential in any discussions of the futures of literacy to look at where we have been, and why the forms of literacy are as they are.

At the moment all the discussions around futures of literacy are focussed on the new technologies and their likely effects. At the same time there is a strong element in public debates which quite stridently advocates a return to traditional forms of literacy. All this ignores the fact that in Europe the traditional forms have grown out of a long era which has now come to an end.

The feudal agricultural economies, which were still recognisably present into the middle of the last century in many parts of Europe (and depicted in the paintings by the Breughels, elder or younger), could be seen alive and active well into the 1950s in central Europe, and into the 1970s in southern Europe. However, that era has been overlaid, during the last two hundred years, by the period of industrialisation, of the industrially-based nation states, of mass societies with their mass bureaucracies, of mass-production, and of mass communication and the mass media. That era has bequeathed us the forms of literacy which we have come to regard as natural, as normal, as essential. These are forms of literacy which served the era well, which were essential for the successes of that era, (and which were implicated in its problems). They are not likely to serve us well for the demands of the new era that we have entered.

So what were the shapes of that past, and what are the discernable shapes of the near future? It is clear, as I have suggested, that we are moving out of an era of relative stability of a very long duration. We tend to be focussed on the industrial revolution and its effects in so many ways in our debates on literacy. But we also know that the invention of the printing press predates the early period of industrialisation by a good two centuries; and indeed it is often held that the invention of printing using moveable type represents the first stage in the process of mass industrial production. The significant point however is that when the printing press became commonly available, and replaced the medieval scribe – whether at the court or in the church – it was the forms of writing of the medieval scribe which came to dominate the new technology. And when some one hundred years later English had become the language of public life (at the court, in the judiciary, in commerce, and in some areas of intellectual life), it was the spoken forms of the educated elites and of their public forms of speech, which became the basis for written forms. The page-long paragraphs of the writings of Hobbes or Newton or Milton, with its paragraph long 'sentences', were the result of the mixing of the learned grammars of Greek and Latin and the structures of public oratory.

Below is a brief example: it comes from John Milton's tract against censorship, *Areopagitica*. Its sentence structures and cadences are influenced by the structures of written Greek and Latin, both of which an educated person was expected to be competent in; but these sentence forms come from the forms of speech as public oratory.

Extract from Milton's Areopagitica

Good and evill we know in the field of this World grow up together almost inseparably; and the knowledge of good is so involved and interwoven with the knowledge of evill, and in so many cunning resemblances hardly to be discern'd, that those confused seeds which were impos'd on Psyche as an incessant labour to cull out, and sort asunder, were not more intermixt. It was from out the rinde of one apple tasted, that the knowledge left forth into the world. And perhaps this is that doom which *Adam* fell into of knowing god and evill, that is to say of knowing good by evill. As therefore the state of man now is; what wisdome can there be to choose what continence to forbeare without the knowledge of evill? He that can apprehend and consider vice with all her baits and seeming pleasures, and yet abstain, and yet distinguish, and yet prefer that which is truly better, he is the true warfaring Christian. I cannot praise a fugitive and cloister'd vertue, unexercis'd & unbreath'd, that never sallies out and sees her adversary, but slinks out of the race, where that immortall garland is to be run for, not without dust and heat. Assuredly we bring not innocence into the world, we bring impurity much rather: that which purifies us in triall, and triall is by what is contrary.

That vertue therefore which is but a youngling in the contemplation of evill, and knows not the utmost that vice promises to her followers, and rejects it, is but a blank vertue, not

> a pure; her whitenesse is but an excremental whitenesse; which was the reason why our sage and serious Poet *Spencer*. whom I dare be known to think a better teacher than *Scotus* or *Aquinas*, describing true temperance under the person of *Guion*, brings him in with his palmer through the cave of Mammon, and the bowr of earthly blisse that he might see and know, and yet abstain. Since therefore the knowledge and survay of vice is in the world so necessary to scanning of error to the confirmation of truth, how can we more safely, and with lesse danger scout into the regions of sin and falsity then by reading all manner or tractats, and hearing all manner of reason?

But the printing presses had by then become ubiquitous in cities like London: one of the reasons why the government of the day attempted to control their use. This medium could now be available to the lower classes. as a means for their expression, representation and communication. Here is another brief example, a religious rather than a political tract this time, roughly contemporaneous with Milton's text.

Am *Anna Trapnel*, the daughter of *William Trapnel*, Shipwright, who lived in *Poplar*, in *Stepney* Pariſh; my father and mother living and dying in the profeſſion of the Lord Jeſus; my mother died nine years ago, the laſt words ſhe uttered upon her death-bed, were theſe to the Lord for her daughter. Lord! Double thy Spirit upon my child; Theſe words ſhe uttered with much eagerneſs three times, and ſpoke no more; I was trained up to my book and writing, I have walked in fellowſhip with the Church-meeting at *All-hallows*, (whereof Mr. *John Simpſon* is a Member) for the ſpace of about four years; I am well known to him and that whole Society, alſo to Mr. *Greenhil* Preacher at *Stepney*, and moſt of that ſociety, to Mr. *Henry Jeſſe*, and moſt of his ſociety, to Mr. *Venning* Preacher at *Olaves* in *Southwark*, and moſt of his ſociety, to Mr. *Knollis*, and moſt of his ſociety, who have knowledge of me, and of my converſation; If any deſire to be ſatisfied of it, they can give teſtimony of me, and of my walking in times paſt.

Anna Trapnell

However this is written by someone not trained in Greek and Latin, and not versed in the elite forms of public oratory.

It is clear that just like John Milton, Anna Trapnell draws on the forms which were available to her: they are all forms which she knows as speech (though she would also have *read* the Bible):

- *the speech of her parents* (a lower class English): "my mother died nine years ago . . . I was trained up to my book and writing";
- *the speech of the preacher*: "my father and mother living and dying in the profession of the lord Jesus", heavily influenced by the bible;
- *the speech of the religious community* of which she had become a member: "the last words she uttered on her death bed were these to the Lord for her daughter . . .".

The point that I wish to make here is that we always draw on the resources which we have available to us, for the purposes of making the representations that we wish or need to make. In the process the existing resources are transformed, reshaped in the direction of the requirements of the environment of communication and by the interests of the maker of that message/representation. That applied to Milton as much as to Trapnell, even though the resources available to them were very different; and the social valuations of these resources unequal. Nevertheless, the processes were the same for both, and are still the same for anyone engaged in communication In both texts, that of Milton and that of Trapnell, we see the emergence of the written sentence, out of the resources of speech of different kinds (and of the grammars of Greek and Latin in the case of Milton).

But there are two other points to be made here. One concerns technology and its effects and influences. The printing press, with its moveable type had superseded the scribe and his practices: but the traditions left by the scribe and the forms of both the elites and the non-elites immediately colonized the new medium. The other point goes somewhat in the opposite direction: yes, the old resources colonized the new medium; but at the same time the possibilities offered by the new medium – its affordances (what the medium makes it possible to do, or what it makes difficult) reshaped the resources. The printing press had its effects on writing and on the sentence. The written sentence is as much an effect of the affordances of that medium in interaction with the users and the environments of use, as it is an effect of the resources which were brought to writing for print. These are essential points to be borne in mind in thinking about literacy at a time when the effects of technology are again overwhelmingly present.

Framings of literacy

The forms of literacy which I know come from that past era. In their precise and specific form they come from the requirements of the European nation state, which in its western European form and structures has lasted for at least one hundred and fifty years. They are being brought forward into an era of instability. The frames which had provided stability for social institutions and for their forms of literacy are dissolving or have already dissolved; more strongly, by and large, in anglophone societies than in non-anglophone societies. These were the stabilities of the nation state, of its economies, of its structures, and of its values.

These framings have all begun to weaken, to soften, to dissolve; and many factors are involved in that. The migrations of the last century have produced societies which are insistently plural, culturally. Changes in forms of transport – whether transport of information or of goods or people – have eliminated distance as an overwhelming fact of life. Changes in the global economy largely brought about by that change in the possibilities of transport have meant that capital is both global and free of the restrictions of time: capital can be anywhere at any time. Culture has also been freed of geographical and social place by the possibilities of transport: I can receive the products of any culture in the place where I am, provided that I am a part of the affluent section of society. Culture has, in that sense, become severed from locality; it too has become global. Technologies are involved in all of these though technology is never the single cause for any of these changes.

Forms of literacy which had developed around the requirements of 'fixing' things around the requirements of stability, and the human dispositions fostered by those requirements, will become increasingly irrelevant and even dysfunctional. One clear indicator is the increasing changeability and fluidity of genres. Genres are the shaping features of texts, arising from the characteristics of the social structures and settings in which texts are made. An interview is a clear example: in a social structure where someone has the right to ask questions of someone who has information to provide in response to that question, the genre of 'interview' realizes that social arrangement. But we know that interviews change, and are changing constantly: just as we know that advertisements existed in the past in magazines, as texts that were formed out of the social relations of seller and buyer, which led to texts in genres of a particular kind. Some twenty years ago or so there were quite clear demarcations between texts in which the reader of a magazine was addressed as the reader – of a feature article, let's say, or of an advertisement. These distinctions now hardly exist: it is difficult (or one should say, really, pointless) to look for that distinction. The social structures have changed in ways which no longer make that distinction, and so the generic difference is going also.

THE TIMES

Newspaper front pages – circa 1959.

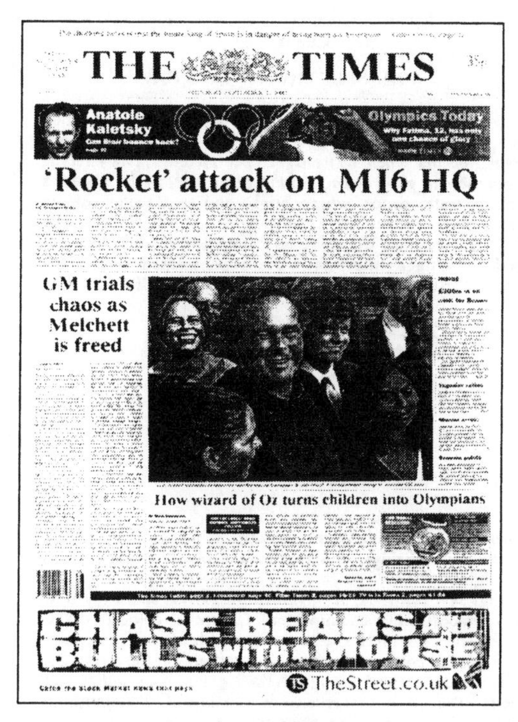

THE TIMES

Anatole Kaletsky

Olympics Today

'Rocket' attack on MI6 HQ

GM trials chaos as Melchett is freed

How wizard of Oz turns children into Olympians

CHASE BEARS AND BULLS WITH A MOUSE

TheStreet.co.uk

circa 2000.

This brings with it severe challenges for those whose modes of (literate) being were formed in that former era. In that era, writing of quite specific kinds occupied a central place in hierarchies of communication.

In the newspaper of 1959 writing is the central mode of communication: it was simply impossible to think that writing might be other than *the* mode of communication. Class was not an issue in this: even the papers which were for the working class had pages filled with writing: indeed from that point of view there was seemingly less of a difference between classes.

The forms of writing were closely related to the social and economic structures which they served and in whose service they had (been) developed. One can think here of the heavily formal styles which so typified bureaucratic writing, scientific writing, academic writing, but also journalism of a 'serious' kind, etc. These forms of writing mirrored the social, economic and political structures of that

CHAPTER 34

THE SPECTRUM. COLOUR

34.1. Analysis and synthesis of white light. Mention has already been made, in 31.9, of the dispersion of colour which occurs when white light is refracted. The effect is noticeable in glass prisms, in cut gems such as the diamond and in water drops, where bright and spectacular colours seem to have been added to the light. It was Newton who first showed that the material concerned in this does not add anything to the light. He experimented with a triangular glass prism, allowing a shaft of sunlight to pass

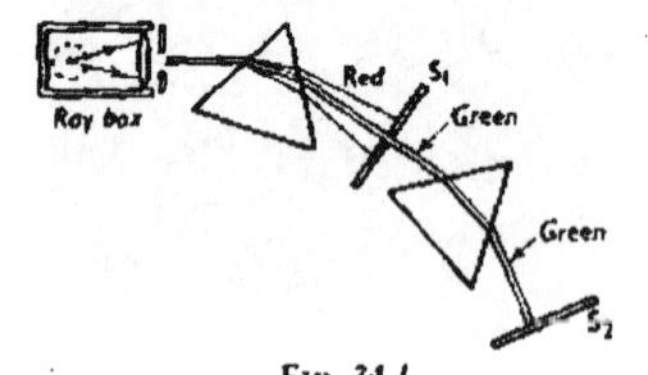

FIG. 34.1

through a small hole in the shutters of his rooms in Trinity College, Cambridge, and placing the prism in the path of the light.

Fig. 34.1 shows an arrangement of prisms and a ray-box, for demonstrating an important feature of coloured light. The first prism gives rise to a coloured beam which appears as a band of colour, known as a **spectrum,** on the screen S_1. The colours are red, orange, yellow, green, blue, indigo and violet—though not everyone is capable of distinguishing between blue and indigo, the latter being a navy-blue. If a

507

Book content pages: circa 1950s

era, and in their form they encoded knowledge in specific ways: as stable, as orderly, as sequentially unfolding, as tightly framed. The book was the medium which quintessentially enshrined both the form of writing and that form of knowledge. The organisation of the book – in its chapters, its subsections and paragraphs, organised around *the potentials of the page, reflected* that organisation in all its aspects.

The structure of the book, with its chapters, each of which contained a discrete part of the 'body of knowledge' set forth in the book, each following in an essential and necessary form on the preceding one, realised the structure of the knowledge which was being 'set forth': orderly, sequenced, coherent.

The genres of writing supported these arrangements. In a narrative (whether as short story, or as a novel, or as the report of an experiment), the *setting* clearly precedes the *complication,* which

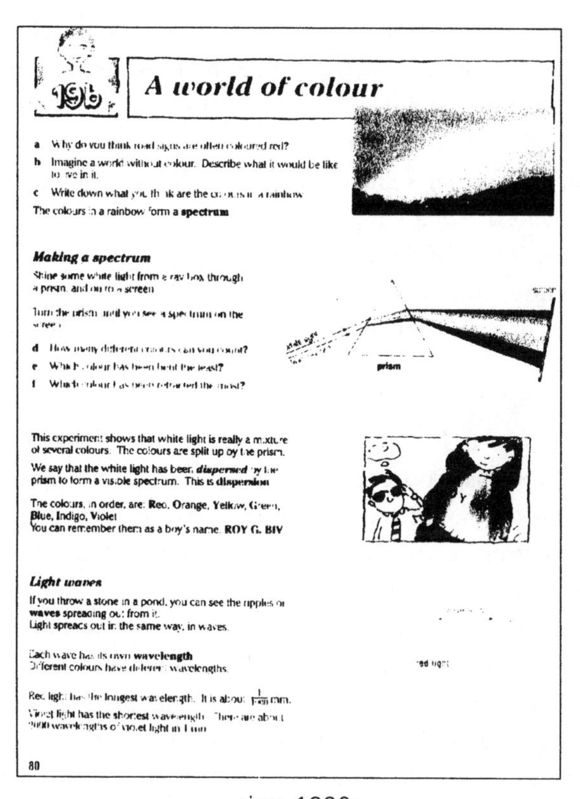

A world of colour

a Why do you think road signs are often coloured red?
b Imagine a world without colour. Describe what it would be like to live in it.
c Write down what you think are the colours in a rainbow.
The colours in a rainbow form a **spectrum**

Making a spectrum

Shine some white light from a ray box through a prism, and on to a screen

Turn the prism until you see a spectrum on the screen.

d How many different colours can you count?
e Which colour has been bent the least?
f Which colour has been refracted the most?

This experiment shows that white light is really a mixture of several colours. The colours are split up by the prism.

We say that the white light has been ***dispersed*** by the prism to form a visible spectrum. This is **dispersion**

The colours, in order, are: **Red, Orange, Yellow, Green, Blue, Indigo, Violet**
You can remember them as a boy's name. **ROY G. BIV**

Light waves

If you throw a stone in a pond, you can see the ripples or **waves** spreading out from it.
Light spreads out in the same way, in waves.

Each wave has its own **wavelength**
Different colours have different wavelengths.

Red light has the longest wavelength. It is about $\frac{1}{[illegible]}$ mm.

Violet light has the shortest wavelength. There are about [illegible] wavelengths of violet light in 1 mm

80

circa 1990s

clearly precedes the *resolution* of the story, just as clearly as Chapter 1 preceded Chapter 2. Genres were (relatively) stable, and the stability of the textual/ linguistic form underpinned, guaranteed nearly, the stability of the structures of knowledge. These kinds of order are now under the severest challenge, and with them, the organizations of knowledge which they had supported.

2 Practical and theoretical considerations around the stuff of literacy

In this section I will look at some of the specific notions, categories, concepts and forms which need to be considered in serious discussions of literacy. I will also comment on the kind of theory which will be necessary for new thinking in this domain.

I will start with some theoretical issues. Let me start with the issue of grammar, and contrast it with the notion of resources. The term grammar has a range of meanings:

- the language user's knowledge of the rules which underlie competent use of a language;
- the description of these rules by a linguist;
- the codification of these rules;
- the conventions which are insisted on in relation to such rules; etc.

The meaning which I will focus on here gathers up the first two 'definitions', that is, it treats grammar, broadly speaking as that knowledge which underlies what people do when they speak or write, and which may be relatively adequately represented by a linguist's description of that knowledge, as a set of elements and the rules of their relation to each other. The commonsense view during the last century had been that this knowledge gets internalized in the processes of language learning, (although in Chomskian theories of language that knowledge is already *there* in the heads of children when they are born. as innate knowledge, which is activated by the exposure to and experience of specific languages). The knowledge is social knowledge: it is not individually produced, and it is not alterable by individuals in their use of that knowledge. It is also knowledge which is 'policed': conventions of language use mean that divergence from the rules brings with it sanctions of various kinds.

The 'system' of grammar (or of language) is therefore seen as stable, and beyond the influence of individual interaction. The metaphor which has been dominant in relation to the learning of language has been that of 'acquisition': you acquire language as you might acquire some other possession, a car perhaps or a sofa. There is no interaction between object acquired, and the acquirer. The object or phenomenon is there, and you take it into yourself, or adapt yourself to it.

In contrast, I want to propose an entirely active conception, one in which the learner changes the stuff which (s)he uses in learning, and in the process changes him/herself. So when a child of three, struggling up a steep grassy slope, says: "that's a heavy hill", the child is not simply "acquiring" a new term, *heavy;* or, in another approach to this example, 'just making a mistake'. Rather, the child is shaping for himself the potential uses of 'heavy'. He uses the resources which he has available to him (he does not know the word 'steep' at this stage) to express the meaning: 'it takes real effort to get up this hill!' The best available word-resource he has at that point is the word 'heavy', and so it serves the purpose which he has. I want to say that this is not an unusual, or a merely childish example, but that it is the way in which language is learned, always, and meaning is made, always.

But this is a highly interactive notion both of language learning and of meaning making (and one which makes, at that level, no real distinction between them): it treats the learner and the meaning-maker as transformative of both the stuff that is learned, and used, in meaning making, and of his own 'inner' resources. It makes grammar into a resource which is constantly remade/reshaped/transformed as a result of the actions of the users/transformers of the resources. That reshaping is not accidental, anarchic, or arbitrary: the child's use of "heavy" was not an accident, nor was it arbitrary. It reflected his 'interest' in the situation, and an attitude towards the activity he was engaged in, an expression of what he felt: the effort required. The interest reflected his 'take' on the world at this moment, his position in the world. And that too is characteristic of all meaning-making which always reflects my interest at *this* point, at *this* moment: an interest which is shaped by my social history, my social place, and the characteristics of the environment in which I am making my message/meaning.

Grammar as resource is therefore the record of the interested and transformative action of all language users. The stuff of language records the history of the interests of those who have shaped it. That applies to literacy as much as it applies to all aspects of language use. In the case of literacy I am speaking of the shaping of the resources of writing in exactly the same way. This is totally different to the common sense 'use of elements without change to the shape of the elements' of which I wrote above. The resource of grammar is constantly in flux, constantly changing, because those who use it as a

resource for communication constantly change it in their use. *There is no stable grammatical system, and there is no fixed grammar.* At times there are attempts by those who have an interest in doing so to slow the pace of change, but that always remains an attempt: it can never succeed because it goes against the basic principles of human meaning making. For myself I now prefer to speak, in a somewhat Hallidayan fashion,
of the resources of language-as-writing (and the resources of language-as-speech), rather than of 'the grammar' of writing. Of course these resources are organized, and of course they display regularities, 'rules', which come from their histories of use, and are understood – even if in quite different ways – by those who use these resources. But I find it difficult now to speak of 'the grammatical system' as that seems to me to imply stability, rigidity and boundaries which I know are not there.

But if we think of grammar as a resource which is constantly remade, then some other things follow: for one thing, as I said, the individual 'language user' is agentive and transformative; individuals are constantly innovative and creative in their relation to the resources of 'their' language. That realization has profound consequences for education, because it requires a change in our view of what learning is, and of what learners are and do. Building a curriculum of language on that notion and seeing students as constantly innovative and creative places them in a very different position from one based on an 'acquisition' theory of learning: the former moves away from a focus on insufficiency, on incompetence and error, to a focus on the students' interests, their motivations for the uses of *these* forms rather than *those* others, a sense of all students as creative and innovative in their linguistic action.

This different sense of learning and of learners, which places transformation, creativity and innovation at the centre, applies to all language-learning contexts, to all contexts of linguistic activity, no matter what the age of the learners or the meaning-makers. That is one, and for me essential, consequence. The other is somewhat different. It concerns the issue I raised above, that of the pluralizing of the term 'literacy' to 'literacies'. The motivations for that move are clear enough: to avoid reifying the notion of literacy, avoid seeing and describing it as a stable thing, to place emphasis on the innumerably different forms of literacy in all the contexts in which it appears. And yet I believe that despite its correct identification of a political problem, as in literacy (and "illiteracy") programmes, it is based on the very same mistaken notion of language, and of grammar as a stable system, which has given rise to the problem in the first place, and which the pluralizing form then attempts to correct.

The move, it seems to me, is somewhat like this: we can see that in all the contexts in which we look at the uses of literacy, people do all sorts of different things with 'it'. In order to give recognition to these recognizably distinct things we name them, we give them different labels, thereby saying: this is a distinct form from that; let us recognize its distinctiveness in order to recognize the value of what this group is doing (rather than allowing some extraneous version of literacy or language to be imposed on them). Each of these forms has its own 'grammar', its own logic, its own histories, its own uses and functionalities. But if we realize that language and therefore grammar *are* always and constantly reshaped by those who make use of these resources, then there are no different grammars, there are no different literacies. Rather there is the fundamental human fact of meaning making: the constant transformation of resources in line with the interested action of those who use the resources to give shape to their meanings, in this form here and in this other shape there.

The pluralization of the term *literacy* entrenches,without wishing to do so, the very theory which had given rise to the problem which is to be counteracted, solved, overcome. This is not an uncommon strategy in social and political matters; but for me the solution is to make the theory fit the practices.

Two examples of 'transformation'

The first is a brief extract from a history of Australia, written by an Australian historian, Geoffrey Blainey, *A Land Half Won* in 1972.

Extract from A Land Half Won

In Central Australia . . . the Pitjantjatjara were driven by drought to expand into the territory of a neighbour. Several of these invasions might be partly explained by a domino theory: the coastal invasion of the whites initially pushing over one black domino which in turn pushed down outer dominoes. But it would be sensible to believe that dominoes where also rising and falling occasionally during the centuries of black history. We should be wary of whitewashing the white invasions. We should also be wary of the idea that Australia knew no black invasions. Even when Aboriginal tribes clung to their traditional territory, fatal fighting within the same tribe or between members of hostile tribes was common. It is possible that many tribes suffered more deaths through tribal fighting than through warfare with the British colonists in the 19th century.

Blainey was at that time regarded as a politically progressive historian, and so he did not wish to perpetuate the earlier formulation that had been used of the process of colonization, which had been that of "settlement" with its implication that there had been an empty land waiting to be "settled".

The term which had then come into use as an attempt to name the reality of that history was that of "invasion": a hostile, violent, aggressive act against people and their land. Blainey softens that term in two directions (apart from implicating black Australians in the process also), by introducing the much more neutral terms "expansion" and "expand" and by altering the more usual syntax of the verb *invade,* and of the noun *invasion.* In its more usual usage, *invade* takes as an object-noun a word that stands for a political or social entity of some kind. I can invade something which belongs to someone politically or socially: I can invade someone's property, land, territory, privacy, or the pitch at a cricket game, etc. But I can't, in the usual usage, invade something that is open, not owned, that belongs to everyone, something that is natural rather than social. But that is what Blainey does: ". . . the coastal invasion of the whites . . ." is changing the grammatical/syntactic scope of the verb *invade* and of its derived noun *invasion;* it is now an act ("the whites invaded the coast") which can take place in relation to a new category: something which is merely a geographical, a natural thing, not a political/social one. Similarly, to use a descriptive adjective such as *white* with the word *invasion* has the effect both of obscuring real agency (not: "the British invasion"), and allows, of course, the ideologically convenient act of equating white and black invasions (where "black invasions" had been caused by *natural* events, such as droughts, implying, perhaps, a natural cause for the "white invasions" also).

The syntax of a small part of the language has been changed; and the interest of the transformative agent, the historian, is quite clear. A historian who wished to appear as politically progressive, but whose instincts/interests were deeply against such a move, transforms the existing resources of the language to serve the directions of his interest. The fact of the normal invisibility of this process should not deflect us from recognizing its usuality, its utter normality.

My second example goes into a somewhat different direction. It consists of two linked texts: a *Position Description* and a *job application* based on that description.

Position Description and Job Application

Position Description

Responsible for: The supervision of office staff providing administrative services to the academic staff. The provision and coordination of all student and student related activities within the Institute.

1 Co-ordination and supervision of the office staff providing administrative support.
2 Supervision of the attendance/flexitime system for all Institute staff.
3 Co-ordination of student enquiries and related activities.
4 Assist in organisation of student admission/enrolment/registration/assessment.

5 Prepare documentation for submission to the Institute's Admission and Progression Committee and act as an Executive Officer to the Committee.
6 Ensure the accurate maintenance of student records.
7 Preparation of correspondence relating student records/ progression/transcripts.
(*Numbering of paragraphs as in the original*).

The points I wish to make here are 1) about the productive potential of the resources of the language; 2) the reproduction of the resources in the form in which they have been received, and what that might mean; and 3) the effects of power in this.

To take the first point first: The PD is an example par excellence of the heavy, nominalized (that is, verb-based, dynamic forms made into noun-based static, frozen forms) language, with nouns of inordinate length *and* complexity. The opening nominal is the best (or worst) example:

"The supervision of office *staff providing administrative services to the academic staff" is* one single nominal, that is, an element which acts as a single noun. Other examples abound:

> *"co-ordination and supervision of the office staff providing administrative support", etc.*

These are noun-entities formed by a productive linguistic process, in response to certain social processes and structures. These social processes and structures occur regularly and frequently over a period, so that they come to seem not like new *events* each time, but as the existence of a stable *phenomenon* outside time – which is always like that, best represented by a noun-like linguistic entity.

A more event-like representation would be something like this:

> *"The successful applicant will supervise those staff who work in the office, who provide services to the academic staff. "*

This description is not particularly elegant, but nor is it very much longer than the nominal form: so efficiency of space, a usual reason given, cannot be the reason. But the nominal form has much more authority, because it is the name of something which exists as a thing (rather then in the more verb-like form which is a description of events that happen); above all. as a stable thing it *can* be administered, something difficult to do with events.

The productive potential of the resources of writing here lies in being able to turn event-in-time into object-out-of-time; to turn events with human participants into entities with no real overt trace of human presence; to turn the world of human social endeavour into the world of general categories. We might feel that the word "productive" is somewhat misapplied here: but the point is that that would not be the bureaucrat's view of it. This is the technology that enables the bureaucrat – and many others: the scientist, the policy- and the law-maker, etc. – to turn the messy world of events and actions into the stable, unchanging, orderly world of entities. It is and has been an essential technology in the era of the industrialized economies and of their social and political structures.

Job Application

In all positons held, good oral and written communication skills have been essential in satisfying job requirements. Communication at all levels from students to company executives, to College Principals has required clear concise expression together with attention to confidentiality and sensitivity. Supervisory and management skills have been developed over my career. Most recently in Student Administration, it has been my responsibility to form work teams, oversee work flow and set short term goals to meet deadlines. Immediate responsiveness to client/student enquiries has been required in previous positions, whilst the planning and organisation of day to day business was carried out. Skills of resource organisation and decision making were quickly acquired. Experience with computer systems has been gained whilst working with . . .

The applicant for the job has a *difficult task;* she recognizes, from the language of the PD, how this institution represents itself, and feels that she needs to approximate to that language in her application – to show that she understands and is in sympathy with this institution. At the same time she needs to show some signs of 'individuality', which is, on the face of it, one of the characteristics looked for: nobody wants (nobody wants to admit that they want) an employee who is merely a clone of the institutional structure. She must therefore show some signs of this individuality while giving the impression that she will fit in without causing a ripple. Her strategy is to move a little way back towards the more event-like structure. So she too uses the resources of the language innovatively: she unmakes some of the heaviest of the bureaucratic language. She doesn't name herself directly; nor does she make herself the agent of any action. She *suggests* her own agency by the use of the possessive pronoun *my,* as in the first sentence of the second paragraph: ". . . skills have been developed over *my* career . . ." (rather than: "I developed skills . . ."), and the use of the active verbal form but with an inanimate agent-noun: "Communication at all levels . . . has required . . ." (rather than: "Communication at all levels has required from *me* that . . .", or "I responded to the demands of communication by . . .").

When I saw this text some years ago (it comes from an academic institution in Australia, circa. 1988) I felt that it showed the power of language, of bureaucratic *discourse,* (that is, the world organized and represented through the lens of the bureaucratic institution). I thought that the applicant felt constrained to replicate the language of the institution and to fit herself to it. When I look at this text now – through my now somewhat changed lens of a view of grammar I still see the power of the language, of the discourse, and of the institution, but I also see the attempt by the 'applicant' to show herself as somewhat, somehow distinct from the institutional forms, as speaking in her own way, through the use she makes of the resources of the language. She has changed the heavily nominalized language of the original into a direction where she can appear, however faintly, in the way I have outlined.

In this text I now think there is both the power of the language and of the institution from which it comes and at the same time there is the power of the individual to transform the resources in the direction of her own interests. These interests, as always, include a complex of issues: and after all if one wants a job that one is applying for, it includes an attempt to present oneself as though one was already shaped in line with the purposes of the institution. This job application seems to be saying: "*I am already the position* you are seeking to fill".

3 The emergence of image: the shift from page to screen

Some of the earlier examples have shown what I regard as the major issue in relation to literacy at the moment: the emergence of image into the centre of the communicational landscape in many areas of public communication. Image has co-existed with word for a very long time: whether that was the spoken word of the priest who explained the meanings of the frescoes on the walls of a medieval church to the parishioners, or the written word in the environment of a page of print. The change which marks the recent period is that image has become communicationally equal in its function with writing. Whereas in recent European history, writing has been the dominant and most highly valued mode of communication, image is now emerging as frequently coequal and at times as dominant. Let me show what this change has been about. I will use the book content pages from pages 72–3 to make the point.

In the first of the two illustrations, a page from a book first published in 1936, the relation of word to image is the 'traditional' one: word is the dominant mode, and image serves as illustration. To put it simply, the student-reader (the text was for 13 to 14 year olds) could read the writing alone, ignoring the images, and get access to the curricular content. (Of course, that ignores the fact that students would have been expected to know the visual description of a bar magnet and of a magnetic field: but that information was also present in the writing). In the second example, the page from 1988, the curricular content is no longer communicated via writing: if the student wishes to know what a circuit is, (s)he needs to look at the images, the diagrams and the other images. That is where the curricular content is. Here image is communicationally dominant, or central.

There are many questions to be answered around this change. One concerns the affordances of writing and image, what the potentials for communication of each of these two are. Here it is important to think about the deep logics of the two modes of communication: writing and image. Although writing is *graphic,* that is, it is mode of communication inscribed on a surface, in its structures writing does lean on the structures of speech to some extent. And the structure of speech is given by its deep logic, namely that of *sequence in time.* The sounds of speech have to be produced one after another, as do words; and as do clauses – at least in informal casual speech. This logic underlies the syntactic/grammatical organization of speech: which is one of the sequence of events (represented by sequence of clauses). That is the logic which has given rise to narrative in its various forms, even if narratives work with and against this logic in a variety of ways.

In literate societies, that is, societies in which writing has a history of a reasonable length, writing develops a grammar which is distinct from that of informal casual speech (or indeed from speech in general). That is, writing ceases to be merely the transcription of speech, but develops linguistically in directions which are a result of its distinct social uses and functions, as well as of the possibilities of the graphic mode. One big difference between language-as-speech and language-as-writing is that the syntax of speech is close to the logic of temporal sequence: one clause tends to follow on another, linked by conjunctions of various kinds (and, but, so, when, therefore, etc) as well as by the linking device of intonation. The syntax of writing by contrast tends towards the logic of hierarchy: clauses are syntactically subordinated to another. Temporality and temporal sequence dominate in speech (the question is: what happened? what happened next? and what happened then? . . .), whereas hierarchy – a kind of virtual spatiality – tends to dominate in writing (the question is: what is the main thought and idea? and what are the ideas and thoughts which are subordinated to it?).

The sentence is, quintessentially, the unit of writing, while the "clause-chain" – clauses linked in a relatively light fashion – is the unit of informal speech. The sentence is not a unit of speech, though it can of course be imported into spoken language from the resources of writing. The sense of the close connectedness of speech and writing, as 'language', stems in the main from the fact that both share the clause as the major grammatical/textual/semantic unit. The clause is the linguistic entity which encodes one event, or relation, or state of affairs in the world

The affordances of writing are thus a mixture of the possibilities offered by sequence (deriving from the structures of speech) and the possibilities offered by hierarchy. Image obeys the logic of space, not of time; and of simultaneity rather than of sequence. Its orderings therefore depend on the possibilities offered by space, and the space of the (framed) image. The frame may of course have the most diverse forms: the contemporaneously most potent frame is that around the screen of the computer. Within the space of the framed image cultures have developed ways of assigning meaning to the different areas of the image-space. The centre of the image space may have special significance, as may the lower and the upper sections of the space, the left or the right. Elements may be placed in the various 'zones', and may be related/linked to other elements in other zones.

In the image overleaf (drawn by a six year old) there are two clear elements, at a first look: the girl-author in the foreground, and the display of toys in the 'background'. At the next level of analysis we can then see that each of these has a further structure: the figure of the girl is placed absolutely centrally; she is, she has placed herself as "the centre of attention". The display of the toys in the 'background' has its own structure: elements rising in size, from left to right, and equidistant from one another. They are 'ordered' in size; but their equidistance shows that they were all equally important: on the display-shelf each takes the same amount of space.

This *is not* a 'story'; this is not *telling us* 'what happened'. This is a display; it is *showing* us 'what was there'. The 'take' on the world is entirely different; not "what events happened, and in what sequence", but "what were the significant elements, and what was the relation between them". My contention is that in contemporary forms of messages in communication we are increasingly being asked to make a choice about the 'take' that we have on the world, and therefore about the aptness of our representations: is image better than speech, or writing, or is writing better. Or, if, as is increasingly

My visit to the Toy Museum.

normal, the two, writing and image co-occur, the question is: what part of my message is best given in image, and what part is best given in writing?

In the meantime another revolution has taken place, which has led to a profound change – even if only in the dominant metaphors of public consciousness – in the unit we now regard as the unit (and in the medium) of public communication. The change in medium is from book to IT, and the change in the unit is from page to screen.

The slogan: *from page to screen is* no longer new, but its implications for literacy specifically and for communication more generally have not been fully realized. In the former era of communication, the page was the material unit and the site/space of communication. The logic of the organisation of the page was derived, as I said above, from the deep logic of writing: a logic of sequence in time through which the meanings of writing could be represented.

This temporal order (which could of course be made hugely more complex and abstract) was represented by the *graphic means* of succession of letters in a word, of words in a line, of lines in a paragraph, and of paragraphs on the page. Where images occurred – and relations of image and word have been a constant feature of the era of writing and of printing – image fitted into the logic of the unfolding of writing on the page; and image did not disturb the logic of writing.

In the new communicational arrangements, the dominant space/site of communication is the screen. The logic of the screen is not a temporal but a spatial logic. Whereas in writing meaning derives from sequence – what is first, and what is last (in a sentence, or in a paragraph or in a chapter), on the screen, meaning derives from the arrangement of elements in spatial relations to each other. Meanings attach to what is central and marginal, foregrounded or backgrounded, connected – spatially or not – to other elements, made salient by placement or by colour or by size, etc. Writing has to fit into the logic of the visual arrangements of the screen; and that

CHAPTER 34

THE SPECTRUM. COLOUR

34.1. Analysis and synthesis of white light. Mention has already been made, in **31.9**, of the dispersion of colour which occurs when white light is refracted. The effect is noticeable in glass prisms, in cut gems such as the diamond and in water drops, where bright and spectacular colours seem to have been added to the light. It was Newton who first showed that the material concerned in this does not add anything to the light. He experimented with a triangular glass prism, allowing a shaft of sunlight to pass

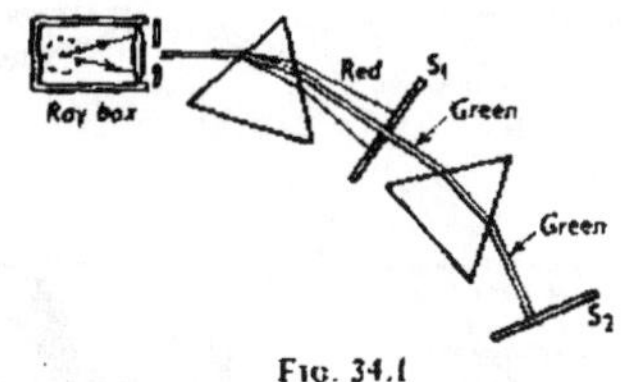

FIG. 34.1

through a small hole in the shutters of his rooms in Trinity College, Cambridge, and placing the prism in the path of the light.

Fig. 34.1 shows an arrangement of prisms and a ray-box, for demonstrating an important feature of coloured light. The first prism gives rise to a coloured beam which appears as a band of colour, known as a **spectrum,** on the screen S_1. The colours are red, orange, yellow, green, blue, indigo and violet—though not everyone is capable of distinguishing between blue and indigo, the latter being a navy-blue. If a

507

Image and writing in older science textbook.

subordination to the logic of the visual changes the communicational value, function and force of writing decisively. My prediction is – and indeed some of the 'pages' I have used as examples here already show it – that the structures of writing will change in response to that.

Look at the relative length (use as a measure the number of clauses per sentence) of sentences in the pages of the science textbooks from 50 years ago or so, and those of the contemporary era. In the counting that I have done, the sentences of the writing of the former era can be between 6 and 8 clauses in length; the sentences of the contemporary page tend to be between 1 and 2 clauses in length.

Of course, 'pages' continue to exist; but increasingly now pages – whether the frontpages of newspapers, or the pages of science books, are organized to resemble screens. The screen is becoming the dominant mode of organizing and structuring messages. And the

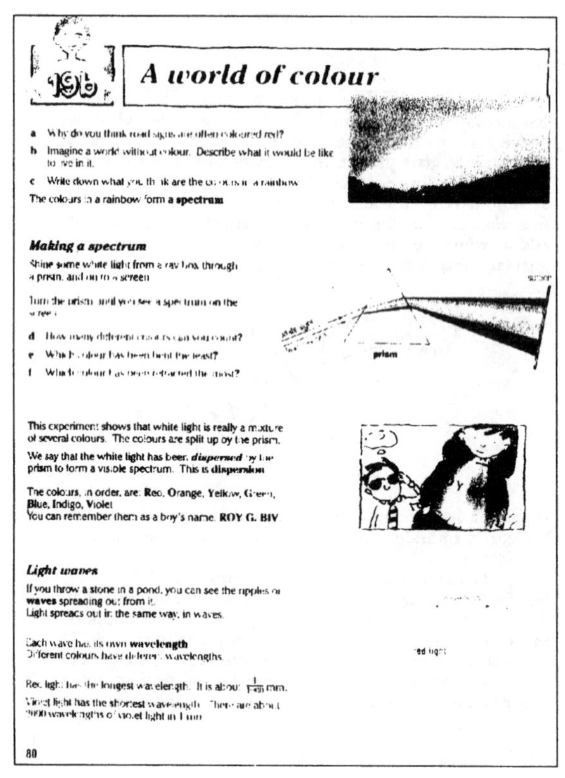

A world of colour

a Why do you think road signs are often coloured red?

b Imagine a world without colour. Describe what it would be like to live in it.

c Write down what you think are the colours in a rainbow.

The colours in a rainbow form a **spectrum**

Making a spectrum

Shine some white light from a ray box through a prism, and on to a screen.

Turn the prism until you see a spectrum on the screen.

d How many different colours can you count?

e Which colour has been bent the least?

f Which colour has been refracted the most?

This experiment shows that white light is really a mixture of several colours. The colours are split up by the prism.

We say that the white light has been ***dispersed*** by the prism to form a visible spectrum. This is **dispersion**

The colours, in order, are: **Red, Orange, Yellow, Green, Blue, Indigo, Violet**
You can remember them as a boy's name. **ROY G. BIV**

Light waves

If you throw a stone in a pond, you can see the ripples or **waves** spreading out from it.
Light spreads out in the same way, in waves.

Each wave has its own **wavelength**
Different colours have different wavelengths.

Red light has the longest wavelength. It is about [illegible] mm.

Violet light has the shortest wavelength. There are about [illegible] wavelengths of violet light in 1 mm

80

Image and writing in new science textbook

screen is organized by the logic of the image. It is the logic in which meaning is expressed through the relations of elements to each other shown spatially. So writing increasingly fits into that logic. That appears in little things, such as the bullet point, which has replaced the former unit of the paragraph; or in the boxed areas of writing which are visual entities functioning in visual designs of layout first, and are written elements only at a second level. This is also disrupting the formerly seemingly unshakeable ordering of left to right and top to bottom sequencing of the written page. Writing can now go in all directions. In power-point presentations letters fly in from all directions to assemble themselves into words. The visuality of the screen is bringing the visual characteristics of writing newly to the fore.

The always present graphic features of writing will become more emphatic, and will become a new resource for meaning.

The effect of all of this on the structurings of knowledge (knowledge which is represented in the form of a narrative is entirely differently organized than "the same" knowledge organized in visual form) is difficult to assess at the moment. It is nevertheless clear that this will have the most profound consequences in all domains of social, economic and cultural life.

4 Frames and genres: from text to connectivity

The communicational world has already changed in ways more profound than we are prepared to realize. To me the most important questions are these:

- what are the real needs of the world of tomorrow, and how do we prepare the young or the not so young for that world?
- what of the world of the former era do we wish to retain, what do we feel is too important for us to give up in order to have the forms of cultural and social lives that we desire?

This is a very different agenda to that either of the cultural pessimists, who see the end of civilization as they knew it, and have merely regret and nostalgia to offer; or of the political and social reactionaries who insist that the stabilities of the former world must be insisted on: and because they can't be recovered in the political and economic sphere, they must be recovered in the domain of culture. Both of these miss the point, in quite different ways, and with differing consequences.

It was the political, social and economic stabilities of the era of the nation state and its economies of mass-production which had led to the production and persistence of strong frames of values, of knowledge, of institutions, of forms and media of representation. These frames existed in the texts which characterize this era: strongly framed genres, with no slippages between genres at the margins, not all that much movement in the dynamic of genres. With the dissolution of the frames in all areas, the texts which are produced no longer have these frames as their anchorage. Genres which were once stable in their form, secure in their anchorage in social relations and structures, are beginning to change in response to the changes in the world.

This changes the conditions of the learning of literacy, and of the production of texts. It is now less a matter of describing the forms of genres and making these forms the content of the curriculum of literacy education; it is now more a matter of making knowledge available about the generative principles which underlie the production of texts and of their generic form. From a situation in which the task was one of learning the forms and how to use these, the situation now is that of learning the potentials of the resources, and the possibilities of their use in specific social situations. The 'work' of text-making has changed.

I think this is what lies behind the slogan of "from text to connectivity". Textual forms are in flux. The task now is to understand what resources are available, what the potentials of these resources are, and how they can be assembled to form the textual designs which correspond to my interests in this environment at this moment. The question is no longer: "what is the appropriate form to use in this context?", rather it is "what are the appropriate resources for this task, and in what arrangements will they best realize my interests vis-à-vis this audience in this social environment?". The task, the work, has changed from 'use' to design. Of course I am overstating the case somewhat; the social is not chaotic, it still has shape, power still exists in telling ways, and the manifestations of power in the form of class, of gender, of age, of profession, and so on, are tangibly there and still affect the everyday of all our lives. And so I would still teach about genres and generic forms, and about the principles of their constitution. The stabilities, however, are less or have gone in many situations where they formerly existed.

5 Multimodal communication and the design of texts

The facts of multimodality pose entirely new questions in the domain of literacy. They intensify the need for the recognition of the category of *design* as a central notion in literacy. Where before the question might have been: "In this context, what are the appropriate linguistic resources, and in what generic form?", now the question is "Given these meanings that I wish to communicate in this environment to this audience, what are the best modes to use for which aspects of my meaning? With this question we have moved entirely away from 'competence in use' to 'best design". This requires a full understanding of the potentials of each mode – what can it do? what does it do best? what is best for this audience, given their cultural, ethnic, class background? etc. 'Design' makes entirely different demands for the production of text; it sees the maker of the text as much more in command (not as controlled by the grammar. but as in control of the deployment of the resources). 'Design' is founded on the 'interest' of the designer: what does (s)he wish to achieve in this situation, arising out of the social histories of the designers and of their affective dispositions.

Each instance of design changes the potentials of the resources: what had not been done before is now a possible thing to do; it extends the possibilities of representation for the next instance of communication. Each instance of communication transforms the resources, makes them into what they had not been before, and does it as a reflection of the sign-maker's interest at that point of communication. This is how the social histories of individuals – and therefore of their society – gets into the resources. Given that the resources always used are socially shaped, it also accounts for how the social and the cultural 'gets into' the individual. Each act of communication, each instance of the work of transformation changes the individual: the resources that (s)he 'has' have changed, have been transformed. The individual's inner make-up and the potential for action are different.

6 Educating for instability: literacy curricula for unknowable futures

Given the stability of this now past era, the goal of competent use in relation to stable modes of representation, and a conception of literacy as a stable system of resources, was sufficient. It encouraged and produced personalities with dispositions towards stability, dependability, reliability. It left innovation, creativity, and even critique, to the very few; those designated individuals to whom society had given these tasks: artists, intellectuals, and maybe a few others, such as rare scientists. Creativity was the privilege of the few: the many were asked to perform the same tasks over and over again, reliably, dependably, uncomplainingly. Such dispositions were highly prized in the era of mass-institutions, and of mass production. The task of education was to prepare those who were educated for stability: the stability of social systems which, whatever changes took place on the surface, remained stable in their fundamental outlines. Education was a device for reproducing the culture, by producing the young in the image of the old.

Literacy had its role in that. Forms of communication were stable, or held stable – at least relatively speaking. There were tight frames around speech and writing, with the structures of each not permitted to penetrate the other or at least not in occasions of formal public communication. The formalities of social structures, themselves of course realising the realities of distributions of power, were entrenched and reproduced in the formalities of writing (and of speech also): in grammar and syntax, as much as in textual form. The persistence of stable social arrangements over long periods meant that the genres of public communication, and especially so in writing, took on the appearance of being naturally so: after all, no one had known them to be different, so that their origins in the social structures became quite invisible.

These arrangements and these dispositions will not serve the demands and needs of the new era. We have entered an era of quite radical instability, and the dispositions required for this period will be configured around innovation, creativity, ease with change, feeling comfortable with difference. The modes which we have for representing the world to ourselves, and to others are absolutely crucially implicated in this. If I believe that in my representations I must adhere to the known, the conventional, the tried, the well understood, I will be ill equipped to deal with constant and if anything, accelerating, change. The dispositions we need to foster are precisely those needed most in relation to the new demands for communication: full understanding of the potential of the resources, confidence in the ability to translate one's own interests into designs which fully embody these interests in the environments of communication.

Let me say that I think that over the next two or three decades writing, even of a quite formal kind, will remain the preferred mode of communication of the political and certain cultural elites. For that reason alone, for reasons of access and equity for all the young people in schools, writing will need to remain absolutely strongly present in a curriculum of communication, and writing will need to be taught much better than it is now (and here I do not have in mind the narrow agenda of the present Literacy Strategy). But writing will not be the sole or even the relevant form of communication for many sectors of the economy, of society, or of public communication generally. For that reason schools will need to develop curricula of communication which go entirely beyond writing, in which writing is one element among many. The newly dominant screen with its logic, superseding the formerly dominant page, means that this is inevitable. Of course this is not due to the effects of technology alone, or even to its powers in the first place. Many other factors contribute to that shift: changes in the economy, changes in the cultural make-up of contemporary societies, the globalization of culture, and many other factors, all play their part in this.

An information-based society and economy cannot make do without the image. The information explosion of which we hear so much had been produced by the written word: and it cannot be solved by the mode that produced it in the first place. Knowledge in many domains will be represented in images of the most diverse forms, often in combination with writing, and often not. But equally, while a society and an economy based on information cannot rely on the word as the central form of communication, we cannot at the moment do without the word – spoken or written – either.

So what of imagination, what of the affective side of humans? Will these be lost if written narratives become less central, if the word is not the central means of representing the culture to its members? My answer is that of course there will be a change; but it will be a change in which the forms of imagination which have always existed for painters, sculptors, dancers, musicians can come to the fore, in which the vast range of human potentials which have been largely ignored or even suppressed by the dominance of writing, can find their essential place in the new economies of knowledge and information, where culture and cultural difference are valued as essential elements in a society founded on innovation. But much effort will be needed to realize this potential. Other, darker futures can easily be imagined. I think that the contribution of the old to the world of the young is not to burden them with the nostalgias, anxieties and fears that we have inherited, but to open for them possibilities that will allow them to be more fully human than were the societies reflected in the now still potent and dominant forms of writing.

A2 Policy

Further RaPAL references about policy

RaPAL Editors. 'Literacy Research Hits the Headlines', *RaPAL Bulletin* Vol. 3, 1987.

Moss, W. and Lobley, G. 'ILEA Goodbye!', *RaPAL Bulletin* Vol. 11, 1990.

Dalton, N. 'Adult Literacy – The Educational Challenge of the 1990s', *RaPAL Bulletin* Vol. 12, 1990.

Gardener, S. 'Basic Education and the Further and Higher Education Reform Bill', *RaPAL Bulletin* Vol. 16, 1991.

Burton, M. 'The Stigma of Illiteracy', *RaPAL Bulletin* Vol. 47, 2001–2002.

Keeping alive alternative visions

Mary Hamilton
RaPAL Bulletin No. 36, Summer 1998

Mary Hamilton works in the Department of Educational Research at Lancaster University, UK, where she is a member of the Literacy Research Group. She is also a founder member of RaPAL.

1. Introduction

The last decade has been a time of radical change in the political, social and educational landscape of the U.K. and a crucial time for adult basic education (ABE). We have moved from the literacy campaign of the mid-seventies to a permanent and increasingly formalised system of learning opportunities for adults, but with this has come a redefinition of what counts as literacy, the goals of literacy learning and the means for achieving them. In some ways this has been progress, but in others, especially for those of us committed to ideals of community-based literacy practice, it has been a dispiriting journey in which our particular visions of literacy have moved further into the distance.

The main focus of the ALPHA97 project is on the relationship between policies, institutions and local actions concerning literacy, and the extent to which these reflect the literacy practices and expressed needs of local communities. I take this to be, at base, a practical focus, about the points at which ideals meet reality and what it is possible to achieve in this translation. As suggested in the overall project outline, I will explore the issues under three main headings:

1. How has the relationship between policy, institutional settings

and local literacies changed in the UK over the last 10 years?
2. What are the constraints or antagonisms which may interfere with communication between these three levels?
3. How can we make policy and institutions more responsive to local needs and marginalised communities so as to develop sustainable literacy projects?

I have been connected with literacy work since the adult literacy campaign of the early 1970s. I am currently based in a university where I am involved in a range of research projects, including documenting vernacular literacy practices outside of educational settings. I am a founder member and current chair of the Research and Practice in Adult Literacy network (RaPAL) which promotes participatory research, and makes links with community-based projects and learner centred practices wherever they occur.

For this article I have drawn on a number of documentary sources, including government policy documents and research reports. I have also consulted a range of projects in the UK that are working responsively with local needs and agendas. I have seen this process of consultation as a kind of action research which could help us begin to articulate the current concerns of local communities and formulate some new directions for policy. Sanguinetti (1995) suggests that action research of this kind can provide the opportunity for participants to develop alternative discourses which can be used to visualise and argue for change. This is especially important in the current UK context where one of the profound ways communities and literacy projects are disempowered is by the lack of alternatives to a strongly dominant centralist discourse which legitimates only one version of the work that is going on (Hamilton 1995).

The process of consulting has already been an education for me. It has revealed many creative and determinedly resistant responses to current constraints and important regional differences in the policy context (especially in Scotland and Northern Ireland) and the research has generated conversations in which people challenged the prevailing language and definitions of "market led" literacy as well as some of the assumptions underlying my own thinking.

2. The Last Ten Years

The policy and institutional context of literacy work in the UK is multi-layered. To understand how it has been shaped we need to consider the European, national, regional and local levels, all of which determine the actual learning opportunities available to particular adults. We also need to know something about the broader context of political and cultural change during this period which has both local and international dimensions to it.

In terms of the key issues of ALPHA97 it is worth mentioning that developments in adult literacy programmes have been taking place within the bleak context of increasing poverty and social exclusion for a significant minority of the population. For example, the Child Poverty Action Group and others have documented the increasing differentials in wealth that have lead to nearly a quarter of households with children living under the poverty line in the 1990s (Oppenheim and Harker, 1996). Particularly worrying from the point of view of literacy are Department for Education and Employment figures showing that more children than ever are now being permanently excluded from school because they are judged to be disrupting classroom life. The figures have increased from 11,000 in 1993 to 15,000 in 1996 (national Guardian newspaper report 23/4/96). In her analysis of a series of riots that took place across the country in the summer of 1991, Bea Campbell describes the ways in which whole communities have been abandoned by employers, by their official political representatives and by the police, living in a state of emergency, without resources, and treated as beyond the pale, criminalised but not protected (Campbell, 1993).

National political changes have included the erosion of local democracy and corresponding mechanisms for planning and allocating resources at the same time as tightening control from the centre. Many elected and accountable bodies have been replaced by unelected "quangos" disbursing money from central sources. All public services – transport, health, education, social services, prisons – have been scrutinised. Government funding has been reduced through privatisation and the introduction of a contract culture and an "internal market", whereby different parts of a public service buy goods and services from one another or elsewhere in a system of competitive tendering. The notion of public service has been replaced by the metaphor of the market place. This has been accompanied by a conscious change in the vocabulary of social relationships. So for example, users of public services are now referred to as "customers" or "consumers".

> One of the most significant developments has been the dismantling into smaller units of the larger, metropolitan district authorities such as the Greater London Council and, most recently, Lothian Regional Council in Scotland. These co-ordinated provision of many resources for local communities and have tended to be generous and frequently creative in their suppport for community education and development activities.

Recent changes in the funding arrangements for social welfare and community development programmes have also had knock-on effects to learning opportunities, for example, for members of linguistic minority groups, refugees and asylum seekers. The introduction of integrated social funding through a "single regeneration budget" has been important for urban areas that qualify for such funding but many smaller towns and communities do not have access to it. Such funding can have divisive effects and, as Rutledge (1996) points out, it

> *"encourages a peculiar form of local politics. The structure is headed by an unselected quango, with a number of subsidiary groups seeking to gain resources (often sorely needed) for their own constituents, and in direct competition with each other".*

As another writer puts it, groups are compelled to compete with one another "on the basis of their level of misery or the respective skills of bidders". (Reeves et al 1993, p.98). The short-term nature of much project funding also leads to a huge burden of paperwork, insecure staff contracts and the impossibility of forward planning.

Funding through European Community is now an important source for many projects, and comes with its own set of criteria, target group definitions and paperwork headaches. In general it has been useful in allowing developments that could not have happened with national resources and political priorities over the last 10 years. Women, and communities with high unemployment, have particularly benefitted from this handing. Special funding has been available to Northern Ireland through the "peace and reconciliation" programme. Exchanges and visits are often part of funded programmes, as a powerful way of creating a new European identity.

Changes in culture and community life have been brought about by the emphasis on individual competitiveness, rather than cooperative strategies, in all areas of public and private life. This has led to fewer collective resources and more fragmentation in community initiatives. The government now advocates individual solutions to social problems within the market place, rather than the welfarist policies that framed the original literacy campaign in the early 1970s and which are now denigrated as both inefficient and patronising (see Hamilton, 1996; Withnall, 1994). The National Lottery has become a major new focus for the financial hopes of both individuals and the community groups who bid for grants from its profits.

New media developments are changing the role of literacy in people's day-to-day lives for example in the widespread use of video libraries and home computers. These changes have been strongly encouraged by a government that places a high value on new technologies and the consumer goods that accompany them -home videos and computers, mobile phones and so on -such that these have become symbols of status and prosperity as well as offering new possibilities for communication.

Changes in the organisation of the workplace and the structure of paid employment have led to the weakening of traditions of collective action through Trade Unions and a corresponding erosion of working conditions. "Flexible working" has led to a huge increase in temporary and part-time jobs which do not offer access to training (See Frank and Hamilton, 1993). There is significant unemployment, including many people who are out of paid work permanently or on a long-term basis. (see Unemployment Unit. 1996). These changes are the product of global economic trends combined with regressive government policy on employment and industrial relations. We are encouraged to blame unemployed people and those who are low-skilled for the country's economic problems. International comparisons are not argued through in terms of the global market and the power of international business. Neither do they explore what "flexible" working means in terms of peoples daily lives, health and the security of their families.

Broader reforms have taken place in the education and training system within which adult basic education and literacy are minor players. Changes in ABE programmes are motivated by top-down policy decisions enforced by financial incentives or penalties.

In many instances, ABE workers are having to deal with the unintended consequences of legislation designed for other constituencies or purposes, rather than a coherent policy aimed at developing and supporting literacy and adult learning (see Fullan, 1991; Derrick, 1996).

Most notable among these changes has been the Further and Higher Education Act of 1992 which separated vocational, qualificationbearing courses from non-vocational courses for adults. In England and Wales ABE was placed within the vocational sector so that the institutional setting of literacy programmes has changed, away from community settings towards the greater formality of Further Education Colleges.

> The separation of "vocational" from "non-vocational" is particularly unhelpful to ABE. Neither category is adequate to define the range of fuctions that literacy programmes should be able to fulfil and the division contradicts a long tradition of viewing liberal or community education as having many functions-from cultural, community development and citizenship roles, to making early steps back into the employment market (see Tuckett, 1991 and also the Scottish Community Education model which currently preserves this integrated view). The Act has made this questionable distinction into a solid legal entity and further redefined "non-vocational" as "leisure". Thus any activity that cannot be linked to direct emploment outcomes is classed as a luxury commodity and therefore expendable (Derrick, 1996). This particularly discriminates against older adults who have retired from paid employment

and poses problems in resourcing courses for adults with serious disabilities or learning difficulties. It results in a two-tier system- one highly resourced and closely monitored and the other fragmented and marginalized.

Adult literacy educators have now forged better links across different levels of the education system (for example links with Higher Education through concerns for students who find academic writing difficult and links with schools via family literacy programmes). There are stronger links with employers and vocational training. But these educators have also lost many of the local and regional networks that used to be an important part of staff development opportunities. The many locally responsive projects keep in touch by informal means, but often feel isolated or invisible.

There is rhetorical support for the notion of lifetime learning, both at national and European level but considerable reluctance to invest public money to improve opportunities for adult learning (National Institute for Adult and Continuing Education, 1996). England and Wales has a government-funded central agency, the Basic Skills Agency (BSA) which formulates policy and this has resulted in a higher profile for adult literacy issues than in Scotland or Northern Ireland where there is no similar agency.

Through the BSA, and by controlling finding, the government exerts strong central pressures on programmes and projects for accountability, quality control and accreditation. Outcomes are measured by nationally defined standards and a body of skills rather than in terms of individual profiles or the impact of projects on communities. Expansions in recruitment are motivated by money and maximising the number of student enrolments, rather than rights to education.

The internal market in education has taken the form of local management of schools and colleges (see Tuckett, 1991) whereby each institution is funded directly from central government and operates as an independent business. Democratic control of programmes has been eroded at many levels, including the representation of student and staff voices and control over curriculum and assessment (Smith,

1994). Collective action and representation has been replaced by the idea of "consumer power" at the individual level.

The research base of adult literacy has grown but has had a limited impact on practice. The national Basic Skills Agency (BSA) limits itself to "research into the scale and characteristics of basic skills need and the effectiveness of basic skills programmes" (BSA, 1995) and has tended to fund quantitative, policy driven research that assumes, rather than explores the needs and interests of local communities. Academic research in higher education on the "new literacy studies" has been developing alternative notions of literacy as social practice (e.g. Mace, 1993, 1995; Barton, 1994; Street 1995). These ideas have underpinned much community-responsive literacy work, especially within non-government funded organisations in the UK and internationally. This cultural approach to literacy emphasises the significance of local contexts and purposes for literacy but it has not found a sympathetic climate in government policy circles. Gaining a positive platform for these ideas has been hindered by the entrenched views of the media reporting on literacy and by the traditionally distant and mistrustful relationship between Higher Education and ABE practice in the UK.

3. Examples of projects working in a responsive way to local and marginalised communities

This section offers some examples of responsive projects that have been developed, over the last ten years. These projects show how responsive literacy work can still be carried on despite the difficult climate outlined above, and the principles on which it is based. I have included examples of different kinds of institutional partnerships that span the possible range in the UK and illustrate different strategies and priorities.

Example A: Community Projects Funded by Non-Governmental Sources

Pecket Well residential college for adults opened in 1992 in a rural location near Halifax in the North of England. Courses are run for students and by students and the college welcomes adults with disabilities or learning difficulties. It has been developed by a small group of committed people and is based on principles of democratic learning which show themselves at all levels in the running of the college: learning methods, course design, relationships between workshop leaders and participants (who at different points in time, may be the same people), management of the college, researching local community needs through employing outreach workers and through students' activities on courses.

Pecket Well raises money for its activities from a range of sources including trusts and the European Community. So far, the college has been able to operate in an independent way but even a small, independent project like this cannot escape the effects of bigger changes in policy. Increasingly it is under pressure from funders to target, label and measure the people who come onto courses:

> *"It is very difficult, the constraints the funders put on us. They try to turn us more formal, make us less democratic. Certain funding restricts many people, for example age, time to complete courses, background, culture, areas of residence etc. Accreditation is also "forced" upon us"* (interview notes, July 1996)

Pecket Well has made use of the media to publicise itself as a unique project and to explain the importance of residential courses -frequently a life-changing experience for the adults who take part. It uses these opportunities to challenge mainstream language and assumptions about literacy and put over students' views.

One of its strengths is the emphasis it places on developing a public voice for students at the college through collectively documenting and publishing the writing that is done in courses (Pecket Well, 1994; 1995). In this, the college is supported by a national network, the Federation of Worker Writers and Community

Publishers (FWWCP) which offers national events and exchanges with other projects that promote student writing, contacts and ideas.

Example B: Local government funded community education services

In local areas where there are a large proportion of bilingual speakers, a number of innovative initiatives have developed within the statutory sector, despite the fact that only English and Welsh are recognised for support within literacy programmes. There is no state funded heritage or community language programmes, either for children or adults and this a deliberate policy decision. This means that the language needs and literacy practices of bilingual ethnic minority groups are not generally acknowledged within literacy programmes (see Hamilton, 1996, for details).

A notable exception to this is the Sheffield Black Literacy campaign that started with the Yemeni community in 1989 and has continued to the present, adding new community strands to its activities (see Gurnah, 1992). The campaign is organised by the Sheffield Unified Multi-cultural Education Service (SUMES) within the Local Education Authority, in partnership with the black communities and a local university which trains literacy assistants as part of a year long access course. The literacy assistants are young, unemployed members of the black communities who recruit and teach learners as part of their outreach and development work in the community. At the same time they themselves get a training which offers access to further study and employment. There are other examples of responsive developments in local government organised community education services that have circumvented the prevailing models (see for example Lucero and Thompson, 1994). Key factors in the successful development of these programmes seem to be a political will and committed and informed decision-makers within local councils to respond to community needs; involvement of younger members of local communities; partnership funding from a range of sources and creative use of existing funding criteria.

Example C: Further Education Colleges with a strong community focused service

There are also examples of further education colleges developing a wide range of community learning opportunities. For example, Bilston College in the West Midlands of England operates with a strong held view of itself as a community college engaging with "popular education" (Reeves et al., 1993). Based in an area of former steel works with high levels of unemployment and poverty, resources in the local community are scarce and those offered by the college are valuable. It sees its brief as "not merely to visit community groups to tell them about the college, but to participate on an equal basis in community organisations and where relevant, to identify with their objectives".

It has gone to great lengths to develop courses in community venues that respond to local needs, including courses for women and older adults, and separate centres for Afro-Caribbean and Asian Education and Training that teach in community languages. A Basic Skills Partnership links the college with local community organisations and a well resourced "Key into Learning" centre is a main focus of the college effort where adults from the local communities can obtain help with reading and writing as well as computer skills. The college organisation reflects its underlying commitments. It has a number of directorates including a "community" directorate and a "quality and equal opportunities" directorate which enable these goals to be properly planned and linked in with all areas of the college's work. Bilston College is unusual in having such explicit, well-thought through and high level management support for community access. Having set out its values and vision, it is then able to re-interpret and take control of the demands that fenders make on it, enthusiastically develop new technologies in ways that do not isolate and disempower students, and argue for definitions of "quality" that take into account local conditions and resources. Bilston College is still limited by its role as a government funded tertiary college but it offers impressive evidence of how a strong and explicit community mission can be used to challenge the barriers resulting from funding constraints and open up the institution to local community involvement.

Example D: Partnerships with University-based Continuing Education

In a climate where local education authorities and voluntary groups have struggled to support community education initiatives, sometimes partnerships with higher education have produced innovative responses. Coare and Jones (1995) report on a writing project that was set up between the Continuing Education Department at the University of Sussex in Brighton, England and a local organisation for homeless adults. This project built on writing activities associated with the BIG ISSUE – a national magazine written by and for homeless people and sold on the streets to finance shelters and as a means of livelihood for the sellers. The magazine distribution offices also function as advocacy and advice centres for homeless people (see Guardian newspaper report 2/10/96). The BIG ISSUE has become a familiar part of street-life in all major cities in the U.K. and was originally started with help from a business. In itself the magazine is an example of a creative literacy response to a social need that has grown to affect the lives of thousands of people.

The Brighton project used a local day-centre for homeless people and was able to be very flexible in what counted as "attendance" on the course and outcomes from it – important because of the chaotic and unpredictable nature of participants lives, and their changing needs. Such flexibility is not easy to obtain through more mainstream funding but is an essential part of responsive literacy work with extremely marginalised groups such as people living on the streets.

Example E: Network supporting Community-Based Literacy

There are many other initiatives of the sort mentioned above, which seem to form a strong but largely invisible thread running alongside the mainstream, official version of what is happening in literacy work. It is a problem that such projects do not have a louder public voice because their contribution is not properly acknowledged and they are unable to contribute directly to shaping the future through policy debates and decisions. In such a context, national networks that can help draw together the contributions of local projects and work that is not recognised by the mainstream ABE providers are potentially very important.

Few networks have survived the changes of the past decade. One of these has already been mentioned – the Federation of Worker Writers and Community Publishers (FWWCP), which was founded in 1976. It is almost entirely invisible in the mainstream of ABE which does not highly value the activities of student writing and publishing, despite the fact that it still exists in some form in many programmes (see O'Rourke and Mace, 1992; Mace, 1995).

Another network is the Research and Practice in Adult Literacy Group (RaPAL) which started in 1985. RaPAL links practitioners and researchers together, publishes a regular newsletter and supports participatory research activities which include students. A self-funded network for workplace trainers has also recently been started, reflecting new developments in workplace basic skills programmes.

The National Literacy Trust is an umbrella organisation, launched in 1993 and supported by a number of different trusts and corporate sponsors. It sees its role as encouraging access for all citizens to a literate culture, and to this end it works across all sectors of the education system, children and adults. It encourages partnerships with other cultural organisations, such as libraries, publishers, newspapers, the British Film Institute, arts councils, and with charities and businesses to find ways in which all of these can put their resources to work for local communities. It works with the media and offers its services as a consultant and broker to policy-makers, schools and colleges, parents and pressure groups. It holds a database of literacy initiatives in the UK from all sectors and it is currently publishing a position paper on the future of literacy drawing on the perspectives of all these groups. One of its main strengths is that it can work across the boundaries imposed on many literacy-oriented groups. One of its key notions is that of literacy for *pleasure*. In a climate where all programmes to develop literacy have to justify their activities in terms of the functional and vocational, insisting on the importance of pleasure to literacy learning is a gesture of challenge! However, the NLT is not a political organisation does not identify itself with partisan views. It will be interesting to see how far it can maintain this neutral stance in the longer term.

4. Constraints on Developing Community Responsive Literacy

Over the last ten years, adult literacy and basic skills has become a permanent and accepted area of provision in the UK. However, the more these programmes have been accepted into the mainstream of post-school education and training, the more pressures there are on them to become standardised and therefore less responsive to the needs and priorities of local and marginalised communities. There is consensus about how this has been achieved. The mechanisms have been clear, open and brutally applied: the introduction of outcome funding and quality criteria, standardised accreditation, devaluing of outreach and development work, a vocational slant on courses which marginalises those who are not looking for direct employment outcomes.

> Those of us committed to community responsive literacy do not believe that this is an inevitable effect of the process of "mainstreaming". More creative and responsive ways could be found of developing a system of securely funded and high quality learning opportunities if the political will existed. The projects described above show some of the many ways this can be done done. The very different way in which Scotland has developed its ABE service within a strong community education framework offers another model.

The factors that contribute to the current "gap" between the perspectives of national policy-makers and local communities are summarised below.

Firstly, there is a lack of real knowledge about the priorities and practices within local communities by those making decisions and setting up programmes to develop literacy. Restrictive funding mechanisms limit the possibilities for the necessary research and consultation to be carried out and for project workers to develop their knowledge and feed it back to policy makers. Time and again this is mentioned by project workers as a frustration and an essential component of more measurable work they do. It is also a key issue in literacy in other countries, where it often goes unrecognised by policy makers.

Secondly, centralised control of policy decisions, and the corresponding erosion of local democracy and fragmentation of collective organisations contributes to the "gap". A description in the Adult Literacy and Basic Skills Unit (ALBSU, now the BSA) Annual Report of how the recent family literacy initiative was developed illustrates the policy process that currently takes place. There were discussions between ALBSU and various government ministers; ALBSU staff carried out research into family literacy elsewhere in Europe and visits to programmes in the USA which were considered as models for developments in the UK. Finally, ALBSU visited "programmes and projects that work with children and parents and talked with several experts" (Adult Literacy and Basic Skills Unit, 1993, p. 10). This looks like a very top-down process. Importing social and educational policy ideas from the USA has been a common strategy in recent years (see Finegold et al, 1993). Within this strategy there is little room for consultation with local communities or participants in existing ABE' programmes and there are certainly no rights for such groups to be involved in the decision-making process.

Thirdly, hierarchical, undemocratic learning institutions (mainly further education colleges) work in isolation and in-competition with one another within a "market driven" economy, vying for students and unable to share information and referral systems that would put the interests of students first, rather than the interests of the institution itself.' As we have seen, under these conditions, change and expansion is determined by financial incentives rather than developed in partnership with project workers and participants. But, as Reeves points out

> *"against a background of thinly-spread educational resources, widespread social deprivation and educational underachievement, the idea of encouraging competition between educational institutions makes little sense"* (Reeves et al 1993, p.10.)

Similarly irrational, even in terms of the government's own priorities, is the decreasing support given to the voluntary sector in terms of the partnership funding which in the past has helped many projects keep afloat. This wastes cost-effective resources that already exist in local communities to support literacy.

Finally, we suffer from narrow and elitist assumptions about what is of cultural value in terms of literacy and other popular media; simplistic, utilitarian views of literacy and of what counts as "the community interest"(i.e. the business community first and foremost). As I have discussed above, the separation of "vocational" from "leisure" is particularly restrictive and the legal redefinition of ABE is one way in which the discourse of literacy has changed since the early days of the campaign. A powerful feature of the policy process is that it labels and categorises groups of people, institutions and activities and then gives weight to this classification system through selective allocation of funding. Politicians and the media legitimise these categories and de-value alternatives views through what Ball (1990) has called "a discourse of derision". This process constructs a strong notion of "the other" to exclude some people and target others for support, and is extremely oppressive if not accompanied by consultation with those affected by it (see Stuart and Thomson, 1995). Narrow and rigid views of how literacy is important to adults' lives that mean consultation with local groups is not seen as important – answers are known in advance, provision is offered rather than negotiated.

Even more fundamentally, experience from international adult education work suggests that as long as ABE remains solely within the domain and discourse of education and training it will deny community aspirations, focusing on individuals rather than on group concerns and co-operative activities (see Rogers, 1992).

There is at present a lack of opportunities to develop and make public *alternatives* to this dominant, legitimated, government discourse.

Alternative discourses are needed that can voice the concerns and priorities of local and marginalised communities in order to regain control of the literacy agenda (see Wickert, 1996).

5. What should we do?

Recommendations for action at three different levels suggest themselves from the discussion above. They correspond roughly to the levels of local communities, institutions and central policy and they are based on some general understandings. Firstly that we need to view literacy holistically, in the context of peoples full lives and community context. Secondly, that strong democratic consultation and decision making mechanisms are essential to developing truly responsive community literacy. Thirdly, that it is also essential to have the desire for and practical possibility of co-operation and partnership between all interested groups and organisations. Fourthly, we must create opportunities for open debate about the local uses and meanings of literacy and develop alternative public discourses of literacy through publishing and other mass media.

1. Networks: In our current context we lack the networks that need to be in place to ensure a flow of information and debate between actors at different levels and to ensure the visibility of community concerns.

There is a role for a co-ordinating body to link non-government funded organisations, as exist in some developing countries. This could act as an alternative focus for policy discourse and a critical voice, as well as being a resource and information exchange.

Such a body could encourage the active support and involvement of students and their communities in lobbying the media and policy-makers. Equally, they could support strong professional networks of teachers and community workers across the whole range of E literacy related programmes.

Research and development skills and strategies need to be developed with resources devolved to local communities for this purpose. A possible model for this can be found in Australia's research networks which fund practitioner-based research projects through links with local Higher Education Institutions. The Higher Education institutions are vehicles for these activities but do not own the resources.

Finally, international links and exchanges develop understandings of the global influences on local decisions and the range of solutions discovered in other countries. These can all inform a national policy.

2. Institutional Systems are the bridge between community networks and policy, and as such they can open the way for mutual learning and dialogue, or deter this process.

We need institutional systems which can support a range of learning opportunities on the spectrum from informal to formal learning as stepping stones for those who need them. As part of this, we need to work out a productive and complementary relationship between statutory and voluntary organisations. Institutions should also aim to support the general development of a culture of literacy and enabling access to it for all. This is not just an educational issue but many cultural institutions should be involved in partnerships for cultural action – for example libraries, arts projects, community publishing and other media.

To achieve these things we must have democratic management in the institutions themselves so that community concerns can shape the curriculum and ensure dialogue between informal and formal systems.

Finally, institutions should critique sterile notions of competency based education and training and insist on alternative forms of assessment and measures of quality including input, process as well as output. A great deal of this knowledge already exists, for example in the widespread experience of Open Colleges around the country that have found flexible ways of accrediting courses, tailored to local needs.

3. Policy changes. Policy is about planning and allocating resources and designing effective mechanisms for doing this. Top level policy can dampen or stimulate energy at the level of institutions and communities to develop literacy. Policy can also have unintended consequences for groups who are not the main focus of the initiative and we need action to protect their interests.

We need to make it possible for representatives of local and marginalised groups to have the right to input into policy decisions, at all levels, not just to be occasionally consulted.

It is essential that literacy initiatives be coordinated with other social policy initiatives so that they are not happening in isolation. Literacy is not just an educational issue but must be linked with the broader culture and socio-economic conditions of peoples' lives which may be working against life-long learning and community development – such as unemployment and insecure jobs, poverty and homelessness, fragmentation.

We need staff development to be taken seriously as a policy issue, with programmes that aim to produce reflective professionals and a real sense of professional community and voice, to input seriously into policy. There are number of higher education institutions ready to contribute to this, and the Research and Practice in Adult Literacy network also has a potential role to play.

Positive action programmes should be put in place that resist the imposition of mono-cultural, monolingual approaches to literacy education and recognise the resources that are available in bilingual communities.

Finally, we need to review the role of a central agency for literacy taking into account the contrasting experiences in England and Wales, Scotland and Northern Ireland, and examining the solutions adopted in other countries. We should consider other mechanisms that could be put into place which might better meet the current demands of the field.

Gaber-Katz and Watson (1991) in their discussion of community based literacy, talk about *building* community not just responding to what is already there. They point out that poverty and stress limit peoples' aspirations as they struggle with the day-to-day tasks of survival, r and part of the role of community literacy is to enlarge people's own vision for their lives and strengthen the resources available to them. This seems very relevant to the depleted communities of turn-of the millennium Britain and is recognised by many working at the local level.

The consultation I report in this article reveals that many of us in the UK are well-practised in the art of working in the cracks, finding the spaces to do what we know to be important. Alternative visions of literacy are alive but out on the margins. Finding ways of moving them into the centre is the challenge we face. What we have not learned to do effectively is to engage with the central processes of policy formation and decision-making, to use the powerful institutions of the media to put over these visions. Many of the resources available to us to do this have been removed during the last 10 years by a government that is very adept at using these institutions for its own purposes. It is time for us, too, to focus on the big picture and how to redraw it.

Bibliography

Adult Literacy and Basic Skills Unit (1993) Annual Report 1992–3.

Barton, D. (1994) Literacy: The Ecology of Written Language. Blackwells.

Ball, S. (1990) Politics and Policy Making in Education. Routledge.

Basic Skills Agency (1995) Annual Report 1994/5, Spreading the Word BSA London.

Campbell, B. (1993) Goliath: Britain's Dangerous Places. Methuen.

Coare, P. and Jones, L. *Inside-Outside: A Homeless People's Writing Project*, in Adults Learning, Jan 1996. Vol 7/5, pp 105–106.

Derrick, J. (1996) Adult Learners in Colleges and the FEFC. Paper presented at the NIAE Annual Conference, Warwick University, March 1996.

DfEE (1996) Lifetime Learning: A Consultation Paper. HMSO Publications.

Dept of Education for Northern Ireland (1996) Lifetime Learning: A Consultation Paper. Training and Employment Agency.

Finegold, D, McFarland, L and Richardson, W (eds) (1993) Something Borrowed, Something Blue? A Study of the Thatcher Government's Appropriation of American Education and Training Policy. Triangle Books, Oxfordshire.

Frank, E and Hamilton, M. (1993) Not Just a Number: The Role of Adult Basic Education in the Changing Workplace. Report to Leverhulme Trust June 1993.

Fullan, M. (1991) The New meaning of Educational Change. London, Cassell.

Gaber-Katz, E. and Watson, G.M. (1991) The land that we dream of: a participatory study of community-based literacy. OISE Press, Toronto, Ontario.

Guardian Newspaper *New School option for boy in strike row* Guardian front page report, 23/4/96 (quoting DfEE statistics).

Gurnah, A. (ed) (1992) *Literacy for a Change: A Special Issue on the Black Literacy Campaign in Sheffield*. Adults Learning. Vol 3, No. 8. Leicester.

Hamilton, M. (1992) *The Development of Adult Literacy Policy in the UK. A Cautionary Tale* in the International Yearbook of Adult Education. Bohlau Verlag, FRG. Sanguinetti, J. Evaluation as Discursive Resistance. Paper Presented at LERN Conference, June 1995 Australia.

Hamilton, M. (1995) *Networks and Communities in Adult Basic Education* in Savitsky, F. (ed) Living Literacies: Papers from a conference on Multiple Literacies and Lifelong Learning. London: London Language and Literacy Unit.

Hamilton, M. (1996) *Adult Literacy and Basic Education* in Fieldhouse, R (ed) A Modern History of Adult Education. MACE.

Lucero, M. and Thompson, J. (1994) *Teaching English Literacy Using Bilingual Approaches* RaPAL Bulletin No.25 Special Theme Issue on Bilingual Literacy.

Mace, J. (1993) Talking About Literacy London: Routledge.

Mace, J. (1995) (ed) Literacy, Language and Community Publishing: Essays in Adult Education. Multilingual Matters, Clevedon, England.

National Institute of Adult Continuing Education (1996) LifeTime Learning: Response to the Government Consultation Document, NIACE, Leicester, UK.

Oppenheim, C. and Harker, L. (1996) Poverty: The facts (3rd edition) Child Poverty Action Group, London.

O'Rourke, R and Mace, J. (1992) Versions and Variety: A Report on Student Writing and Publishing in Adult Literacy Education, Avanti books, Stevenage.

Pecket Well College (1994) *Forging* a *Common Language, Sharing the Power* in Hamilton, M. Barton, D. Ivanič, R. (eds) Worlds of Literacy, Multilingual Matters/OISE Press: Clevedon, Philadelphia, Adelaide, Toronto.

Pecket Well College (1995) Changing Lives. Annual Report 1994–5. Halifax, England.

Reeves, F. and colleagues (1993) Community Need and Further Education: The practice of community-centred education at Bilston Community College. Education Now Books in partnership with Bilston Community College.

Rogers, A. (1992) Adults Learning for Development. Cassell.

Rutledge, H. (1996) *Empowerment, education and Citizenship: A Local Study* in Adults Learning March 1996. Vol 7/7, pp 172–4.

Sanguinetti, J. (1995) *Evaluation as Discursive Resistance.* Paper presented at LERN Conference, June 1995, Australia.

Smith, M. (1994) Local education: community, conversation, praxis. OUP.

Street, B. (1995) Social Literacies: Critical Approaches to Literacy in Development, Ethnography and Education. Longman London and NY.

Stuart, M. and Thomson, A. (1995) (eds) Engaging with Difference: The "Other" in Adult Education. NIACE.

Tuckett. A. (1991) *Counting the Cost: Managerialism, the market and the education of adults in the 1980s and beyond* in Westwood, S and Thomas, J.E. (eds) Radical Agendas? The Politics of Adult Education. NIACE.

Unemployment Unit Working Brief Issue 76 July 1996. UU: London.

Withnall, A. (1994) *Literacy on the Agenda: The Origins of the Adult Literacy Campaign in the United Kingdom.* Studies in the Education of Adults. Vol 26, No 1, pp 67–85.

Wickert, R (1991) *Maintaining Power Over the Literacy Agenda* in Open Letter. Vol 2, No I.

Response to Moser

RaPAL Working Group
RaPAL Bulletin No. 37, Autumn 1998

In 1998 members of the RaPAL Working Group formulated a response to the government's consultation document on Lifelong Learning, *The Learning Age*. A copy of the response was sent to the Moser Working Group advising on effective post-school basic skills provision. Our letter to Sir Claus Moser is shown here followed by a summary of our response to the consultation document.

Sir Claus Moser, Chairman of the Basic Skills Agency, 7th Floor, Commonwealth House. 1–10 New Oxford Street. London WC1A 1NU.

Dear Sir Claus Moser,

RaPAL would like to respond to the request by the Moser Working Group advising on effective post-school basic provision.

The Research and Practice in Adult Literacy Group is an independent network of learners teachers, managers and researchers in adult basic education and literacy across the post-16 sector. It was established in 1985, and encourages collaborative and reflective research that is closely linked with practice. We produce a newsletter three times a year, other occasional publications and organise an annual conference.

We have already submitted a response to the government's Lifelong Learning consultation document "The Learning Age" directly. Below is a summary of our position, together with our more detailed response, a leaflet summarising the aims of RaPAL and recent examples of the Bulletin we produce.

Among our recommendations are three particular areas within which RaPAL has considerable experience to contribute:

* Promoting more collaborative and qualitative based research especially on student learning experience and the development of good practice.
* Reviewing professional initial and in-service teacher training for basic skills; developing a more thorough foundation level programme training that addresses group as well as individual teaching and a second level which encourages reflection and enquiry as well as demonstration of practical competence.
* Development of networking for those basic skills in response to a current lack of support, networking and opportunities for staff to update their skills across institutions.

We trust that you will find our recommendations useful.

Yours sincerely,
Wendy Moss (RaPAL chair)

Summary and Recommendations from RaPAL in response to Government green paper "The Learning Age".

RaPAL Bulletin No. 37, Autumn 1998

1. General Points

1.1 RaPAL welcomes the green paper the "Learning Age" especially the broad, integrated look across post compulsory learning and the fact that basic education is specifically identified as an area to be addressed in all the initiatives outlined.

1.2 However, we would want to see a stronger response in terms of legislation and funding. We would also want to see a shift in language and vision of the paper so that it is less focused on individual skills and their exchange in the vocational market place. We would wish the government to take a broader and more realistic view of adult learning; to recognise and support the development of literacies in all aspects of people's lives and to build on the diversity of cultures and languages which learners bring to their learning.

1.3 We would promote the development of a leaming democracy in which learners genuinely participate in developing education provision. Genuine consultation and responsiveness to local communities and learners needs will inevitably involve recognising and responding to the diversity of literacy and language practices that people use and need.

2. Specific Recommendations

2.1 A government policy supporting a broader view of basic education as essential for democratic and community participation, parenting and family life, as well as for the workplace (as in the US initiative "Equipped for the Future" and the Australian National Reporting framework).

2.2 The development of a coherent ABE strategy within Lifelong Learning so that this area is not split and marginalised as in the past, but has a strong presence and effect in shaping wider initiatives within education. (The recent FEFC curriculum area survey report on Basic Education (April 1998) reports that this is now the 6th largest subject area in FE colleges)

2.3 Open connections between literacy and key skills – ABE tutors and learners have much to contribute to how key skills are addressed in different domains. For example, within our own network we make connections with those working on development of academic literacies in access courses and higher education.

2.4 Emphasis on learning support in FE is important, but not at the expense of the provision of high quality options for ABE to cater for adults with goals other than immediate formal education. More support and resources for a wide variety of ways of learning at home, in educational institutions, in different community settings and at work are needed. Systematic outreach and mechanisms for consulting learners and potential learners about the nature of their learning needs and interests.

2.5 While partnership is welcome, the government should consider ending small funds for ABE which lead to "bidding fatigue".

2.6 Participatory structures within learning institutions that allow democratic learning to take place alongside subject knowledge and which provide a model of participation that is then carried into the wider community.

2.7 A broader literacy curriculum recognising the wide variety of contexts in which people use literacies and communication skills eg. women's relationship to paid and care work within the home and community; programmes for those with learning disabilities.

2.8 Funding made available for the development of alternative forms of support and provision eg. residential weekends, student conferences and other activities which develop participation and basic skills outside the more traditional framework of education.

2.9 Urgent attention to be paid to ESOL, particularly the wide variety of histories and educational experience of students who need English language skills. We need programmes which recognise and build on bilingual literacy and community languages.

2.10 Recognition of IT as a basic skill and attention to how IT is best integrated into basic skills provision.

2.13 Provision of flexible, progressively staged and understandable qualifications designed for a range of learner styles and capabilities. Alternative and flexible ways of recognising achievement that are not solely accreditation based.

2.15 Financial and other incentives to employers to support workplace learning, and an entitlement of paid release for employees to develop their skills, particularly for temporary and part-time workers.

3. The following recommendations are areas within which RaPAL has particular experience to contribute:

3.1 Promoting more collaborative and qualitative based research especially on student learning experience and development of good practice.

3.2 A review of professional initial and inservice teacher training for basic skills. A more thorough foundation level programme training that addresses group as well as individual teaching and a second level which encourages reflection and enquiry as well as demonstration of practical competence that is internally taught and assessed.

3.3 Development of networking by those teaching basic skills. The recent FEFC inspection report has pointed to the lack of support, networking and opportunities for staff to update their skills across institutions.

The Millennium Dome approach to adult literacy: A practical and theoretical approach to the recent adult literacy strategy

Sue Erlewyn-Lajeunesse and Zoe Fowler

RaPAL Bulletin No. 47, Winter 2001

Sue Erlewyn-Lajeunesse is the manager of ITWorks Harlesden, an adult basic skills outreach centre, and she is currently setting up a sister centre in Kilburn. She is also a part-time MA student at the Institute of Education. Zoe Fowler is a Ph.D. student at the Institute of Education looking at adult literacy and recent government policy. She has also been involved in adult English teaching in FE and is involved in some basic skills teaching with the Big Issue.

The Millennium Dome analogy began one evening in April when we were discussing the emergence of the adult basic skills strategy and considering the impact that this strategy would have on both learners and educators. What was originally a throwaway remark subsequently became the basic structure for our presentation to the RaPAL conference. As with any metaphor or analogy, it offers some potentially useful insights but doesn't map comfortably onto all aspects of the strategy. We are using the analogy as a useful frame through which to observe recent policy, but we realise that this perspective isn't going to work for everyone!

The Millennium Dome approach offered 4 main areas for thought for our RaPAL presentation:

- The foundations
- The architecture
- The zones, or the content
- The Mandelson effect

Returning to these areas, we intend this text to provide an overview of the structure of the strategy, a consideration of its strengths and weaknesses, and some insight into the future of adult literacy provision.

The Foundations

In terms of the recent policy documents on adult basic skills, literacy is defined as "the ability to read, write and speak in English... at a level necessary to function at work and in society in general" [DfEE, 1999]. The strategy documents explain that literacy is valued because it provides the necessary foundations of a learning society, promotes social inclusion, and enhances the competitiveness of the workplaces, businesses and the economy. It is almost as though "literacy" is being seen as the first rung of a ladder which must be climbed in order for the UK to exist as a socially inclusive, economically profitable and well-educated society.

The international surveys of adult literacy, which were carried out during the late 1990s, suggested that the UK has a particularly large number of individuals who have not managed to attain this first rung of the ladder – according to the official statistics, nearly 1 in 5 adults have basic literacy problems. In UK policy documents, this statistic is combined with the belief that low literacy skills are expensive both to the country and to the employer. Long-term surveys of random groups of individuals within UK society suggest that low levels of literacy correlate with various social factors including unemployment and ill health (Bynner and Parsons, 1997): therefore, low levels of literacy result in the state having to pay more money out in the form of unemployment benefits and NHS upkeep. Economists suggest that changing work practices and the use of ICT in the workplace mean that workers with higher literacy skills become more productive within the workplace – "Higher levels of literacy make learning more efficient and thus allow workers to more easily adapt to changing job requirements" (OECD, 1997: 46).

These foundations are wobbly and, it can be argued, do not provide sufficient stability for this strategy to be built upon. There are some problems with accepting the statistics from the international literacy survey. Certainly Mary Hamilton and David Barton argue that they rest upon "an impoverished view of the roles of literacy in society." (Hamilton and Barton, 1999). Whilst the international surveys may have been meaningfully measuring something, that does not mean that they have provided a meaningful measure of national literacy levels. Additionally, the correlations between low levels of literacy and other social factors' are too often appropriated as causal: literacy levels are inferred as the cause rather than the consequence of social factors related to poverty. Furthermore, the economic models of thinking which suggest that there is national profit to be made across the economy by increasing basic levels of literacy are insufficiently explored or justified. Too often these arguments rest upon plausibility and the attractions of persuasion, rather than upon hard evidence:

> *"As a country we are around 20% less productive than the Germans (measured by real hourly wage), and we estimate that around two-thirds of this shortfall (13% of the 20%) is due to our weaker literacy and numeracy skills".* David Blunkett's speech launching Skills for Life, 1st March 2001.

Conflicting theories expressed in the work of Frank Coffield or Peter Robinson, amongst others, would argue that the national economic performance of the UK can best be enhanced by targeting the skills of those at A Level standard, rather than those who lack basic skills.

The Architecture

The government strategy for improving basic skills in the UK is impressive in its scope. Architecturally, we are being offered a strategy which seeks to address all aspects of improving adult basic skills from delivery and funding through to the promotion of learning opportunities. The most influential of the architects involved is Sir Claus Moser, whose working party's recommendations have been embraced almost completely in the final strategy. In 1999, the Moser report, A *Fresh Start*, recommended that "it requires better planning, an emphasis on local partnerships, clearer criteria for funding, better quality and more diverse learning opportunities, and improved and sustained promotion and recruitment" (DfEE, 1999). These thoughts can certainly be seen in the structure of the recent strategy.

However, whilst the structure of this strategy may be as impressive as the structure of the Millennium Dome, we feel that attention should move beyond the architecture and rest instead upon the actual content. (We recognise a slight irony in this – as we travel home tonight we will see the silhouette of the Millennium Dome marking the London skyline. Impressive as this sight is, the building expensively stands without content or visitors whilst its future is to be decided.)

The Zones, or the Content

The RaPAL conference offered a valuable opportunity to discuss with other practitioners what works in the adult literacy classroom, and this was a key element of the discussion following our presentation. We shared with our colleagues the concern that the content of the adult literacy strategy was too strongly focused upon accreditation of the learner's skills, rather than upon the process of the individual's learning experience – we are perhaps being offered "one amazing day" of accreditation rather than a longer-lasting experience. The focus on accreditation extends beyond the introduction of national tests; the restrictions of FEFC funding prompt an almost obsessive preoccupation with quantitative outcomes and there are fears that pedagogical creativity will be lost if learning increasingly becomes a controlled process of skills acquisition. Through discussion, we felt that one means of balancing the processes of learning and the demands of accreditation was to contextualise accreditation within the learning process rather than offering tests as the final hurdle of a basic skills course.

We do, however, recognise why measurable aspects of the strategy, such as the numbers of learners achieving qualifications, are being prioritised. Substantial investment has been made in this strategy and there is the need to have some form of measure by which the success of the strategy can be evaluated. It will prove difficult for the government to calculate the extent to which the adult literacy strategy has improved social inclusion, promoted lifelong learning, and affected national economic effectiveness. Therefore, a focus is being placed on that which is easily measurable – for example, the number of teachers who have received training, the numbers of students who are participating on courses, the number of students who have reached different rungs in the national targets. These facts are readily available and are assumed to be able to demonstrate, for example, whether the target of 750,000 adults improving their basic skills by the year 2004 has been achieved. But there are further problems with this thinking – Greg Brooks' recent research demonstrated that some students make little progress on adult literacy courses and their literacy levels can go down as well as up (Brooks et al., 2001). In this case, whilst statistics may suggest that 750,000 adults have improved their basic skills because 750,000 adults have passed basic skills tests – this does not mean that 750,000 adults have continued to demonstrate that they had better basic skills than when they took the tests. This led our discussion into the more general area of how achievement can be effectively measured, and has provided us with an area for future consideration.

There has been an increased flexibility of learning opportunities through the increased use of ICT and the variety of learning centres which are available to learners: this is good but there does need to be some kind of balance. Open learning is valuable – it offers the learner the opportunity to improve their skills at a time and place to suit him or herself; however it can also be an alienating and isolating experience. Through offering group learning opportunities alongside open learning, the learner gains additional benefits such as a sense of mutual support within the class and a shared identity as a learner. Using class-based activities is not as cheap or efficient as delivering courses through, for example, the Internet, but it may have wider benefits which are being neglected within the national strategy.

Although the core curriculum places some emphasis on group work, we are concerned about the balance which will be possible between open learning resources and class based teaching. If we truly wish the adult basic skills strategy to meet its aim of providing the foundations of a learning society, then we need to devote time and thought to the value of the learning experience as well as offering an overview of the intended outcomes in terms of tests and curricula.

The Mandelson Effect

This strategy has been effectively marketed and promoted through the use of the media and we welcome the fact that substantial attention has been given to this area of adult learning. However, we do have some reservations about the ways in which the language used to describe adult literacy has been manipulated. Both A *Fresh Start* and *Skills for Life* associated low levels of literacy with 'stress', 'losing out', worry' and 'uncertainty', amongst other things. Individuals with low literacy levels were described as fatalistic and inert. When you are marketing a strategy it is presumably tempting to use language to advertise your cause, but we feel uncomfortable with this use of language to describe real people within our society. The effect of this emotive language is pernicious; it encourages the reader to think of the individual with low literacy skills as being somehow inferior to the person with literacy skills. This effect is further accentuated through the presentation of correlations as causations, as we discussed earlier in this paper. The recent debate on the basic skills internet discussion group suggests that the recent 'Gremlins' television campaign continues to promote this morally loaded view of the individual with low basic skills. There seems to be a certain contradiction in the simultaneous attempt to promote basic skills courses to the learner whilst alienating and patronising the same individuals. If we want the strategy to attract high numbers of learners, we need to positively address the learners themselves as well as marketing the strategy.

Conclusions

Overall, we feel that much is to be admired about this strategy. It is a serious attempt to address the area of basic skills and it has attracted substantial attention and funding to this area. The strategy is overwhelmingly based upon the government using promotion and marketing to attract the learner, although the possibilities of using economic force are being piloted with job seekers in the North Nottinghamshire area. However, we question the foundations upon which this strategy rests and we are concerned by the amount of attention and responsibility which is being placed upon the learner – as Ann Hodgson suggests, placing the focus of the strategy upon the learner "opens up the potential for placing blame for social exclusion and poverty on non-participating individuals rather than looking to broader societal trends and issues". (Hodgson, 2000). The focus of this strategy risks placing the burden of society onto the learner, rather than the burden of the learner onto society. Our other main concern is that if the strategy continues to be developed *for* rather than *with* learners and educators, then when government attention is removed from this strategy, only the redundant architecture will remain.

Bibliography

Brooks, G., Davies, R., Duckett, L., Hutchison, D., Kendall, S. and Wilkin, A. (2001), *Progress in Adult Learning: Do Learners Learn?* London: Basic Skills Agency.

Bynner, J. and Parsons, S. (1997), *It doesn't get any better: the impact of poor basic skills on the lives of 37yearolds*. London: Basic Skills Agency.

DfEE (1999), *A Fresh Start: Improving Literacy and Numeracy. The report of the working party chaired by Sir Claus Moser*. Sheffield: DfEE.

Hamilton, M. and Barton, D. (1999), *The International Literacy Survey: What Does It Really Measure? Lancaster: Centre for Languages in Social Life.*

Hodgson, A. (ed.) (2000), *Policies, Politics and the Future of Lifelong Learning.* London: Kogan Page.

OECD (1997), *Literacy Skills for the Knowledge Society: Further Results of the International Adult Literacy Survey*. Paris: OECD.

The contribution of the mass media to adult literacy, numeracy and ESOL policy in England, 1970–2000

Iffat Shahnaz and Mary Hamilton

RaPAL Journal No. 50, Spring 2003

Iffat worked from 2002–2003 as a Research Associate on the ESRC Changing Faces Project and Mary, based at the University of Lancaster, is currently co-manager of the project.

Introduction

The research study on which this article is based is called "Changing Faces of Adult Literacy, Numeracy and ESOL: A Critical History of Policy and Practice 1970–2000". The study is funded by the Economic and Social Research Council and is a collaboration between Lancaster University, City University and the Centre for Longitudinal Studies, Institution of Education London. The fact that this study has been funded by the main national research council for the social sciences is an indication of the importance placed on the field of Adult Literacy in the UK at the present time.

The aim of the project is to track policy initiatives stemming from the 1970s adult literacy campaign up to the launch, in 2000, of a new Adult Basic Skills campaign (*Skills for Life*). Writing an historical account – like any other act of literacy – privileges some voices and marginalizes others. It draws upon fallible memories and incomplete records. We did not want to create simply a "famous person" view of history, but an oral and documentary history that would include the perspectives of different groups. We wished to explore how people position themselves in relation to policy, their beliefs and values and responses to it. We are therefore using documentary, longitudinal and statistical data alongside oral history interviews to integrate the perspectives of three key interest groups involved in the field:

- decision-makers in a range of government and national agencies,
- practitioners engaged in teaching and organizing within ABE programmes,
- adults with basic skills needs.

We are treating adult literacy, numeracy and ESOL (from now on referred to in this paper as ALNE) as a case study that illuminates the bigger changes that have taken place in post-16 provision in England and Wales and explores this experience in the context of parallel international developments.

We have chosen to report here on a specific aspect of our project: the role that the mass media have played in developing policy and practice in the field of ALNE. In this paper we describe our efforts so far to create a public timeline for these media events and some of the issues that have arisen as we did so. We discuss some of the difficulties of piecing together comprehensive data and documentation of media events from public sources and indicate the ways in which personal recollections and public records can work together to produce a history.

The role of the mass media

The involvement of the mass media in the field of ALNE is important for at least two reasons. First, as in any field of contemporary social policy, we believe that the mass media play a significant role in framing policy issues and promoting public awareness and debate. Within the field of educational policy this role is generally acknowledged but very few research studies address it directly. Of particular interest to the Changing Faces Project are the ways in which images of the learners and teachers are constructed in media messages, how the teaching-learning process is represented and how the goals of adult literacy are articulated. Since "illiteracy" has frequently been perceived as a stigmatised state, presenting positive images of potential learners and breaking down negative stereotypes has always been a challenge for those creating publicity in the field. In our project, we are gaining access to some accounts of the debates around this that have taken place within media organizations themselves.

Second, in the original adult literacy campaign in the mid-1970s, the British Broadcasting Corporation (a state-funded agency) played a major role. Published accounts describe how apolitical campaign, led by The British Association of Settlements and championed by an individual member of parliament, worked closely with the BBC to bring adult literacy into the mainstream of local authority adult education provision (see BAS, 1974; Withnall, 1994, Hamilton 1996; Hamilton and Merrifield, 1999). These developments did not take place in a vacuum, but came out of wider concerns and innovations of the time, including the introduction of the Open University in 1969 and optimism about the possible role of public and educational broadcasting in widening access to education.

When the BBC developed the television series *On the Move* in 1975, it pioneered media strategies that have since been used more widely. These include the idea of linking broadcast programming to a "hot-line" referral service and follow up support, including paper-based materials (Hargreaves, 1980). The *On the Move* programmes, starring Bob Hoskins, were shown at Sunday teatime with day-time repeats. A parallel set of radio programmes aimed to recruit volunteer tutors. By December of 1975, 10,000 people had contacted the national helpline. By 1978, 75,000 volunteers had been trained to take part in the adult literacy campaign from national referrals. Many worked on a one-to-one basis in the home.

On the Move aimed to break down the stigma and stereotypes of adults not being able to read and write. David Hargreaves has written a detailed diary account of this process (Hargreaves 1980) and describes how it took time to arrive at the formula that was eventually broadcast and how the process of piloting the scripts was fraught and difficult. The final result was the 'everyman' character embodied by Bob Hoskins, televised at peak viewing times, rather than in the "public education" time slots. Testimonials from adults with literacy difficulties were increasingly broadcast alongside humorous episodes of the programme. Viewers were encouraged to think "that's me!", not to feel embarrassed or ashamed and also to request and receive support as discreetly as they wished. The importance of this approach is confirmed in the interviews and group discussions we have carried out. The images used in this campaign were very different from the "horror" tactics used in the current Gremlins campaign publicising the Skills for Life strategy (See Barnes, 2003, for the advertising agency's account of this approach). Perhaps *On the Move* represents a different, more innocent era of public broadcasting and advertising.

However the issues are approached, the social construction of the man or woman who needs help in improving reading and writing remains full of contradictions. There is an unproven assumption that if you could read and write better then your life will improve. The idea of an individualistic, meritocratic society, where you can improve everything by your own efforts is very strong. In the present day, information and communication technologies are seen as being the main conduits for these efforts and are a major focus of funding initiatives, for example, UK Online and Learn Direct (Shahnaz, 2001). The government regards a more literate population as leading to a more economically and knowledge-rich society. However to have a literate and digitally literate population is not a guarantee of a more equal society.

Putting together a public timeline for media campaigns since the 1970s

Our timeline so far is shown in Table 1 on pages 136–41.

This demonstrates that media campaigns have been carried by adverts, storylines in popular soaps, specially created "educational" programmes, news items and documentaries. The original "*On the Move*" series was clearly a campaign because it was tied to a policy agenda and was explicitly trying to rally support, both from volunteers and potential learners. But there have been many media initiatives that are less-clear cut. For example documentary programmes such as those shown periodically by *"World in Action"* or *"Dispatches"* could be very influential in shaping public perceptions of literacy issues, but were isolated programmes with no follow up or co-ordination with policy. Other programmes such as *"Parosi"* (Neighbours) or the family literacy initiative *"Read and Write Together"* were offered more as direct teaching/learning resources than having a campaigning orientation. Regular representations of literacy, numeracy and ESOL appeared incidentally in newspaper and television coverage and may have unintentionally promoted existing stereotypes and run counter to the deliberate campaigning messages. For example, a recent episode of the soap opera *Brookside* featured a character experiencing difficulties with an uncaring and bureaucratic Further Education college.

Accessing information

We have not been able to identify any central place where information about media programmes is kept. We have identified a number of related archives: the BBC Written Archive, the Workers Educational Association archive, the National Institute for Adult Continuing Education, and the Write First Time archive at Buskin College. We are also discovering individual organizations with their own archive records and people with large personal collections, some of whom are deliberately saving this material for the historical record.

Personal recollections gathered from our interviews are another source of information about media programmes especially from the people who were directly consulted or involved in the making of them. The most popular programmes remembered without prompting in both individual and group interviews are *On the Move* and *Parosi*. Occasionally someone recalled a programme that we did not have on our timeline. Then we were able to check public sources and add it in. For example a very early "World in Action" documentary programme entitled A Best Kept Secret was mentioned by an informant and then identified in the article published by Alex Withnall on the origins of the adult literacy campaign (Withnall, 1994). When *On the Move* was mentioned this was remembered more in terms of raising awareness about the issue and overcoming and challenging the stigma of "illiteracy".

The programmes are also remembered fondly by people recruited as volunteers at the time and the accompanying tutor handbook – "that green book" – is also recalled as one of the very few resources available at the time. The campaign made one of the interviewees want to get involved in this area of work in which she has remained until this day:

> *"I had never seen anything like that before. I thought it was really well done".*

Parosi tends only to be recalled by people active in the ESOL movement. Although it was the first of its kind, it did not have the same impact culturally and politically. Its impact was felt more in areas of the country with high concentrations of ethnic minorities at the time such as Birmingham, Coventry and London. Some ESOL practitioners interviewed have mentioned *Parosi*'s impact on the Local Authority funding of home tutor schemes. In other areas, the needs of these groups were not considered to be important. For example, one of our case study areas, Norfolk, had very little ESOL during the 1970s and 1980s. This has now changed due to the more recent policy of dispersing refugees around the country.

As for numeracy campaigns, with prompting we have found some awareness of the *Count Me In* Campaign by the BBC. This seems to be the one best remembered over the decades. The use of celebrities in these programmes clearly affects peoples' memories. For example, one London based interviewee commented:

> *"That was Carol Voderman, was it Carol Voderman?" . . . I remember their names but I don't actually remember what they looked like and what they did . . . they obviously didn't make a major impact on me did they?"*

This illustrates the power that well-known personalities can have, but perhaps also the limitations?

1970s

Date	Event	Themes	Source
17th of July 1972	Granada World in Action documentary: *A Well Kept Secret*	Public awareness-raising	Article by Alex Withnall and adult learner informant recalled it in a group discussion. He remembered and identifies with a milkman who couldn't read or write. First time he had seen it in the public domain
Pre-BBC Campaign	ITV Thames TV Item on magazine show *Good Afternoon* with feedback possibility	Public awareness-raising, referral for volunteers	Interview with informant involved at the BBC
Early 1974	Teaching Adults to Read (Radio 4, 8 programmes)	Recruitment and Pre-training for volunteers in preparation for the *On The Move* campaign programmes	Interview with two informants involved at the BBC
October 1975 (repeated the next year)	Launch of *On the Move* Produced by David Hargreaves	Media awareness raising and referral of potential learners	Interview with informant involved at the BBC Participants in 3 Group Interviews all mentioned *On the Move*
October 1976 (repeated the next year)	*Your Move* Radio 2 20 x 25 min programmes	Tutor Support	Interviews with two informants involved at the BBC

1976-7	*Next Move* – Radio 2	Readings aimed at tutors	Interviews with two informants involved at the BBC
	Move On – Radio 2	Readings aimed at tutors	Interviews with two informants involved at the BBC
October 1977	*Parosi* BBC TV series produced by Robert Clamp	ESOL (Hindustani)	Documents (Workbook) Interview with two Key informants, one involved with ESOL practice, one from the BBC/practitioner interviews
1978-ish	*Figure It Out* BBC Series	Early ancestor of numeracy programming	Interviews with key informant involved at the BBC
1987	*Make It Count* YTV, produced by Peter Scroggs	Numeracy Series, with back up materials from NEC	Documents: Sargant, 1990, p.13
1978-9	*Speak for Yourself* BBC produced series	Aimed at wider range of community languages	Documents (Paper written by practitioner) Interview with informant at the BBC.

1980s

Date	Event	Themes	Source
1983	*Make It Count* Ch4 (updated from original YiV programmes)	Numeracy Series, with back up materials from NEC	Documents: Sargant, 1990, p.38
1983	Numbers At Work Ch4	Employment related Numeracy series	Documents: Sargant, 1990, p.38; gMRB research report, Sept 1983.
1983	*Counting On* Ch4, series of 10 programmes	Next stage numeracy	Documents: Sargant, 1990, p.38
1984/1985 (3 seasons)	*Switch onto English* Programme aired on Sundays on BBC1 at 10.30-10.55 am (13 January-17 March) Wednesdays, on BBC2, at 12.55-1.20 p.m (16 January-27 March)	Supporting ESOL teaching	Documents (Workbook) Interviews with two informants, one at the BBC, one practitioner.
1985-6	*Write Now*	Supporting Writing	Interviews with two informants at the BBC
1986	*Write On*, YTV /Channel 4 Series of 10 programmes	Supporting Writing, in association with the NEC and ALBSU	Documents: Minutes of planning meetings/Informant from Ch4; Sargant 1990:p 66
1987-8	BBC TV: *Spelling It Out*	Supporting Spelling	Interviews with informant at the BBC, one practitioner.

1986 or 1987	Granada/ITV World in Action Documentary *Starting at the Bottom*	MORI poll on extent of need in Rochdale. Also drew on National Child Development Survey work at Lancaster	(Documents: video) Member of project team was involved.
1989	Two BBC series, *Stepping Up* and *Step up to WordPower and Number Power*, with linked publication.	Linked to new student accreditation initiatives validated by City and Guilds	Interview with informant seconded to the BBC to work on this.

1990s

Date	Event	Themes	Source
1992 10am on Sundays	BBC *Inside English*: 8 programmes plus video and trainers' notes Direct teaching of ESOL	Linked to HMI Report Bilingual Adults in Education and Training Aimed at learners and teachers	Documents: ALBSU report 1992
February 1995	*Read and Write Together* One week organized by ALBSU and BBC Four programmes plus series of short ads?	Linked to ALBSU Family Literacy Initiative Raising public awareness and to train teachers to understand family literacy	Documents (ALBSU Family Literacy Newsletter; Paper written by practitioner Informant); interview Newsletter; Paper written by practitioner with informant at BBC
1997-8	*Count Me In* – BBC. Programmes launched with accompanying CD Rom	Numeracy support, BSA involved	Interview with ex-BBC informant who developed the programme/practitioner interviews
1998 Feb	*Brookie Basics*, Ch 4 Soap storyline in Brookside (Also similar one in East Enders on ITV around same time) With referral line and follow-up materials	Part of the activities for the National Year of Reading organized by the National Literacy Trust	Documents (workbooks and video) Pilot Interview with learner. London LSERN Group member

2000s				
2000		Gremlins Campaign promotional campaign *"Get On"* by DfES. Part of Skills for life Strategy	Raising public awareness, referral to Learn Direct hotline	Documents (publicity materials) LSERN comment "horror of how this would help things move forwards". However some pilot interviewees thought it was good but wouldn't use/call up the helpline.

Undated: Write Away; Help Your Child with Reading programmes 1–8; Sum Chance – 8 (maths); A Way with numbers, 5 and 17; and Summing Up – Numeracy

Over time, media campaigns appear to have become more sophisticated and wider in their "reach". Recent campaigns such as the *"National Year of Reading"* and the campaign to promote the *Skills for Life* policy have used multiple media outlets to saturate the audience with the main message. The *"National Year of Reading"* in 1998 was perhaps the most co-ordinated media literacy campaign to date, covering children's literacy as well as adults and a broad range of media messages. It used ads, soap-lines in East-Enders and Brookside, newspaper articles, posters and linked commercial campaigns such as *"Free Books for Schools"* initiated by Rupert Murdoch's company, News Corporation, and also tied into the promotion of snack foods.

In the *"Get On!"* campaign launched as part of the English Skills for Life policy, television and radio broadcasting have been complemented by the use of posters, postcards, desk calendars and a range of other publicity. A parallel campaign in Wales has its own separately designed broadcast publicity. The use of help lines and referral services has now been linked into the promise of a whole world of on-line learning through Learn Direct, in line with current excitement about the possibilities of new Information and Communications Technologies (ICTs).

Issues emerging about the use of mass media

1. The Resourcing Issue: Co-ordinating Demand and Supply

The interviews to date have highlighted resourcing issues. Right from the start of the *"On the Move"* campaign tensions emerged from the challenge of needing more services in place to meet the anticipated increase in demand. The evidence from the 1970s clearly shows that many practitioners were starting from scratch in terms of teaching materials and resources. Although there was expertise in voluntary organizations, especially in the field of ESOL, provision was ad hoc and practitioners were isolated. One interviewee talked about having set something up in her living room to deal with the demand. Another interviewee, a practitioner at the time, described her experience in Kirby, Liverpool. She received a consultation document on what became "*On the Move*" and also a phone call from the new national Adult Literacy Resource Agency (ALRA) asking:

> *"'Are you ready to meet this demand? We are expecting a huge response, can you cope?'*
>
> *People responded: 'I don't know whether I can cope or not, but we should take it anyway we'll have to develop systems that result from that.' It was a wake-up call, there was going to have to be more provision because there would be huge demand through the broadcasts"*

There was very little infrastructure to support the practice that was being developed on a large scale for the first time. Resources had to be created. New partnerships, training and ways of working were developed. Over the years the BBC campaigns were considered to be too successful and agencies could not cope with the demands. David Hargreaves has suggested that this was the issue that eventually slowed down the national media campaign, replacing it with more limited and carefully targeted local publicity.

The thrust of current *Skills for Life* policy initiatives is that many of the adults most in need of basic skills help will be difficult to reach and that it will be important to "drive up" demand for courses. This seems to be in contradiction to the experience of past media campaigns where the response overwhelmed available provision and publicity had to be fine tuned to appeal to particular target groups that the government wanted to reach.

2. The Impact of Changing Media Technologies

Huge changes have taken place in available information and communication technologies since the 1970s. There are several implications of this for our study.

First, we encountered the changing modalities of storage, reproduction, circulation of information in very graphic and material ways – the quality of reproductions of teaching and learning materials, the purple ink of "banda" machines in the early 1970s, unevenly typewritten memos, old fashioned seals, signatures and crests on official documents, grainy videos and black and white photographs.

Second, storage issues. It is not just a matter of archiving documents, but of having the equipment to access them, especially where these are electronically stored. There are also visual materials such as film and video.

Third, the definitions of "literacy" itself are challenged by the technological changes that have taken place. Are electronic literacies an extension of print literacies or are they a different area of knowledge and expertise? Is e-mail communication or text messaging a new "hybrid" form of communication, not identical with either writing or speaking? How should the different technologies be linked together in teaching and learning practice? Current policy in ALNE places great store on the new ICTs as the conduits for improved literacy and numeracy. But does this involve the inclusion of electronic based literacies as part of reading and writing as well as being considered as a vehicle for teaching traditionally defined skills?

3. Assessing Need and Impact

The *increase* in the estimated numbers in need of ALNE over the years from 2 million in the 1970s to seven million in the present raises some interesting questions in relation to media coverage. Has the unique work in publicising and addressing adult needs in ESOL, Literacy, Numeracy been successful? In particular, how should the impact of media campaigns be monitored and evaluated? Perhaps the increase indicates the success of the media in raising public awareness and breaking down stigma around literacy enabling the extent of ALNE to be ever more visible. Or perhaps these estimates are moving in response to some other, political agenda, in a more complicated relationship to media activities.

Apart from the unusual evidence assembled by David Hargreaves (1980), we have been surprised at the lack of evidence in the public domain about the impact of media campaigns. This lack of evidence points to a bigger issue mentioned in the interviews: that there has been little systematic effort to track learners in the field to identify their progression routes, where they have come from and where they move on to.

The most common monitoring activity associated with media campaigns is to estimate numbers watching programmes via audience panel surveys, and to report the numbers using referral hotlines as a gauge of public response. Much of this evidence is collected by broadcasters themselves. The numbers responding are typically of an order many times the rate generated by local outreach activities. As an example of this, we know that the first series of the "Gremlins" advertisements resulted in a huge demand for information. ("50,000 people have phoned the hotline in the one year since Sept last" Source: TUC Learning Services website, accessed Sept 2002).

Beyond this basic monitoring exercise, little effort appears to have been made, even during the current campaign, to pursue the research opportunity offered by this surge of interest and find out more about who the callers are and where they end up. Which groups of people respond and who do not? Do phone calls translate into people taking up learning opportunities? Where do people go if local service providers are not able to cope with the demand generated by media campaigns? And should we call a campaign "successful" if existing provision cannot cope with the demand it generates? This is a question of co-ordinating the media campaign and the policy and practice response. Our research, therefore points to the need to think through the assessment of media impact in general: on whose terms and within what parameters and policy agenda can we consider a public media campaign in the field of ALNE to be successful?

Conclusion

We are still at the point of filling gaps in our timelines and of thinking through the issues that are emerging from the accounts we are collecting. Even so, we have been amazed so far at the richness of the information generated by our research especially when we put together both documentary evidence and oral history interviews. The archive that we build from this project will be a resource for the future, enabling researchers and others involved in professional development courses to explore these issues.

Bibliography

Barnes, S. *Why Gremlins?* St Lukes Advertising Agency. http://www.dfes.gov.uk/geton/docs/11gre10.doc Dfes website, accessed Jan 2003.

Barton, D and Hamilton, M. (1998) *Local Literacies: Reading and Writing in One Community* Routledge

British Association of Settlements. (1974). *A right to read: Action for a literate Britain*. London: Author.

Bynner, J. and Parsons, S (1997) *It Doesn't Get Any Better: The Impact of Poor Basic Skills on the lives of 37 year olds*. London: Basic Skills Agency.

Hamilton, M. (1996) "Adult literacy and basic education". In R. Fieldhouse (Ed.), A *Modern History of Adult Education*. Leicester: National Institute of Adult Continuing Education.

Hamilton, M. and Merrifield, J. (1999) "Adult Basic Education in the U.K. Lessons for the US." in National Centre for the Study of Adult Learning and Literacy *Annual Review of Adult Learning and Literacy*. Jossey Bass.

Hargreaves, D. (1980) *Adult Literacy and Broadcasting*. 1st Edition. Frances Pinter (Publishers) Ltd Organization for Economic Co-operation and Development (1997) Literacy Skills for the Knowledge Society. Paris: OECD.

Ozga, J.(1999) *Policy research in Educational Settings: Contested Terrain*. Open University Press. Perri, 6. and Jupp, P. (2001) Divided by Information. Demos

Shahnaz, I. (2001) *UK Online: An Answer to Social Inclusion*. City University

Withnall, A. (1994) "Literacy on the Agenda: The Origins of the Adult Literacy Campaign in the United Kingdom". *Studies in the Education of Adults*, Vol. 26 pp.67–85

A3 Practitioner roles and identity

Further RaPAL references about practitioner roles and identity

Ivanič, R. and Richmond, J. 'We Need a Clearer View', *RaPAL Bulletin* No. 3, 1987.

Herrington, M. and Clayfield, A. 'Guidance in Adult Basic Education: A Report on an Action Research/Staff Development Project in Leicestershire', *RaPAL Bulletin* No. 20, 1993.

Tobias-Green, K. 'Counselling and the Adult Learner', *RaPAL Bulletin* No. 20, 1993.

Hillier, Y. 'So What is Good Practice?', *RaPAL Bulletin* No. 30, 1996.

Gardener, S. 'Student Writing in the 70s and 80s: What We did, Why, What Happened', *RaPAL Bulletin* No. 40, 1999–2000.

Peutrell, R. 'Literacy in the FE Mainstream: Good Practice and Our Class (room) Struggle', *RaPAL Bulletin* No. 47, 2001.

Talk into text: Reflections on the relationship between author and scribe in writing through language experience

Wendy Moss

RaPAL Bulletin No. 40, Winter 1999

Wendy Moss is a member of the RaPAL Production Group and works for the City Lit Training Unit in London. This article is based on research done in 1986 as part of her MA.

Navel String

When all my baby born
I collected the navel string
and send it home to Nigeria,
to my mother in law
for them to bury it at home.
It could be buried
next to a coconut tree or orange tree.
After that,
the tree belongs to the child.

(Ngusi Thompson in *Every Birth it Comes Different*, Hackney Reading Centre and Centerprise Publications, 1980)

This moving account of an African tradition is taken from a book produced in Hackney Reading Centre; many of the texts were produced by women writers talking on tape or dictating to a scribe, then discussing and editing their writing. This technique is often termed The Language Experience approach. This article looks at this process, particular writing with a scribe.

In adult literacy teaching, there are two reasons why language experience writing has been used: firstly, as above, to give a voice to those whose written language skills are early but whose oral voices are rich with content, and secondly as a technique to provide early readers with adult based, and predictable so readable simple texts. The technique derives from a particular view of literacy learning that stems back to Freire.

My research looked in detail at tapes and transcripts from student-tutor pairs in adult literacy groups in the mid-80s. I was on a language studies course and I was particularly interested in the transition from speech to writing. I asked several pairs of tutors and students to try using language experience (the student-writer dictates and the tutor scribes) and then analysed the ensuing tapes and texts. The rest of this article is a summary of my findings. I should note that I was concentrating on issues of transcription; I didn't look at later stages where student-writers might re-read or re-listen to their texts and edit them.

Written and spoken English are different

'I can say everything but I can't write it' – every writing tutor has heard this. The cry of frustration of the less confident writer. At one level the barrier might be the mechanics of writing: spelling, or use of standard English for more formal written genres. Oral language also carries much meaning through placing of words, intonation, gesture, accentuation which cannot be transcribed. Finally, there are many common written genres that are very different from those of everyday speech, and the new writer must learn to recognise and reproduce these. Given this range, I expected some of these differences to impact on transition from talk to text using a scribe.

My first example from my research tapes is from a student-tutor pair who were discussing a royal wedding that had been on television that week. Helen was telling Jennifer (the scribe) what she would like to write. It was the first time they had tried this approach to writing.

Jennifer: Well then, now do you want to include there the blessing of the rings?
Helen: Oh yes. And then they blessed the rings and then they got married.
Jennifer: Well yes. How would you say that?
Helen: While they were getting married they blessed the ring.
Jennifer: Yes but how did it affect you, that is really what we want to know?
Helen: (interrupts) It was very nice.
Jennifer: How did it affect you? Something you hadn't seen before?
Helen: Yes it was something I hadn't seen before.
Jennifer: Ah, you were impressed with the blessing of the rings by the archbishop?
Helen: That's right.

Written text: *I was impressed by the blessing of the ring by the archbishop.*

In the second line of the extract, Helen says what she would like written down: *And then they blessed the rings and then they got married.* Instead of scribing this, Jennifer spends the rest of the dialogue trying to negotiate something different. Jennifer was new to language experience. She felt her job was to try and encourage Helen to write in a 'literary' rather than conversational style (as she told me later). She has therefore somewhat missed the point of the process. However it is worth looking at *why* Jennifer tried to change Helen's text as it illustrates some of the issues that arise when speech is transcribed into writing.

Jennifer may have not liked Helen's first offer of text *'And then they blessed the rings and then they got married'* because of the repeated 'ands'. This is very typical in the texts of emerging writers because in everyday informal situations we do not *talk* in formed sentences, but in bursts of ideas of seven or eight words long, generally linked with ands, erms or buts. Speech is generally more *fragmented* (See Chafe (1982) if you want to read more on this). In speech we don't notice these 'ands' because speech is gone in a flash. But overused in writing they sit and stare at us from the page and can sound repetitive. So skilled writers use other devices for connecting ideas together. Nevertheless, there is no reason why Jennifer could not have recorded this sentence – Helen might have changed it later if there were too many 'ands' in the whole piece.

Helen's second offer of text: *'While they were getting married, they blessed the ring'* is more integrated (using 'while' to link the sentences together), but this is still problematic as far as Jennifer is concerned perhaps because its not clear who the 'they's' refer to. Jennifer may feel Helen needs to be more *explicit* for a non-present reader (another distinguishing feature of writing). So Jennifer quietly inserts 'by the archbishop' at the end of the final written sentence.

Finally Jennifer seems to feel the writing lacks expressive vocabulary, characteristic of good 'literary' writing. She ignores Helen's suggestions 'It was very nice'. The final text: *'I was impressed by the blessing of the ring by the archbishop'* is not really in Helen's words at all.

Jennifer hadn't, yet, grasped the importance of giving control to the learner-writer in language experience writing. However, my tapes showed that even where a scribe is trying faithfully to do this, she may still intervene and control the text in a myriad of ways.

For example, in the sequence below, Carol and myself were talking about Carol's childhood in South East London:

Carol: Well they always knew who was in and who was out, you know. All the neighbours knew which neighbour was in and which neighbour was out and they used to come and knock at me mum's door, come in and have a cup of tea. Chat.

Wendy: (writes) who was in and who was out. They used to come in and knock at my mum's door and ermmm come in and have a chat?

Carol: Yes
Wendy: (reads while writing) and come in and have a chat.

Written text. *All the neighbours knew who was in and who was out. They used to come in and knock at my mum's door and come in and have a chat.*

I have kept closer to Carol's words than Jennifer but still omitted erms and repetitions and turned what she has said into two complete integrated sentences. Carol was a lively and skilled oral storyteller, but fluent oral speech is more fragmented (see above); I've used my writing skills to compress her story for writing. But I've not really checked it with her first. I've also changed 'me' to 'my' ('me mum's' to 'my mum's'.) This subtle shift to more standard language by scribes was quite typical. We often left in some regional vocabulary but selected out or made minor adjustments to other aspects of dialect. (My guess is that we did this almost unconsciously).

Learner-writers and Tutor-scribes in dialogue

The other area I explored in my research was how scribes intervened orally in the writing of the text. I found that some confusion often happened because there were two conversations happening at once: one conversation was about what should go into the writing, the other was an oral conversation about the topic itself. For example:

Wendy: . . . seven boys and the girls. We can change this after. Now where to go next. Do you want to speak about your street first?
Carol: er how we lived in New Cross
Wendy: Yeah
Carol: Okay then
Wendy: We lived at New Cross?
Carol: sorry

Carol seems to have some difficulty following my meaning. This is because throughout this piece I was talking in what I called 'Discourse Field A' – I was talking about what should go in the piece of writing, but Carol was talking in 'Discourse Field B' – she thought I wanted to make conversations about her experiences. This conflict of conversational goals – ie talking at cross purposes – was quite frequent in the tapes. Tutors, on the whole, where more often operating in Discourse Field A (talk for text) than students.

Tutors controlled the text in other ways too. There were regular interventions by tutors that directed the development of the text. For example, sometimes we *asked questions*:

Jennifer: Would you like to say anything about the bouquet?
Wendy: Do you want to speak about your street first?
Sue: Lets get back to this set pot thing. What was the first thing you had to do?

Sometimes we *reformulated* what students said in 'better' words:

Terry: I hate it (Laughs). The whole thing you know
Janet: You feel it is something that . . . It's something you feel very strongly about . . .
Janet: Do you want to put that at the end? That you feel very strongly about it?
Terry: Yes can do.

Sometimes we offered the *first part of a sentence* for the writer to complete:

Wendy: You said it was a dead end street because it had . . .
Carol: school at the end of the road.

So, in many parts of the tapes, tutors were using dialogue to control the shape and content of the emerging text. I noticed also that they were doing this using many of the techniques used in client-centred counselling and I discuss this more below.

The role of the tutor-scribe: controlling or disempowering

At first sight I thought the tutors interventions (apart from in the case of Jennifer) were intended to keep control with learner, but were still, in the end, empowering. However, discussing the issue with Sue Gardener (see her article in this issue), I modified this position.

We noted that if a learner was *new* to language experience (as in most of my samples) they would naturally be more hesitant and find the switch between discourse fields more difficult. However, we felt that, in our experience, students did increasingly direct text more and more as they became used to the process. One of the student-writers in my tapes had experience of dictating exam answers to a teacher at school. She was more confident and seemed to be operating in Discourse Field A much more with her tutor.

I also noted (and expected) that tutors were likely to intervene more with some genres rather than others. For example, in a narrative (story telling), the organisation and sequence are easier because they follow the sequence of events in the world. Carol and myself, writing about Carol's childhood struggled where to start at first. However, once we moved on to a story about her having to wear her brother's trousers, Carol was able to dictate her text with more confidence.

Sue's interventions with Margaret, writing about Wash Day were largely about where to start, and making terms explicit. Although there is a natural sequence to describing Wash Day, the difficulty was setting it in a context at the beginning:

> Sue: Let's see if we can start with . . . You can think of just a particular washday?
> *Margaret:* No well (that/it) depended on the weather like
> *Sue:* Alright, that's a good start. It used to depend on the weather. Let's call it 'Wash Day' . . . it used to depend on the weather. Right. We'd better say why. It's obvious in a way that if it was wet . . . What did you do . . .

Perhaps the hardest transition from speech to writing is expressing a point of view. Terry dictated a piece on 'child abuse'. Until Terry decided on a theme to develop, she seemed to struggle from line to line, listing points disconnected to each other, even though she was able to be quite fluent in dictation.

So, although the whole point of language experience writing is that the learner-writer is in control of the text, my research suggested that it is perhaps simplistic to think that any new writer can dictate *any* text to a scribe (or on tape) with ease: there was some justification for sensitive intervention. I also noted that most scribes, when they did intervene, used similar techniques to those used by client-centred counsellors. For example, the reformulation 'So you lived at the end of the street' both validates and affirms a speaker (a counselling technique) but also directs the conversation and the text.

What is to be learned for practice

Nevertheless, scribe-tutors need to find boundaries between facilitation and a control that is disempowering. My suggestions are:

- The learner writer and scribe need to agree clearly at specific points that the talk will become writing – ie to ensure both are operating in Discourse Field A (talk for text). As the learner writer become more confident much more of the talk will be operating in this field.
- Tutors can make their role clearer when they are using strategies to elicit text. For example: 'I'll ask you some questions if you get stuck because it may help you work out what you want to say'.
- If the writing is non-chronological, learner writers may have particular problems in getting started, with ordering and with ending. It may be helpful to explain why it is difficult and discuss possible ways it could be ordered. I could have said to Carol 'What is the most important thing you want your reader to know first?'
- It is, of course, very important that issues of language awareness are discussed with student-writers. There has been much discussion about the importance of encouraging student writers to use their own language variety in writing rather than

a standard form (see Sue Gardener's article in this issue). However, the differences between oral and written forms of language do not only lie in dialect. For example, in oral story telling, it is common to switch to present tense in the middle of a story, at its crux. This heightens the tension and the 'realness' of the story. This is a sophisticated oral story telling strategy but generally seen as 'tense inconsistency' in writing. As learner writers explore and choose written styles, it's important they see themselves as learning to be bilingual (or multilingual) in speech and writing. The language skills we use orally are not always easily transferred to writing and new techniques have to be devised.

In summary, if tutor scribes working with new writers are as explicit as they can about the process, in the ways described above, and the initial transcription is followed by a period of discussion and editing, then student writers should be able to take increasing control and see the final text as genuinely 'their' writing.

References:

Chafe, W. L.(1982) 'Integration and involvement in speaking, writing and oral literature'. In D.Tannen (ed) *Spoken and Written Language: exploring orality and literacy*. Norwood, MNJ; Ablex

Moss, W. 'Controlling or Empowering? Writing through a Scribe in Adult Basic Education', in Jane Mace (ed) (1985), *Literacy Language and Community Publishing* Multilingual Matters Ltd.

The literacy teacher as model writer

Annette Green
RaPAL Bulletin No. 39, Summer 1999

Annette Green used to work as an adult literacy tutor for the Adult Community Education College in Lismore in Northern New South Wales, Australia. She is now a vocational education and training tutor at Charles Sturt University, Wagga Wagga (a large regional city situated in the Riverina Region midway between Sydney and Melbourne NSW).

We often discuss "modelling" in terms of writing, so I decided to take this a step further and write for my Certificates in General Education for Adults (CGEA)[1] class, who are completing "Module 2: Reading and Writing".

My students had already made a few attempts at writing for self-expression and had not really achieved the required competencies at this level, and in this task in particular. I therefore decided it would be an appropriate genre to personally model for the class, and hoped that the 'ensuing discussion would open up the field of personal writing for them. They seem to feel that functional writing is "more their level", and we have spent a lot of time on procedural texts, letters, writing for public debate and other functional styles of writing. The genres they feel most competent with are the ones which have a clear format which they can follow. They equate personal writing with failures in writing at school – they could never decide what the teacher wanted them to produce. This group is accustomed to discussing and sharing their writing. They carefully offer suggestions as to how their own texts and those of other students could be improved. Although they continually look to me for reassurance, they have strong ideas as to how texts should sound. Their writing style often seems to reflect the fact that they, as people who have done limited reading as adults, tend to gather much of their language from radio and television.

The class was very excited when I outlined what I had in mind for these sessions. Were they expecting a minor masterpiece, I wondered nervously as I approached the first session. I am still trying to quash the idea of "Annette, the master speller" and replace it with the idea that literacy learning is lifelong and that nobody has mastered the spelling of every word in the English language. I consoled myself with the thought that at least they were unlikely to idealise me as "Annette, the perfect writer" after these sessions.

Although I have modelled texts and involved the class quite frequently with joint construction of texts, I had not written "off my own bat" in front of the class before. I drafted, edited and completed the text on a large-screen computer. By printing out and labelling the various drafts, we were able to check the progress of the text. This is a useful technique in presenting student work for moderation, as the issue of teacher assistance can be evidenced more easily. The enormous screen meant that everyone could see what was being written, so I did not always read aloud. Two members of the group have recently bought home computers and we have all been trying to experiment with writing first drafts on the computer (rather than copying the "final version" of texts for publication, an approach I like less than writing/composing directly on to the computer).

We had an extended discussion on the fairness or otherwise of two students using a spellchecker on assignment pieces. We concluded that it was a great tool in many circumstances, and that if our spelling was close enough for the spellchecker to pick up what word we meant, we were getting somewhere.

At the beginning, the students were very interested in helping me. I found it quite disconcerting – in some ways, it was more like joint construction of a text rather than me demonstrating how a person writes. We had done a lot of joint construction, which I found an artificial experience insofar as I was not really at ease composing and constructing in front of a class while trying to keep the writing my own rather than just using it as a teaching tool.

I learned during this experience that the parameters of our set curriculum, the CGEA, made the writing rather constrained. I was always checking back to what the writing needed to demonstrate in terms of criteria and competencies. When trying to decide whether each text achieved the competencies required for each type of text at each level, I became much more in tune with the difficulties experienced by both the students as authors and myself as a marker. After this experience, I wondered if we should change the curriculum so that all the competencies are on a checklist but are not grouped according to assessment piece, so that students may assemble a portfolio of texts which demonstrate the achievement of all the competencies without them being assigned to a particular assessment item. The rewritten CGEA is closer to this idea and consequently easier to use.

The class continually questioned both the material and the style, something I am still not sure that they do with their own texts. They were also very interested in any spelling or typing errors I made. They wanted the text to be perfect from the beginning, a reflection on the type of writing they believed good writers would turn out "first go", rather than editing at a later stage. I wanted the class to focus on the writing process, the ways we get material down, rather than sticking at the surface level and concentrating on a perfect product from the start. Their comments showed they were shocked that I could happily read over the text without continually fixing it up. I think this was very instructive for them.

Beneath the surface

In the second writing session, the class was a lot more vocal and critical. Words were suggested continuously, and I became far more receptive to the class and went along with their suggestions far more readily. Perhaps I was considering the idea of giving up "personal voice" in favour of altering the focus to working together on my text. It seems it *is* possible to retain ownership while accepting comments and suggestions for revision. I also really enjoyed the fact that we came to grips with topics such as editing, being overly concerned with perfection during the first draft and changing tenses while in a text. For example, the idea of present and past continuous was discussed at length during this session.

The strategies that the class used to try to and tighten and amend parts of my writing changed quite markedly between the two sessions. In the second session, the focus was far more on the text itself, and what the writer was attempting to convey. The first session was dominated by content questions and pre-writing activities, combined with an almost obsessive interest in surface details. In the second session, the class jumped all over the text and indicated sections they found strange, obscure or clumsy. They questioned the inclusion of sections they felt were irrelevant and encouraged expansion in other areas.

The total experience was very informative for me as a teacher as I found myself very much in the position that the class must be in when they begin their writing assignments. To some extent, we used a similar format, in that we discussed the topic and criteria for assessing the piece at the beginning, as we do with the students' pieces. We did not examine the model text in this case, as we usually do, as I was hoping my text would be useful as a model. I tried at all times to keep the writing in the kind of language and style I thought would appeal to the class. Mindful that this class have had very limited reading and textual experiences, I didn't want to write in a style which was inaccessible to them. I believe I may have gone too far in the other direction, and that the final text is overly simplistic and thin. I was trying very hard to stick with simple sentence structures, as the main fault with all writing produced by the students is that they attempt enormously complex sentence structures which don't quite hang together.

I found the fact that we do not share metalanguage to describe faults and features of texts very frustrating. At the request of the class, we spent some time early in the term examining traditional grammar. We have not devoted much time to this because it is so daunting and time consuming, but they do have some basic terms in place. They wanted them so they can use them to assist their children with homework. (One Lismore school in particular seems to have bowed to the pressure to go "back to basics" by giving the students "Fill in the noun/Add an appropriate adjective" style of grammar exercises.) I prefer the functional approach to grammar, but once again, the limited timeframe means it is difficult to cover topics in depth.

This was a very challenging and interesting experience as I have always considered writing as an activity pursued in reflective silence, with the writer attempting to capture the perfect text from the deep recesses of the mind while simultaneously filling capacious wastepaper baskets with rejected versions. I deliberately avoided pre-planning the actual writing task before the class, so that it would be more honest and more interesting. I am not sure now that this was not a mistake, but at least I was closer as a writer to the position of my students: because writing is a more permanent record of one's language proficiency than is speaking, the demand for unrehearsed writing is more threatening to the learner" (Brown & Hood 1989). Because I put myself in a vulnerable position, exposing myself as a writer, I became more realistically aware of the thoughts and feelings of a less experienced writer and perceived more clearly how writing can be so daunting to my students. I was also allowing them to see the risks writers take in the mistakes I made, so that they could see that there was more to redrafting texts than merely editing.

I wanted to employ a revision model, where changes are tried and considered at all levels. "There is overwhelming evidence that older and more competent writers do more revising for meaning and make more sentence and theme changes than do younger and less competent writers" (Fitzgerald 1988). One of the main points I was attempting to demonstrate to my students is that very few writers get it right first time. I was attempting, as a model writer, to demonstrate the process of reworking and revisiting a text on many levels. The main objective was to demonstrate that, as writers, the primary task is to "get their ideas down on paper, organise these ideas and develop them without the simultaneous need to satisfy the surface demands of written text conventions" (Soter 1987).

Many adult literacy students may have very wide and interesting experiences of spoken genres of language and yet are at a much earlier stage of development with written texts. If we consider the features of spoken and written language as a continuum (Hammond et al. 1992), the students must come to appreciate that the features of written language vary most widely from spoken language when it comes to encapsulating personal experience and expression of responses to experience. I believe that this may be one of the reasons that this class finds functional texts easier to produce, as these are more closely associate with action, as is speech. I agree with Hammond et al. when they define the literacy teacher's role primarily as "teaching students to shape and organise written texts in ways that are different from speech". We need to develop paradigms for describing the content of "typical" texts, a more comprehensive framework that provides rich descriptions of the content or discourse of texts as distinct from a primary focus on grammatical and linguistic textual features. Many adult literacy students have been disadvantaged in writing on many levels. Not only are they typically unfamiliar with many of the highly valued genres of schools and overly concerned with their lack of ability at creative or self-expressive writing, but they are also dismayed by the difficulties they often experience with the actual tools of written discourse such as spelling, punctuation and grammar.

One feature of this activity was of most interest to both the students and myself was composing a piece of writing on the computer. I am not a teacher of computer skills per se, but I do use computers often in the literacy classroom. I may not remember to teach my students the many things I do automatically on the computer to short-cut the writing process. My transcripts were brief partly because we looked at word processing elements at all times. I did not find this a waste of time, as I believe that using the computer will assist my students to take control of their own writing. A recent study found that "being able to produce a quality final word-processed product has increased student self- confidence in their abilities" (Munro 1990). Some of my students would certainly attest to this experience.

Some of the text types required by the CGEA syllabus do not fit into a particular genre. This is certainly true of "writing for self expression" which could involve simple recounts or quite elaborate narratives. In an analysis of various texts (not just CGEA texts) Butt et al (1995) suggests that this kind of writing needs to combine "experiential meanings, interpersonal meanings and textual meanings".[2] Without examining such complex metalanguage with my students, I attempted to examine the text in terms of these features at the expense of dwelling on surface features. This was difficult, but a pursuit I will continue in future writing lessons. I was interested in the language features of a personal recount which could be summarised as "orientation", a "series of events" followed by a "personal comment" (Derewianka 1990). I am concerned that the prescriptive nature of some of the writing tasks the students attempt mean that, as literacy teachers, we are "presenting ourselves as arbiters of what counts as literacy to a community that has not generally learnt to read our announcements critically" (Freebody 1992).

Lessons learnt

In conclusion, the main thing that I have learned during this experience is that writing with a class involves much more than sharing in text construction or even presenting texts I have prepared beforehand to the students as models. I identified much more clearly with the anxieties and problems of students when they are writing. Although I found there were artificial constraints on the writing, and although the product is not a wonderful example of the assignment task the students are currently engaged in constructing, both the class and myself learned a great many things about the way writing happens. I am particularly pleased that they seem to be starting to give up their obsession with surface features, so that focusing on these will not inhibit their writing. The students are also coming to the realisation that writing is always a struggle, and that constructing, clarifying and revising written texts are the most important tasks.

I was interested to discover that the curriculum I have been using has interesting problematic features which I raced to address in future practice. Providing models of text types is difficult if the analysis of textual features is hampered by difficulties with language to describe these features. I am more aware of the need to combine all elements of the writing process with the need to encourage and support student writing. I would encourage other teachers to take that leap and model a text of their own for and with students.

References

Brown, K. & Hood, S. (1989) Writing Matters, Cambridge University Press, Cambridge.

Butt, D. Fahey, R. Spinks, S. &Yallop, C. (1995) Using Functional Grammar: An Explorer's Guide, NCELTL, Macquarie University, Sydney.

Derewianka, B. (1990) Exploring How Texts Work, Primary English Teaching Association, Australia.

Fitzgerald, J. 1988, 'Helping Young Writers to Revise: A Brief Review for Teachers' in The Reading Teacher. Vol. 42, No. 2.

Freebody, P (1992) Assembling Reading and Writing: How Institutions Construct Literacy Competencies, ACAL Conference Papers,Vol. 1, 1991.

Hammond, J. Burns, A. Joyce, H. Brosnan, D. & Gerot, L. (1992) English for Social Purposes: A Handbook for Teachers, NCELTR, Macquarie University, Sydney.

Munro, J. (1990) Computers as an Aid to Basic Education; The Adult Literacy Action Campaign in NSW, Project Report, Adult Literacy Information Office, Redfern.

Soter, A.C. (1987) Recent Research in Writing: Implications for Writing Across the Curriculum in Curriculum Studies, Vol. 19, No. 5.

Notes

1 The Certificates of General Education for Adults (CGEA) is an accredited, competency-based qualification equivalent to a basic certificate/foundation studies certificate for 16 year old school leavers. It covers literacy, numeracy and 'basic workplace competencies'.

2. These are terms used by Systematic Functional Grammarians. Annette prefers the SFG approach as it is language and grammar orientated and the emphasis is on meaning and the way we exercise choices in language

to make meaning. The three terms she is using refer to the metafunctions of language:
Experiential – This function is important when representing experience (denotation). It refers to the kinds of writing which involve sharing real life experienced knowledge.
Interpersonal – The functional view where language is considered as interaction and refers to communication in the sense of messages to and from the writer(s). The writer/speaker establishes or maintains a relationship with the reader/listener. This includes the connative and expressive functions.
Textual – refers to the meaning conveyed by the text itself, eg, a review will necessarily have various elements such as a brief overview of the plot or opinions etc. Or, a shopping list will have items which need to be purchased.

Struggling with authority: Texts, power and the curriculum

Catherine Macrae

RaPAL Bulletin No. 32, Spring 1997

Catherine Macrae is a Team Leader for City of Edinburgh ABE Team (a post she job-shares) and a part-time research student with the Linguistics Department of Lancaster University.

In this article I will describe research I'm currently engaged in, based on interviews with experienced ABE practitioners. I want to explain how their different conceptions of writing influenced their interpretations of their role and authority as tutors in adult learning. Drawing on the this research[1], I will suggest ways of thinking about written language and the literacy curriculum which I hope will be useful to ABE practitioners striving to work with the complex realities of literacy.

ABE tutors' struggles with authority

In the interviews, tutors recounted stories to me of actual events, involving real people, from their day to day practice. I have analysed five of these narratives, looking at tutors' underlying models of writing, and the things they define as 'problems' in students' writing. I found they were both acknowledging and questioning accepted beliefs and values about the 'good' adult education tutor within student centred learning. Their dilemmas stemmed from their own experiences of tutoring and their different conceptions of writing and learning to write. I will try to explain these here, although I will necessarily simplify tutors' complex efforts to integrate developments in their practice.

In their narratives, some tutors represented writing as self expression. These tutors evaluated students' texts in terms of their originality, creativity or authenticity. In explaining their practice to me, these tutors focused on the decisions they made in suggesting writing activities to students, or discussing problems in students' texts. They identified as problems errors such as spelling, verb tenses or repetition of conjunctions, and questioned when, whether and how tutors should point out and correct these errors. The tutors legitimated their decisions by foregrounding the supportive counsellor aspect of their role (what the tutor may and may not do), focusing on the limits of tutors' authority.

Other tutors, who were involved in accreditation, represented writing as socially constrained. They evaluated students' texts in terms of how far they had achieved a recognised text type or understood the concept of genre. These tutor's explanations also focused on their decisions about writing activities and responding to problems in students' texts. However their writing activities were very strongly defined (with particular audiences and purposes) and the problems in the students texts were 'below the surface' and concerned generic conventions. (One student omitted a personal opinion in a report; another appeared to have drawn on the language of guidebooks in a letter to a friend.) These tutors were struggling to find a way to represent to me their active role in either introducing students to the idea of appropriateness[2] or in expecting students to change their writing in the light of this. The tutors felt their actions were different from the existing norms of ABE work and that they needed to legitimate their actions by foregrounding the effective educator aspect of their role (what the tutor ought to do), stressing responsibilities involved in the tutors' authority.

The nature of written language: three metaphors – mirroring, moulding and casting

Where tutors represented writing in their narrative as self expression,

I think their metaphor for writing was one of *mirroring:* through written language the student commits to paper her experience and perceptions of the real world; the writing reflects what she believes or has experienced and therefore the tutor must not interfere with the text itself, save to sort out surface defects distorting the reflection such as spelling, handwriting and so on. The mirroring conception treats the act of writing as unproblematic once you have thought of what you want to say and you have acquired the skills involved in handwriting, wordprocessing and spelling. In many respects this account of writing is close to the traditional (autonomous) model of literacy[3] and fits well with many students' expectations since the idea of literacy as a set of skills dominates public perceptions.

Where tutors represented writing as socially constrained, they were moving beyond the surface level and problematising writing itself, not as a reflection of the world, but as a representation. This account of writing suggests to me a *moulding* metaphor: the writer needs to be aware of social conventions for written language regarding audience and purpose and needs to use these conventions as socially available 'moulds' to shape her text in terms of language, structure, layout and so on. I think this view of writing as representation shaped by socially evolved conventions is closer to the 'ideological' and complex account of literacies in New Literacy Studies[4].

The tutors I interviewed assumed that school writing practices were unacceptable: copying of texts or essay writing were to be avoided and students were to be moved beyond sentence completion exercises or worksheets of comprehension questions, to composing their own texts. They were however struggling to develop a coherent, negotiated adult literacy curriculum from an account of writing that unhelpfully divides some kinds of writing from others: autobiography or 'community' history, are seen as self expression[5] and letters or reports are seen as socially moulded. I want to argue that we need to rethink these oppositions and acknowledge that 'reality is much more complex and multi-dimensional than we ordinarily suppose it to be, and it *is* contradictory'.[6] To do this, we need a more sophisticated metaphor for writing.

Casting is an ancient and complex process, which is still used, where solid objects (taps, lamp posts, pillar boxes) are made from molten metal. First the manufacturer makes a mould in two halves by packing sand around a pattern, then removing the *pattern* to leave a hollow, a mould. Before closing the two halves together, she props a shaped *core* within the hollow to create the cavity within the object. Next she pours off *molten metal* from a bucket into a hole in the mould. The hot metal runs into the mould, filling the space between the core and the hollow left by the pattern. When the casting has cooled and solidified, the manufacturer breaks the mould and knocks out the now baked and crumbled core from the object. The finished

object bears the shape determined by both pattern and core yet is not identical to either. Each casting requires a new mould and a new core and although patterns may be re-used, they wear out.
It seems to me that writing is like casting in that we each take something from our own experience of *reality* (the molten metal) and pour it into a mould we have made from *written language conventions* (the pattern) and the self or *identity* we present (the core). The particular identity we choose, the specific written language conventions we decide to use and the part of our experience we draw on are not free from constraint: all of these choices are influenced by social and

cultural expectations of what is 'appropriate'. In this way each text we write is both original and yet also 'populated by the intentions of others'[7]. These intentions or 'voices' confer status on certain meanings and identities and marginalise others. With every word the writer writes, she consciously or unconsciously positions herself in relation to the powerful, dominant voices, either challenging, resisting or accommodating them.

The process of writing produces a permanent object which transcends time and place, representing and standing for the writer and moving beyond her control[8]. The intended and unintended readers may reject the writer's text as not being 'real' writing, or not fitting with their perceptions of the writer's 'proper' identity, or not matching the meanings they prefer. I argue that the act of writing is itself a struggle with authority, especially for those with least power, whose capacity to authorise alternatives to the dominant meanings and identities is most constrained.

The literacy curriculum: linking understanding, knowledge and skills

I suggest we think of the literacy curriculum as a triangle; this notion acknowledges the significance of skills and knowledge but places understanding at the apex. By understanding, I mean developing a sense of how the texts we read and write work as representations of the world, and how texts are shaped by dominant voices. (*See Figure 1.*)

I believe[9] the development of students' understanding should be firmly rooted in exploring how the language choices (wording and grammar), in any given text, relate to the readers writers and uses of that text. I would argue that developing, understanding also involves exploring students' interpretations of how this immediate context relates to wider social practices and dominant voices. I will give an example to illustrate this.

Why texts are as they are and *why* you might choose to challenge, resist or accommodate the dominant ways of reading or writing texts: *exploring beliefs and values, identity and power*

Understanding

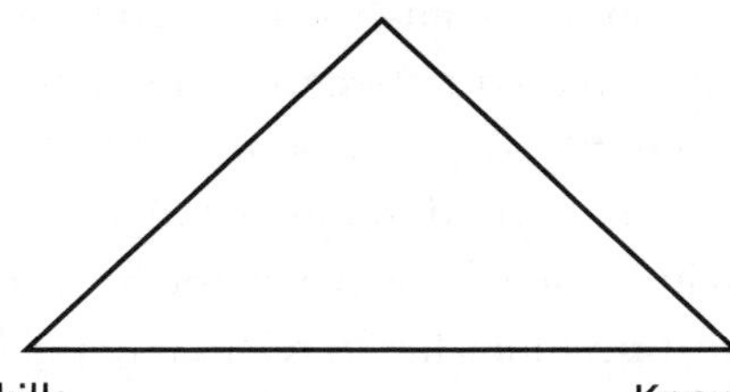

Skills

How to record meaning (words and symbols): *learning to recall and relate shapes meanings i.e. handwriting, wordprocessing, letter/sound and symbol/number correspondences*

Knowledge

What ways there are to make meaning: *learning the conventions for written language (e.g. structure, vocabulary, layout) and learning to relate the conventions to context*

Figure 1

Social workers' struggles with authority

The example is from a workplace course with unqualified social care stall; employed in day and residential children's centres. Their managers wanted them to improve their writing, especially the writing of daily logs or reports summarising these. The 'problem' as seen by the managers, was spelling and handwriting and keeping it brief. They assumed writing the daily log was a simple task: a short (five line) factual recount of the day.

Working from a social view of literacy, we asked 'who reads this text and what is its significance?' It emerged that the text could be read by the young people themselves, their field social workers, independent inspectors, Children's Panel members and social work managers. The text could be used as documentary evidence of a young person's behaviour, of the worker's competence, of the quality of care in the centre and so on.

The uses of these 'simple' texts made them difficult to write and we anticipated that the students would be very conscious of this. We felt our role was to develop a conscious awareness of how issues of authority shape the language choices in everyday social work texts.

Working from an understanding that *the texts would themselves reveal traces of their writers' struggles* with beliefs, values and identity, the tutor asked centre managers to provide examples of 'acceptable' writing and, with the students, analysed and compared these with their own texts. They explored differences and discussed why these might occur. The group found that what makes texts acceptable and valid in this setting went beyond 'professional' jargon. The writer is required to be 'non judgmental' by distancing herself from the 'facts' she presents. Her text positions her in relation to public and media voices about the care and control of children and the voice of social work itself. Significant language choices included: the naming of young people (they compared social work naming with an extract from Bea Campbell's *Goliath* (Methuen 1993)); links between ideas which contrasted very deliberately negative information (via 'however', 'although', 'only once' etc.) with more positive generalisations (via 'on the whole', 'mostly', etc.); frequent use of passive verbs e.g. 'he is often accused of . . .' and mitigating metaphors 'he found himself at odds with . . .'.

The students felt very ambivalent about acquiring the social work 'voice' as a writer. They believed it was not always in the young people's interests to be bland and impartial. Also, they felt their own experiences and perceptions as workers and carers were excluded. However, they did not feel powerful enough to resist or challenge this openly, although they hoped one day to be in positions where this might be possible. On a positive note, the group discovered their managers did not share a consistent view of 'appropriate' writing and by looking at texts in detail, students found some space within which they could make the language choices they prefer.

It seems to me important that as practitioners we represent written language and literacy in all their complexity. I hope I have made a case for doing so here'[10]. I have tried to show that the dilemmas and contradictions in our work signify our struggle with authority. I hope that, if practitioners can continue to find the time and creativity for reflection, then we may shatter those contradictions. I believe we need to move beyond making critiques of current practice, which resist the dominant voices defining literacy work, to actively challenging these. I hope we can do this by developing a positive sense of our own identity as literacy practitioners.

Notes

1. I want to acknowledge the contribution of colleagues in the ABE Team to my thinking, especially Ann, Katherine Ashe, Bridget, Ann Cohen, Corrie, Deirdra, Elizabeth, Mhairi Gilfillan, Vickie Hobson, Margaret Jessop, Gillian Lawrence, Cath Smith, Eleanor Symms, and Paula Whight.
2. For an account of how central 'appropriateness' has become in British language education and a critique see Norman Fairclough's *The appropriacy of 'appropriateness'* in *Critical Language Awareness* ed. N. Fairclough, Longman, 1992.
3. On the autonomous model of literacy, see Brian Street *Literacy in Theory and Practice.* Cambridge University Press 1984.
4. See David Barton *Literacy: An Introduction to the Ecology of Written Language,* Blackwell 1994, p.25, 33–52.
5. Mary McKeever, in *RaPAL* No. 31, discusses the constraints shaping autobiography.
6. From Liz Stanley and Sue Wise, (p. 64) *Breaking Out Again* Routledge 1993, describing feminist consciousness.
7. For a discussion of this (Bakhtin's) idea and intertextuality, see Norman Fairclough, p. 101–105 in *Discourse and Social Change,* Polity Press 1993.
8. See Dorothy Smith, p167–8, *Texts, Facts and Femininity.* Routledge 1993.
9. I am drawing here on the work of Norman Fairclough and Roz Ivanič. See Norman Fairclough, *Language and Power,* Longman 1989, or *Discourse and Social Change* (reference above) or *Critical Discourse Analysis,* Longman 1995. On 'identity', see R. Ivanič, M. Aitchison and S.Weldon, *Bringing Ourselves into Our Writing* in *RaPAL* Issue 28/29, Spring 1996. Also R. Ivanič, *I* is *for Interpersonal* in *Linguistics and Education,* 6.1 (1994).
10. I hope some of my colleagues will write a follow-up article explaining their ideas for practice.

Some reflections on current literacy teaching

Susan Taylor
RaPAL Bulletin No. 28/29, Autumn 1995/Spring 1996

Sue Taylor teaches English and Sociology at Waltham Forest College, Walthamstow. Her background includes literacy teaching and an interest in learning support. She is also a part-time Ph.D. research student at Goldsmiths' College.

Most of this paper is based on a story which I hope will perhaps be useful in furthering an already existing discussion about the roles of both tutor and student in Adult Basic Education (ABE) with particular reference to Adult Literacy teaching.

ABE appears to be one of Adult Education's most closely guarded 'secrets', despite the current political emphasis on improving basic skills in Britain. We frequently hear of ABE practitioners having to explain ABE practice to other Adult educators who, although working in Adult Education Centres where literacy teaching takes place, know little or nothing about the service and what it is that literacy tutors actually 'do'. As a result there has been a drive by ABE practitioners themselves (supported by the Government) to 'professionalise' the service by accrediting their own teaching while at the same time seeking to improve the public image and status of literacy and numeracy work amongst their fellow adult educators.

Some of the changes in the status of ABE have been due to the fact that the current Government seems to be more willing to fund Literacy and Numeracy work than it is to support other Adult Education activities such as cookery, drama or keep-fit classes. This apparent goodwill towards ABE has come with the proviso that courses should be accredited wherever possible (usually by Wordpower, Numberpower, AEB Basic Tests or GCSEs) which presupposes that ABE students come primarily to gain qualifications, but we must not forget that there are other reasons why people come to class and it is important that practitioners recognise these and respect them.

Some years ago I asked a class of ABE students to keep literacy diaries. One woman, I will call her Shazia, returned to class a week later and told me that she could not keep a diary because she had not needed to read or write anything at all that week. Believing that I understood the 'problem' I explained to her that we frequently use literacy but we probably do not realise that we are doing *so. "What about the names above shops, on items inside the shops or the notices in the post office or doctor's surgery?"* I asked.

Shazia replied that she did not think that any of these situations applied to her. She rarely left her home alone, she never went shopping or to the post office because her daughter-in-law dealt with such matters, and the only time that she believed that she was confronted with the written work was when she came to class.

This interested me. To be truthful, I believed that I understood the situation better than she did and I decided to show her some of the ways that she used literacy in her daily encounters.

I began by talking to Shazia about her lifestyle. She lived with her two sons, their wives and a teenage daughter in a large East London terrace. Her life seemed to be spent partly on domestic tasks and partly on various types of work within her community. She cleaned and cooked for other women who could not care for themselves, she minded other people's children and spent time cooking for a lunch club held locally in a disused school.

I asked Shazia to keep a diary of each of her daily activities and when she returned to class we went through her diary together. "How *did you know which door to knock on when you went to . . . ?" "I was told that it was the only green door on the first floor." "How did you know which bus to board . . .?" "Oh, I don't use buses, 'my daughter-in-law takes me everywhere in the car." "How do you know which day to turn up at the lunch club? Is there a rota?" "Oh no, 1 just turn up every Tuesday."*

"Ah" I said, *"how do you know which day is Tuesday?" "That is the day that my daughter-in-law visits her mother."* I did not seem to be achieving anything, *"What do you do when you receive a letter which is too difficult for you to read?"* Shazia was laughing by this time, *"I receive very few but my son deals with all such matters."*

We discussed some other activities and finally, when I could not pin down one literacy opportunity I suggested that Shazia and I could plot her family and social network. I was surprised to discover that Shazia rarely moved further than five roads away from her house except when she visited the local community centre where we held our class. Apart from myself Shazia rarely spoke to anyone who was not of Pakistani origin, although I should stress here that her spoken English was very good. However, it genuinely seemed that she had few literacy needs. Should she need to read or sign a document then there were many people around to deal with the situation.

This experience made me think a great deal. While I had believed that no one living in Britain could function successfully without some level of literacy, Shazia seemed not to need to be able to read and write at all in her daily life.

I wondered why Shazia continued to attend class and asked her about this. She explained that she had originally joined the class to keep her newly widowed neighbour (a second cousin) company and she continued to attend because she enjoyed her visits, having made many friends at the centre.

Up until this point I had believed that my work was valuable and that my skills as a teacher contributed to improving the lives of men and women by enabling them to become fully independent and ultimately 'empowered'. Faced with Shazia's lack of present need (which like her neighbour could change in the future) I subjected myself to a great deal of reflective thought about the role of literacy teaching in particular and more generally, the assumed importance of literacy teaching as a means of improving people's lives.

I began to ask other men and women why they attended literacy classes. The answers were somewhat predictable and initially reassuring: increased chances of employment, to gain independence, to help young children read. Yet, as I probed a little deeper, other reasons emerged: many people, particularly those who had been formerly employed, claimed to be bored at home and were frequently lonely. Literacy classes provided a safe and, in our London borough, free place in which students could come along, meet other people and, at the same time, improve their written (and usually spoken) English for its own sake rather than to serve a functional need.

Certainly, from the Government's point of view, there should be a differentiation made between classes in which people learn 'something' and in which achievement can be measured, and so called 'leisure' classes which serve primarily a 'social' need and which are subjected to different methods of funding. This changing ethos has led to ABE organisers validating everything that moves and enforcing new 'professional' practices upon experienced and competent practitioners; or, as someone said to me recently, *"I know you do it, you just need to prepare some paperwork to show the inspectors that you do"*

The changes enforced by the Government have meant that student centred learning will, if we are not careful, be subordinated to accrediting unsatisfactory tasks in order to retain funding. If ABE students choose to attend classes with a dual purpose: improving basic skills and at the same time using the occasion for social reasons, they should not be forced into joining a Wordpower or such other class which may be inappropriate to their needs. Neither should all ABE classes be so closely allied to the world of work that people of retirement age, or men and women who are not employed or actively seeking work, are either directly or indirectly excluded by the new 'education for work' ethos. Shazia benefited enormously by attending her literacy class, although perhaps not in exactly the way intended by those who provide literacy support. Another former student, on retirement, realised her life long ambition and came to class determined to learn to read.

I taught her for two years and she made steady progress after which I stopped teaching at her centre. I heard much later that she had stopped attending the literacy class and as this concerned me I telephoned her to find out why. She told me that she had been 'encouraged' to enrol on a Wordpower course and found herself having to sit through classes where she was either expected to practice telephone or interview techniques or otherwise describe to another person how to wire a plug. *"All I want to do,"* she sighed, *"is learn to read and have a laugh . . . What do I want with a bit of paper at my time of life?"*

Myths and realities in adult education

Stephen Brookfield
RaPAL Bulletin No. 10, Autumn 1989

Although I work in America at the moment, I am English. I have been in New York for about seventeen years, which is where I'm based at Columbia University. My main experience with literacy was in the '70s in the adult literacy campaign in the United Kingdom which was strongly reliant on training a cadre of volunteers who would work on a one-to-one basis with people in their own homes. I was involved in that as someone who was setting up training courses for literacy courses for literacy volunteers. I was also able to get a more personalised view of what was happening because my mother has been a volunteer literacy tutor for many years. I can't claim any special expertise in literacy. My interest in adult learning is more broadly conceived than learning literacy so it will be interesting for me to see how close the things that I'm going to talk about are to your own experiences.

I worked for about ten years in England as an adult education organiser and teacher and have spent the last nine years as a univesity lecturer and professor in Canada and the United States teaching teachers of adults. Over the past ten years I have experienced some discrepancies between the literature of adult learning and some of the espoused concepts about adult learning and how we should teach adults. As my professional status elevated and I got more professional credibility, I started to feel that my inner voices of disquiet about things that I felt didn't match the image of adult education that I read about, needed to be said. The research project I've been involved with recently has been talking to many learners about how it feels to experience being an adult student in a classroom. I've talked to learners from contexts as varied as adult literacy classes to students who return to university to do a doctorate. It has covered the gamut of formally planned adult learning experiences. I'm interested to find out whether there are any commonalities amongst these groups of adult learners in terms of what they remember as significant points in their learning, what are highs, what are lows, what kind of teachers they warmed to and then looking at all this experiential material and seeing how it corresponds with the general picture of learning that is presented in the adult education literature.

I'd like to start off talking about the myths I see surrounding adult education particularly the practice of teaching adults in groups. Some of these myths, in the American context anyway, seem to hold great sway in the adult education field. As I go through these myths I'll talk about some of the realities which contradict the myths. There are six myths about adult education which cluster together to form what has been a *group think* about adult education. Group think is a nice concept that I picked up a few years ago developed by Irving Janis. Group think is the phenomenon that happens when a group of people in a similar context evolve a set of ideas about what should happen in that context. Those ideas become reified and elevated and assume such importance that to challenge them is also seen as a treasonable activity, certainly a very deviant one. In the field of adult education there is a strong group think about firstly how adults learn and secondly how we should teach adults.

The Myth of Joyous Release

The first one is that adult learning is essentially a joyful, joyous, releasing process, a kind of exercise in bountiful self actualisation whereby people emerge from the crysallis of their distorted perspectives to fly as this beautifully illuminated butterfly upon the stratospheric winds of intellectual discourse. Learning is an unfolding and a release of potential which is experienced as a wonderful adventure by people. I think that that is true much of the time and certainly many of the learners that I've interviewed and the accounts of learning written by learners that I've been collecting, talk about episodes of joyous release. But they also talk about very significant episodes involving a lot of painful self scrutiny. They talk about significant episodes which are marked by anxiety, are perceived as being very threatening, intimidating and that the joyous aspect to it is a working through episodes like that and emerging the other side psychologically intact and reflecting back on having survived that challenging episode. While it's being experienced, it's not perceived as joyful.

I think the reason why this myth is so strong in the literature of adult education is partly because we'd like to believe that all learning is inherently joyful, partly because the people who write about adult learning have tended to spend a lot of their professional lives as continuing education organisers. If you're organising continuing education programs where your main emphasis is on getting people to show up, because that's what your job depends on, then you don't want to portray learning as a sustained experience of extential angst. There's also a misinterpretation of the work of Carl Rogers who's been very influential in the development of adult education in America. Rogers speaks about the pain of learning as much as the joy of it. Somehow in taking his work and translating it into adult education, that painful aspect has been lost. So when I've asked people: tell me about significant learning aspects that have happened to you; what do you remember as being a high point and a crucial turning point in your journey as a learner? Very frequently the things that they talk about are episodes where there was a lot of challenge and threat involved which induced a lot of painful self scrutiny. But they survived to the point where they suddenly looked at themselves in a new way: Well now I really can do this, perhaps there are elements of intellectual or practical pursuits that I can become involved in after all.

The Myth of Self Directed Learning

The second mythological element which is very strong amongst adult educators is the belief that adult learners have a strong tendency to be self directed. This is really enshrined as a cardinal principle in the American scene. There is a belief that all you have to do as a teacher is show up, remove the artificial constraints of curriculum and content and institutional organisation from learners and they will gleefully bound off on these incredibly exciting voyages of self discovery and self directed learning. The metaphor that I use is that people think that adult learners are like lions of self directedness who are roaring to slip the leash of teacher oppression and roar off into the jungle of self directed learning projects.

My experience has been that if I start an educational experience in self direction, removing limits and constraints and form and say that this is a formless exercise in self direction, then for the majority of people, the immediate reaction is one of intense intimidation mixed with puzzlement stirred with anxiety and a feeling of real discrepancy of what they thought was going on and what they're actually experiencing. They don't take to self direction like ducks to water, they don't feel immediately comfortable in that way of working. For a long time I thought that I was not a good enough teacher. Then I thought that maybe the myth that adults are all innately self directed and straining at the leash is itself wrong and much too over simplified. My experience has been that people are at very different levels of self directedness and there's often a period of weaning and initiation into this way of working that's needed. Even someone who is extremely self directed in one area of learning may need to be extremely teacher dependent in another. I can see that in my own life. I'm one of those people who automatically thinks that no-one can teach him anything. I have a real scepticism about teachers. In certain circumstances this belief has been dysfunctional for me and has been a blockage to my learning.

The Myth of Building on Learners' Experience

A third myth is that adult learners' experience is a rich resource for building curricula upon. The good teacher of adults does a quick pooling of the amassed experience of a group of learners and then is able to magically transform the insights that he or she has about the experiences that people bring to their first class and, conjure up a curriculum which is diverse but is tailored to each individual and fits into their past experiences. I think it would be wonderful if anyone could do that. I think it's very difficult. There are a lot of erroneous assumptions about the whole idea that you can build onto learners' experience. The first one is that for people to interpret their own experiences is very difficult and people may think that they're reflecting on experience but in fact they may have a totally distorted view of what their experience is and they may recall it in a very distorted way. Secondly the longevity of experience isn't related to the intensity or breadth of it so it's the old adage you can be a teacher for thirty years but really only have a year's quality of experience. Many experiences that adults bring to a classroom are ones which don't immediately transform themselves to teaching methods and curricula.

The Myth of Responding to Learners' Needs

The fourth myth is the way you recognise a good adult educational event and a good adult teacher is by observing whether or not that event and that teacher surveys learners for what they say they want to learn, asks them about what their felt needs are and then develops a curriculum and teaching methodologies which build upon those learner declared needs. This idea holds a lot of appeal for me because it seems to be democratic and humane and it seems to emphasise equality between teacher and taught, to do away with artificial institutional barriers and to turn adult education into a very collaborative and participatory activity. The problem is that the mythological dimension to it in my view is that many times in my experience, adult learners have articulated needs which in my gut I know are not in their best interest. So I've had to face the dilemma as a teacher: now do I acknowledge my own intuitive feeling that helping someone do this is really a waste of time in my view or do I say no suppress your own prejudices Stephen and listen to what they are telling you and organise an educational experience in the way, in which they've told you they want it and the way in which they've defined it? Earlier in my career I've tended towards suppressing my prejudices. As I've gotten older, either I've got more arrogant in dismissing learners or I've got greater confidence in my judgement as an educator. Usually I prefer to think that it was the latter interpretation. It seems to me that if you always give people exactly what they want then you run the very real risk of serving needs that are not in people's best interests and you also confine people to the ways of looking at the world and the habitual ways of behaving which are comfortable and familiar to them. That is not good education necessarily. It seems to me that some of the best education experiences are those where the teacher prompts us to step outside our familiar and habitual frames of reference and ways of thinking and ways of acting and looking at something and explore some others. Those are very significant periods of growth for learners. If you always meet learners' felt needs as learners define them then you-deprive them of that opportunity. I realise there's a lot of problems in what I'm saying and it can be perceived as enormously arrogant. But it's a very strong dilemma in adult education.

The Myth of an Adult Learning Style

The fifth myth that is prevalent in adult education is that there is an identifiable and unique adult learning style which is not observable in children or adolescents and that you can point to a group .of adults and say the way these people learn is wholly and separately distinct from the way children and adolescents learn. This is a myth that I believed for many years and spent a lot of time trying to prove, particularly in terms of adults being more self directed than children. Other people have tried to find the "holy grail", of adult learning in other areas. I think the truth is that there are probably more commonalities than differences between the way children learn and the way adults learn. The main difference has to' do with the nature of experience. I think it's more helpful to think about learning in adult contexts and learning the context of childhood and adolescence rather than adult learning being generically separate and different from adult learning. Because the way that people interpret the world and make sense of what's going on around them and make links between new experiences and their existing stock of knowledge is essentially; the same between children, adolescents and adults. There are differences in degree but not differences in kind. The reason why there's such an emphasis on trying to find a distinctively adult learning style is that adult educators feel a real sense of marginality and they feel threatened by cuts and by people's sceptical dismissal of adult education as a field of practice and as a field of study. If adult educators could find a domain of learning which was their own, which was only evident in adults and never showed up in children or adolescents, that did show up in everyone past the chronological age of 21 or 24 or wherever you want to set it, then at a stroke we'd have a professional reason for our existence. We could say that the reason we're in the world as adult educators is because there is this distinct and identifiable way of learning which is adult and it has nothing to do with what goes on in schools or at earlier stages of the life span and therefore our unique role in the universe is being facilitators of this form of distinct and separate form of learning.

The Myth of an Adult Teaching Style

The final myth that strikes me about adult learning and adult teaching is that just as there is supposed to be a unique adult learning style, so there is supposed to be a unique adult teaching style. A lot of people have spent a lot of time arguing that certain ways of teaching adults are inherently more adult than others. When you're working with a group of adults you use certain methods and with a group of children or adolescents you use different ones. Again, this is something that I believed in for a long time and have written in support of and I spent some time arguing that the quintessential way of working with adults was to help them become self directed and I spent another part of my career arguing that the quintessential adult education method was discussion. Discussion as a democratic, collaborative, participatory approach to teaching is one which respects adults as individuals, which involves them in the activity and is therefore somehow unique to adult education. Any adults I worked with, irrespective of the context, irrespective of their cultural or ethnic backgrounds, irrespective of the personality and mix and irrespective of the curriculum or the content or the learning task, I would always use discussion because to me that is what an adult educator did. In doing that I feel I did a real disservice at many times to learners because I refused to acknowledge that very frequently before you can have an effective discussion about something you need to have a period of assimilation or some reflection about something. You can't just walk into any domain of knowledge of skill area and teach by discussion. How do you teach psychomotor skills by discussion?

I think there has been a lot of wasteful energy spent in trying to find the one method to teach adults. What has interested me has been to see real luminaries in adult education struggling to find the one method for teaching adults and then giving up and saying well basically you should use what seems appropriate at the time. The best teacher is one who builds on his or her strengths or who adopts an eclectic approach and is looking for diversity of materials, methods and approaches and so on. This has happened with two really big figures in adult education who represent two very different ideological backgrounds. One is Malcolm Knowles who wrote a book in 1970 called *The Modern Practice of Adult Education* and the subtitle was *Andragogy versus Pedagogy*. He was saying that andragogy was a distinctive style of teaching adults and it was totally separate from pedagogy which was for children. Now he's revised his position and says basically pedagogic ways of teaching are very appropriate with adults at certain times and certain contexts. It all depends on what needs to be learnt, it all depends on the learner's previous stock of knowledge and so on. Andragogical approaches are also appropriate for working with children. Since an andragogical approach is one which is a problem solving, small group, collaborative series of activities which builds on learners' experiences, if you talk to a lot of elementary and kindergarten teachers, they'll say, "Well that's what we've been doing all our lives. That's what Dewey advocated essentially as the core of problem based and problem solving education for children so there's nothing new about it."

The other person is Paulo Freire who in the 1960s and '70s, in *Pedagogy of the Oppressed, Education as a Practice of Freedom* and so on, outlined a methodology of conscientisation, of using culture circles to teach reading and writing and to bring learners up through levels of consciousness. He said that lecturing was an inherently authoritarian form of banking education which dominated oppressed learners. I've heard him twice recently in New York and read a new book that he's written with Ira Shore called *A Pedogogy for Liberation.* In those three contexts he said, "Well, actually I think that lectures can critically illuminate contradictions in reality. They can be inspirational. The approach we should keep in mind is the approach of parallel pedagogies. That is of using a mix of different methods and selecting from them whatever seems to make sense at the time with the people that you're working with."

It's been very refreshing for me to see that, because over the years I've regarded my lecturing or giving presentations in the same way that a chocolate addict regards buying a quarter pound bar of Cadbury's Dairy Milk as something which is sinful and needs to be kept private and which I luxuriate in doing but which I daren't tell anyone about because it's so inherently bad. I've used the lecture method more and more after just refusing to do it for many years because it is useful at certain times and with certain groups of learners. I spent years fighting with my classes by always starting a workshop off with a small group exercise and saying I want you to form yourself into groups of three or four, tell each other what your main experiences have been, what you bring to this event, what you can offer the workshop and what your issues or concerns are. Then we'll get to me as the workshop leader in an hour or so into the workshop. Over and over again I had the experience of people groaning either outwardly or inwardly and saying, "Do we really need to go through this again?" Until fairly recently, I interpreted that as an unfortunate habit of teacher dependency which learners have which it was my duty to break them of. I would say to myself that the reason they're resisting this exercise is because they are so used to following authority and so dependent on external instructions. They will thank me for breaking them of this habit at some time in the future. It's only occurred to me fairly recently that maybe I'm operating on an erroneous assumption. Maybe it isn't always the best thing to do. These days I'm much more likely to say at the beginning of a workshop, I would rather go to a small group activity but I know that if we do that you will fight a lot about it. By simply recognising that fact and changing my practice and meeting people half way, it's made a big difference. It is interesting to set forth some of these ideas about my increasing concerns about the uncritical acceptance of certain ideas in adult education and to find that a great majority of people often feel those same misgivings and disquiets but retain them all privately, never go public with them, never come out of the pedagogic closet. Each one of us tends to be overwhelmed by opinion and feel that in the face of expert opinion, what is contextually appropriate for us isn't really right some how. It doesn't really have inherent validity.

Walking the tightrope: Experiences of women part-time tutors

Jan Sellers

RaPAL Bulletin No. 26, Spring 1995

Jan Sellers works at the University of Kent and at Goldsmith's College. She is also halfway through a part-time Ph.D. at Kent.

Introduction

Can work as a tutor in adult education be described as a 'career'? I use the term broadly here to describe a journey, a course of life in and through adult education teaching, that may involve a range of jobs but may or may not lead to other paid work in the same field.

Since 1987, part of my work each year has been as a part-time tutor, teaching creative writing and return to learn courses. In 1992 I began research (now a PhD thesis) on the careers and working lives of women part-time tutors, asking women in a wide range of organisations about their experiences. I would like to focus here on two questions: what are the main satisfactions in women's 'careers' as part-time adult education tutors, and what are the main difficulties?

To explore these questions, I have drawn on questionnaire responses from 82 women tutors and ex-tutors. Their subject areas are basic education, return to learn, and personal development courses such as self-awareness and confidence building; the majority are based in Humberside, Kent, London, Merseyside and Yorkshire. Between them, they have worked for approximately 50 Colleges and Adult Education Services. I quote their comments on the following pages but, because of confidentiality, have not given their names.

What did tutors find most satisfying?

Their answers fall roughly into three groups: students and their progress, teaching well, and being a team member and employee, with the specific roles and working conditions of part-time tutors.

72% of tutors found working with individual students or with the group highly rewarding. They commented on the pleasures of working with highly motivated students who were supportive towards each other and towards the tutor.

60% of tutors commented on student progress. Tutors noted the benefits of learning together; of working with a wide variety of people from different cultures and communities; of sociable classes and the friendships which developed. Many tutors found teaching adults, or specific groups, highly satisfying.

35% of tutors referred to the satisfaction of teaching well and of supporting students in their learning. Tutors appreciated the opportunity to share knowledge, and the stimulation and variety of the work. 34% commented on their own personal development. 18% noted the usefulness of the work, in service to others and community involvement.

20% praised their colleagues and the organisation they worked for. 17% of tutors found flexibility and relative freedom to be satisfying factors. 10% referred to the pay or the opportunity to work and earn.

What of the dissatisfactions?

Most of the difficulties reported by tutors take place outside the classroom.

34% felt adversely affected by insecurity or lack of control over their working situation, including the impact of cuts or takeovers.

26% expressed dissatisfaction with pay or with arrangements for being paid, and 11% noted the lack of holiday and sick pay, pension and general employment rights.

26% commented on issues relating to staff management, including poor communication and lack of support for part-timers; 23% said that they felt isolated or marginalised, and 11% commented on the lack of career progression.

26% had found difficulties with the peripatetic nature of the work, including time management and travel.

10% said they had experienced discrimination because they were women. 10% noted lack of support, or active hostility, because they had children.

Of the seven women answering the questionnaire who identified as Black or Asian, four commented that either they themselves or colleagues had suffered from racism as adult education tutors.

22% of tutors complained of poor resources or premises. 13% noted problems with teaching.

Conclusion

Although 82 women tutors and ex-tutors are not representative of women tutors as a whole, their reflections are illuminating, and the comments quoted may ring bells for many people. Work in adult education can be enormously satisfying; the enthusiasm and progress of students, and the joy of effective teaching, are perceived as especially rewarding. However, there are many frustrations and concerns reported above that are structural in origin, and could be overcome if there was the organisational or political will to do so. Some of these issues could be addressed at a local level by managers such as heads of centres, tutor organisers and personnel staff.

Insecurity and lack of support, one way or another, form the key problems confronting most tutors: instability of employment, financial insecurity, feeling marginalised, a lack of opportunity to be heard within the organisation, or an absence of practical resources. Regrettably, attitudes of some mainstream staff have proved problematic for some tutors who feel that the value of their work is not recognised. It is sobering to reflect on reports of racism and sexism within adult education services and colleges.

Perhaps it is time for some action research into models of good practice. It would be good to hear some answers to the following questions. Who has developed fast, effective pay mechanisms, for example? Which employer, in the light of current changes in employment rights for part-timers, is leading the way in informing current and former tutors of pension options? Who has strategies helping all staff to challenge racism and sexism, to examine and change their own practice, and to ensure that part-timers have full access to mechanisms for support and, if need be, formal complaint? Who has good information networks and support for tutors working in isolated conditions? Who, in the context of insecurity throughout adult and continuing education, is working to maximise security of employment – or failing that, to effectively share information on changes and developments? One final thought: when did your own organisation last ask tutors for their views and comments?

Literacy, community and citizenship: What do we teach, how do we teach it? Reflections from US experiences

Juliet Merrifield

RaPAL Bulletin No. 38, Spring 1999

Juliet is the Director of the Learning Experience Trust based at Goldsmiths College, London. Its aim is to encourage people at work, in education and in their own lives to make the best use of what they have learned from their experience. The following article consists of key extracts from Juliet's speech made at the 1998 RaPAL conference. This means the words were meant to be spoken – not read. The purpose is to present the main ideas quickly and clearly rather than produce the whole speech word for word.

I'd like to talk about literacy, community and citizenship by telling some stories of my experiences in the US. And I want to address what seems to me the central question for adult education right now – what is our purpose? The three stories are part of my history from the last 20 years. I want to look back at them only as a way to look forward – to learn from experience. And that experience has been a very tiny part of adult education's own history of education for democracy.

Highlander Center: Literacy and Social Change

I worked for ten years at the Highlander Center in Tennessee, a community-based adult education centre which for over 60 years has been doing education for community change and community empowerment. Highlander has played a historic role in the United States in adult education linked to the labour, civil rights and environmental movements.

Highlander recognised the connections between literacy and citizenship quite early. In the mid-fifties, a school bus driver from the Sea. Islands off the coast of South Carolina came to a Highlander civil rights workshop and said that what his African American community most needed was a school in which adults could learn literacy in order to pass the voting rights test. Highlander agreed to help start this Citizenship School. It hired no trained teachers – the first teacher was a hairdresser. It used no textbooks or primers – reading materials were the bible, the constitution, and the Sears Roebuck mail-order catalogue (at that time blacks could not shop in the department stores in Charleston, and depended on mail order). From that small beginning Citizenship Schools sprang up all across the south as part of the civil rights movement, and hundreds of thousands of African Americans took part in them.

By the time I came to work at Highlander in 1977 the civil rights days were long gone, and so was the direct literacy work. Much of my work was with communities and workers trying to deal with environmental problems – toxic waste dumps, chemicals and dust at work. I think of it as being a kind of literacy work, though we taught no one to read, because learning to read and speak the language of scientists was an important part of being able to act on these issues. We did a lot of participatory research, and a lot of workshops which brought people together to reflect on experience, learn from each other, analyse the issues and plan action.

The educational philosophy of the Citizenship Schools also underpinned our work on environmental and occupational health. Some of the lessons about literacy education which came from this work are:

- it was linked to a social movement or social action
- education started from community need
- anyone can be a teacher

- learning used real texts
- learning was for a purpose – to pass the voting rights test

In 1989 I started as director of a brand-new literacy research and resource centre at the University of Tennessee. It was moving from the fringes to the mainstream. I wasn't then and am not now any kind of expert on literacy as reading and writing. What I cared about was people making sense of their world and trying to make it better. But I had to figure out how to do that within the system itself.

Center for Literacy Studies: Community in the Classroom – A three year collaboration with ten community groups

The groups in the project were involved in all kinds of community development work, from a second-hand clothing exchange to fighting contaminated water, from health outreach to housing rehabilitation., They were mostly led by women and they had a particular understanding of education. For many years, education has been seen as the "ticket out" of the Appalachian region for many people – if you have a good education you leave to find a job. People who stay have limited education – jobs in mining, forestry and factories do not require or encourage education. The community groups we worked with see education as the "ticket in" in strong communities, the way to develop local leaders. But they knew it has to be a different kind of education – not like schooling, but more appropriate to the community and the people involved.

One of the groups we worked with was the Mountain Communities United (MCU); a community organisation based in two counties in eastern Kentucky – an area of high unemployment, poverty and low education:

- 40% of the population have income below the official poverty level
- 76% of children under 5 live in poverty
- More than 60% of adults did not graduate from high school.

MCU was started in 1974 by parents organising to bring Head Start pre-school programmes to the area. Since then they have worked in adult basic education, neighbour-to-neighbour health outreach, nutrition programmes, childcare, craft-making, volunteering in schools – with 400–500 participants each year.

MCU decided that their activity for the Community in the Classroom project would be to involve students in creating a teen centre. In this rural community, there are no scouts or recreation centres, and few organised activities for young people. Twelve women welfare recipients volunteered to be part of the planning committee for the Teen Centre. They organised trips to other communities which had youth centres. . They researched equipment needs and costs, including a van (as there is no public transport). They discussed and agreed rules (is it all right to discipline other people's children? all right for parent volunteers to smoke?). They wrote press releases, did financial planning, and went to the local council meeting and asked for money for equipment.

MCU was just one of the groups involved in Community in the Classroom – others did a variety of different things, including a community history class, and some never got going with literacy work. Over the three years of the project we collectively learned some things about how to build community while building skills:

- tie curriculum to learners' lives
- teach literacy practices, not isolated skills
- encourage talk and critical thinking
- teach "research" – finding out about community problems and actions
- create links with community organisations and activities

Perhaps the most important thing we learned about how adult education could contribute to civic participation was to create 'safe spaces' in which learners could find, develop and use their voices. We came to understand that classrooms should:

- model relationships of respect, support and equality

- provide opportunities for team work, collaboration and co-operation
- encourage open debate and discussion
- develop leadership skills in small groups and in the classroom

None of this is very groundbreaking or revolutionary, and you all probably know it already. I think it was useful work, good work to do, but it was just a small step. To learn from it we have to pay attention to what the project didn't do, what we didn't figure out.

We learned about how to teach, but less about what to teach, what the content should be when the aim is to contribute to citizenship and community building. We didn't really know what are the broad literacy skills needed to be active in the community: what do adults need to know and be able to do in order to be active citizens? There was always a tension between the official purpose of the class (get a GED – a high school diploma equivalence test often taken by adults, get a job etc) and what was really going on, what MCU wanted in terms of citizenship. In order to really change how and what we teach, we need to confront the accountability system – that is what drives the programmes.

Equipped for the Future: Purposes for Learning

The National Institute for Literacy has embarked on a very ambitious adult education system reform initiative, the *Equipped for the Future* project, which began where national reform efforts almost never begin – by asking adult learners what they think. In 1993, NIFL invited adult learners to write about what Goal 6 of the national education goals means to them. Analysis of 1500 responses from learners across the country identified four common purposes for learning.

1. gain access to information in order to orient oneself to the world
2. give voice to ideas, so that they will be heard and have an impact on the world

3. make decisions and act independently to solve problems
4. build a bridge to the future, by learning how to learn, to keep up with a changing world

The next phase of the EFF project was to construct "role maps" for the three adult roles of citizen, parent or family member, and worker. The role maps create a framework which adult education can use to develop teaching, assessment and accountability approaches. In particular, the citizen role map for the first time begins to identify what active citizens know and do, and what adult education should be teaching. Based on research with a broad cross-section of civic activists and adult learners we found four areas of responsibility for active citizens:

1. become and stay informed
2. form and express opinions and ideas
3. work together
4. take action to strengthen Communities

Conclusion

We have a lot to offer and we can work to reclaim space for adult education for democracy. Adult literacy education is not in campaign mode any more, it's part of the mainstream education system in this country as it is in the United States. Equipped for the Future says to me that there are institutions willing to take a risk to try to redefine what that mainstream means and what our mission should be. It says that powerful ideas can make a difference. It says that even national work can be participatory, can start with learners and involve practitioners. It's also terribly difficult to do, and risky – there's a danger of being subverted, of creating a monster perhaps. But when you move from the margins to the mainstream, you either get swept away, or you link arms and make a dam so that the water flows where you want it to go.

A4 Literacy, numeracy and ESOL learners

Further RaPAL references about literacy, numeracy and ESOL learners

Ivanič, R., Aitchison, M. and Weldon, S. 'Bringing Ourselves into our Writing', *RaPAL Bulletin* No. 28/29, 1995/1996.

Shiers, G. 'What Do You Really Want to Read?', *RaPAL Bulletin* No. 47, 2001–2002.

Burgess, A. 'Reading and Writing in Everyday Life', *RaPAL Bulletin* No. 48, 2002.

Glynn, J. 'Writing and Research at a Student-Led Organisation', *RaPAL Bulletin* No. 48, 2002.

Student control or student responsibility?

Blackfriars Literacy Scheme
RaPAL Bulletin No. 6, Summer 1988

In 1987, students and tutors at Blackfriars Literacy Scheme in London planned and carried out a research project within their scheme. In this extract from their final report, they describe how their research, and the process of doing it, led to discussion and changes in practice within their scheme. Blackfriars is a voluntary literacy scheme, and has its own management group of students and tutors.

Student Control

In January 1987 we decided it was time to look at what we actually meant by 'student control' in our literacy scheme. Workers had talked about it for years. Volunteers were trained in how to achieve it. Students were supposed to be taking it. But what did the term 'student control' actually mean? Were we all talking about the same concept? Were some of us meaning 'student participation' rather than control? Were others thinking 'student control' meant control over learning, not control over the way the scheme was organised? In fact, could students ever have real control of a Scheme, where the workers themselves were answerable to a hierarchy of funding agents?

These were some of the questions we wanted to answer by searching for a clear definition of the term 'student control' – a definition which could be shared and which would make sense to us all. A half-understood term, or one that was being interpreted in lots of different ways by students and tutors, could only create confusion; it might eventually lead to our Scheme developing a meaningless, misunderstood tradition where the orginal concept of 'student control' was completely lost.

We used two questionnaires to help us with our discussions: one asking members to discuss their ideas about student control of learning, and one asking about student control in pinning the scheme. 24 students took part. Each person was asked individually for their views, which helped to involve all 'Scheme members in the discussion, especially newer students who had not yet been absorbed into a Blackfriars 'tradition'. Many of the responses to the written questions were, of course, recorded by means of scribing – this process in itself led to a lot of individual and group discussion.

One of our regular literacy groups took on the responsibility of collecting and analysing the results. There was no complicated jargon or mysterious research technique used in this process. Everyone filled in a large wall chart with the data from the questionnaires. (Students had no problems with the use of the word 'data'). There was a brief moment of uncertainty when we were faced with so much data, until a student took a pair of scissors and cut out each question heading with the completed data record for that section, and distributed these strips of paper to the group. These smaller chunks of data were much easier to handle by individual students, and made the final drawing up of comparisons a straightforward process.

The Results

Control over learning

Most students seemed to feel satisfied with the amount of control they had over their own learning. Only 2 students wanted more control. On the other hand only 3 out of the 24 interviewed thought that students and tutors shared control 50:50. Almost nobody (only 1 student) thought that tutors preparing materials conflicted with the idea of student control.

Good examples of 'controlling your own learning' were being involved in the preparation of materials, helping each other, leading groups and workshops. Bad examples were being 'forced'; 'hurried', bad communication with the tutor, patronising volunteers, and 'being told to write something and then having it ignored by the tutor'.

Interestingly, the term 'control' seemed to be treated with suspicion by some students. The person who wrote 'there is no good way of being in control' obviously did not want any talk of 'control' in a learning situation.

The fact that there were more examples of 'bad' learning situations than 'good' ones was a cause for concern. It also showed that students had to have the space and opportunity to be able to express negative feelings about their tutors.

Control over the running of the scheme

Over 3/4 of the sample said they were satisfied with the amount of control they had over the running of the scheme, and saw the best way of exercising that control as being through the scheme meeting. However, only a few (4) attended these meetings often (6 never did and the rest only did rarely). So this area needs to be looked at carefully.

Effects of the Study on the Scheme

There have been at least three effects of doing this piece of research in our scheme.

Firstly, we now rotate the day of the scheme meeting so that a larger and more varied group of students are involved in running the scheme. Attendance at the meetings has considerably improved.

Secondly, the discussions which took place while filling in the questionnaires focused attention on the whole subject of 'control' in a more objective way than had previously been the case. In the past, 'control' was often referred to in the context of something bad that was going on.

Thirdly, some students who took part in the study suggested that the term 'student control' might be redefined as 'student responsibility': they didn't feel happy with the use of the word 'control'.

This now seems to be just the start of more discussions on the difference between 'control' and 'responsibility'. However, we have all learnt it is possible to find out about how people feel by asking them objectively and anonymously. We don't think we got the study quite right this time, but we are not afraid to go on trying. It has been hard work getting this together, but we feel it has been well worth while.

Autonomy and dependence in adult basic education

Julia Clarke

RaPAL Bulletin No. 16, Autumn 1991

Julia Clarke has been working in ABE since 1974, in many different contexts. She is now the Area Organiser, West Dorset Adult Literacy and Basic Skills Unit. This article is based on an assignment in which, as part of a course at Southampton University, I was asked to conduct a small-scale enquiry into an aspect of my practice in ABE.

ALBSU have recently produced a "Good Practice Document"[1] which refers to the "established principles and practice" which should be viewed as indicators of quality provision. These principles are not stated but we are told they are implicit in ALBSU publications and that they "underpin staff training, guidance and support for teachers, volunteers and individual". I presume therefore, that when my Area Team in West Dorset instructs trainee voluntary tutors in "ABE Philosophy" we are working within these "established principles and practice". In our training we aim to persuade new tutors that practice in ABE should be based on three principles:

1. student-centred/self-directed learning
2. the negotiated curriculum
3. literacy and basic skills as tools for empowerment

The ALBSU "Good Practice Document" does not mention the first. The second is implicit in a reference to a "learning contract". The third principle, however, is made quite explicit in this document:

". . . it is important to remember that basic skills work is about equity and empowerment not about making people more dependent."

While I agree wholeheartedly with this statement, I have always thought that the route to "equity and empowerment" can only be through a learner-centred approach to facilitating learning.

If independence is one of our aims, a learner-centred approach must be about a lot more than asking a learner what, and how s/he wants to learn. Martin Good et al.[2] suggests that we cannot expect an adult who comes for help with literacy and basic skills to be an independent learner:

"The student needs to learn how to learn. This often implies changes in attitudes as well as techniques, and such changes can only be fostered over time, through debate, discussion, reassurance."

These authors also point out that the changes in self-perception, responsibilities and social relationships brought about through successful learning may not always be welcome. Students may need support in coming to terms with a new, "educated" identity. But is the time allowed for such "debate, discussion, reassurance" in the individualized learning programme, competency statements or the Progress Profile of ABE students in the 1990s?

The research

If the process of "learning to learn" is to be validated as an essential part of ABE practice it needs to develop a coherent rationale. This rationale must be informed by the perceptions of ABE students from the point at which they take the step which warrants their inclusion in this particular category of adult learner. Do they want to become independent learners? Are they already self-directing in the manner described in theories of "Andragogy"?[3] Carl Rogers[4] suggests that person-centred education can be threatening to students. Is this true for adults in Basic Education groups? What are the "attitudes" that need changing, according to Martin Good et al.? What does ALBSU's notion of "equity and empowerment" really mean when related to ordinary people's lives, their relationships, their learning, progress, ambitions, dreams . . .?

I decided to explore some of these issues with a small sample of ABE students in the West Dorset area where I work so I formulated some questions on which to base tape-recorded interviews with small groups.

Sharing the issues

My first task was to "translate" a complex concern into a simple text which would have some meaning for all the participants. This could best be achieved by narrowing the enquiry to focus on a central issue. I chose to pursue the suggestion that perhaps the application of new learning could be difficult for an adult who has been "getting by" without basic skills in situations where such skills are required. Ideas about needing other people, "getting by" and breaking habits are understood in a far wider context than the world of Adult Education and lend themselves to the use of analogies which we can all relate to. I also wanted to challenge the idea that dependency is a bad thing in the hope of enabling participants to acknowledge their own dependency in particular situations.

The participants

The composition of interview groups was negotiated with students and tutors, taking account of other demands on the students' time during the session. One student asked to talk to me on his own. One student chose not to participate. I conducted six interviews with a total of fifteen students. All names used here are fictitious.

Deciding to become a student

While offering a variety of reasons, ten of the students made a largely independent decision to join a class. Of the other five, two were "sent" by YTS or ET schemes and two suggested some parental pressure while one reported that it was a decision made jointly with her mother. The responses can be broadly classified under four headings which are illustrated by examples from the students' accounts: (The number in brackets indicates the number of students making such a response.)

INDEPENDENCE (1)
"I didn't like asking people all the time"

FAMILY (5)
". . . as my children get older they expect me to help more"

EXTERNAL CHANGES (5)
". . . if I hadn't gone to college, I wouldn't be here now . . ."

SELF-ESTEEM (4)
"Well, I'll see how much I can remember and come back and see from there . . ."

Two students said that they had been acutely aware of their need for basic literacy skills for some years but they were only prompted to act when they were encouraged by others who had taken the plunge, *". . . she said that it was quite good . . . and you learn, so I decided to come."*

Whatever prompted the initial decision to come to the class, most of these students acknowledged that they were now motivated by other factors. While dependence was only mentioned by one person in relation to the decision to become a student, it is an issue which looms much larger when students talk about how they "get by" and becomes central to the discussion of long-term goals.

Getting by

Twelve students acknowledged they depended on other people to "get by". For example, John described how he coped as a panel-beater when he was asked to take over from the Estimator.

"... l done my own sort of shorthand, then sat back down at the desk and looked up in his estimates, and I got the customer to write their name and address and done it that way and no-one ever knew."

It thus became clear that dependency is, in fact, a key issue for many of the students. The responses to questions about desired changes and anticipated achievements should help to discover the extent to which they share the goals of "equity and empowerment".

Fulfilling goals

In the discussion of long-term goals, most references to acquisition of skills were linked, at some point during the interview, to the issue of independence. For example, Bill, Chris and Doreen all included independence in their goals.

"If someone gives me a message over the phone, to be able to write it down clearly, without having to go and get it checked."

"It would be nice to be around things and not have to count on other people."

"... when I can learn to do my reading and writing and not to ask my Mum every time, or my other sisters."

Autonomy and dependence in an ABE group

Some of the learning goals quoted above were articulated in response to my asking how long the students expected to keep coming to the group. However, for many students, achieving stated goals does not necessarily mean leaving the group.

When I asked Andrew when he will have reached his learning goals, his reply was a challenge to the distinction between teacher and pupil:

"Well, once I've accomplished all the goals and I feel there's absolutely nothing I can learn, then I think that, o.k. if I've learnt all those

> *things that it's important still to come to help those with lesser abilities."*

Three of the students in the sample could be described as self-directed learners and they were quite clear about which aspects of the group they found most useful. For Sister G. it was the tutor's willingness to seek out appropriate books and references; for Anna it was an opportunity to mix with English people from a variety of backgrounds: ". . . and different ideas and different accents . . ." For Sandra, a big attraction was the word processor. However, they also acknowledged the importance of the group. Sister G. said, *". . . I find it's good to be together and everybody has needs, they're all not the same and one can help another one."*

Conclusions

Twelve of the fifteen students acknowledged that they had been depending on other people to do things which they would rather do for themselves: With the possible exception of Michael and Kevin, (both of whom are still very young), it appears that they all want independence *and equity* in relationships with parents, siblings, spouses and adult sons and daughters. Except in Michael's case, there was no evidence that this desire for independence was associated with loss of security or anxiety about separation or that the "habit" of dependency was obstructing the application of new learning.

There was little evidence of the resistance to person-centred education which Carl Rogers describes apart from in the view of the teacher as an expert as opposed to a facilitator. Perhaps this is one of the attitudes which would need changing before the students can become self-directing in the way that Martin Good et al. suggest.

The inter-dependence which many acknowledge in their membership of an ABE group is highly valued not just for safety and security but also in the learning exchanges fostered within the group. There is little doubt, then, that the students in this sample aspire to a greater degree of autonomy in their daily lives. Of the questions I raised at the beginning of this enquiry, the most problematic one is "Do they want to become independent *learners*?" As I have only managed to write this much because I have been asked by a' course tutor to finish it by tomorrow, and as somebody who is looking forward to my weekly meeting with supportive fellow-students at the University I can only ask, Does *anybody* want to be an "independent learner?"

> *I posed that question six months ago. I returned to the University last week to find that the supportive group referred to above has been sabotaged in the name of "modular delivery". I should now like to answer my own question with an unequivocal "No"!*

Notes

1. Adult Literacy & Basic Skills Unit, "Evaluating Effectiveness" ALBSU, 1991
2. Martin Good, Rachel Jenkins, Sheila Leevers & Andrew Pates, "Basic Education 19–99: A Handbook for Tutors"; National Extension College, 1981.
3. see, for example: Malcolm Knowles "The Adult Learner: A Neglected Species"; Gulf 3rd Ed. 1984. Also: Lindsay Martin "Helping Adults Learn: A Theory of Andragogy"; Sheffield City Polytechnic for the Association for Recurrent Education, 1986.
4. Carl Rogers "The Politics of Education" Journal of Humanistic Education Vol. 1 No. 1 1977. (Re-printed in "The Carl Rogers Reader" Eds. Kirschenbaum & Henderson, Constable 1990).

Learning in open learning

Michael Gray and Wendy Moss
RaPAL Bulletin No. 15, Summer 1991

Introduction

People teaching and learning in adult basic education now accept as a matter of course the importance of students being in control of their own learning, and of building on people's own experience, needs and interests. The ideas of Andragogy, developed by Malcolm Knowles in the early 70s, suggest that adults only learn effectively in a self directed way and by 'experiential' methods (learning from experience). Open learning, currently very popular with policy makers, builds on many of the same principles.

Below are selected extracts from a long letter we have received from Michael Gray who was a student in an Open Learning Centre in the North of England between 1987 and 1989. In the letter he expresses a considerable degree of anger about the philosophy of teaching and learning he met with there. He raises many important issues; in particular he queries what we understand by 'self directed' learning and by 'treating the student as an expert in his/her own learning': He argues towards the end that he wants "my views listened to and respected and treated as mine".

Michael . . .

When I first started attending the college back in 1987, I had some really wonderful ideals preached to me. Andragogy being the relevant word. 'We, the tutors, must get away from that old classroom image once and for all. There must be no more projecting at the students as did the teacher at school, all those years ago. Also they spoke about how important it will be for we, the tutors, to always value the students experience and their ideas, we too can learn much from the students'. Sadly these ideas have not been wholly practised.

'Trying to get a clear picture of my writing ability in terms of my real progress, I find immensely difficult to ascertain, as a student in open learning. However, I will endeavour to give you as clear a picture as possible of my experience . . .

Since coming to the college and getting down to the very painful business of learning how to write, almost all of my written work has always been viewed by my tutors. They invariably tell me my written work is excellent and that my writing has improved enormously. And of course these comments always gave me a great sense of achievement: they like my writing content, and yes, content is of prime importance. Besides, I know I've improved, so I've always taken for granted their praise of my written work. However I'm not so sure any more simply because . . . I cannot really determine the actual degree of progress I have made I have often thought to myself periodically, 'How can it possibly be that from 1987 up to 1989 there has been no problem with any of the writing I have completed?'

The main reason for my deeply felt distrust of my tutors is because of their needlessly unpragmatic approach to meeting my immediate needs. Firstly, they will not go over my work with me. Yes they invariably read my written work for content, but will not go over my work with me for assessment purposes. And it's not for the want of asking my tutors. Indeed some of the time one is left to wander aimlessly: Human contact so inadequate. However, I have since found out that it is not the policy of the tutors to point out problem-areas in any of my writing. This I find not a little difficult to understand and confusing.

Initially all of my written work is drafted first, then I start working on the final version of my draft. This involves spellings, deleting, adding context, punctuation, repetition, crudely constructed sentences. Also when words should be of one structure or more, for example, 'somewhere' not 'some where' and so on. All of this I do to the best of my ability before any tutor is allowed to view any of my work. However . . . surely there has to come a time when my tutors should at least come a little way to meeting my immediate needs. If these policies set by the tutors end up utterly frustrating the student what do the tutors do about it. Do they simply ignore it or what. Here is the type of response I've encountered:

> *'Michael we first can't simply point these things out to you. Teaching is very complex Michael. It's not just telling you where you go wrong and then all will be well. Teaching is knowing when to tell how much, it needs to stretch you, challenge, and yet not go so far as to drown the student. In literacy terms you are the perfect student. You put yourself in a position to teach yourself.'*

These remarks made to me as an adult and not as a child I found to be totally unnecessary. I most certainly was not trying to imply that all will be well . . . they never ask me if I feel comfortable with these ideas of theirs. They simply apply them regardless of how I feel. I have no say in the matter. Teacher knows invariably best.

Punctuation!

I decided to turn for help with my punctuation problems to one of our other tutors. This tutor proceeded by informing me. 'Michael I can't teach you punctuation. No one can teach you how to punctuate'. She then suggested I go to my local library and get myself a book that I was fairly familiar with. She told me to read it and that I should look very closely at the punctuation as this was really the only way by which I would be able to learn to punctuate with my own writing.

I must say I was quite taken back by her advice as I felt it unsound: I had to disagree . . . I told her that I'd been reading books for more than twenty years and had not learnt any punctuation from reading books, otherwise I wouldn't be asking her advice . . . I told her that I would much prefer to learn to punctuate in conjunction with my own written work and not someone else's work. And so disagreed with her vigorously but could make no headway.

However I tried her idea out, and found apart from full stops dot

. . . dot . . . dash . . . dash . . . streak . . . streak. Not a little confusing . . . So I decided to consult with one other tutor. I made it quite clear to this tutor that as far as I was concerned this advice offered me was in reality no more than pure hypothesis – what was the evidence to proving this idea. Her reaction to that was as follows:

> *'Michael the idea of reading someone else's published books is a sound one. It offers you an opportunity to look at correct placing of full stops enough for you to abstract the rules and so I would defend that tutor's suggestion. But I also defend your ideas too. As you say it's all hypothesis. We cannot know if our ideas will work. No one can know how someone will learn. We can only try'.*

These two contrasting views constitute a contradiction in terms. Punctuation from reading books did not work certainly not for me. Indeed, you may as well say that the more books one reads, the more proficient one will become at spelling. Well of course my experience has shown me that it simply isn't true.

I was finally able to get some help on asking one last tutor if should could be prepared to look at some of my writing . . . She immediately responded to my request by giving me some direct input regarding my problem She went over some of my work with me while explaining to me the importance of punctuating the full sentence. To complete the full thought. By way of illustration she pointed out to me some incorrect placing of full stops. Also she asked me to read some of my sentences in my work and to think of the complete thought. In other words for me to try and picture in my mind the complete thought and then full stop. I must say I really did feel comfortable with this idea of working one to one with a tutor . . . and should not I be able to request some direct input with my punctuation? And then I can't very well, with a book, ask questions, if I run into a problem while I'm working at home on my own . . . Sadly this tutor's approach lacked sufficient consistency to be of much help to them.

Since then my punctuation has improved I'm glad to say though if you consider the utter confusion I've experienced. But there's still lots of room for improvement. But if I may say this, reading books had nothing whatsoever to do with my improved punctuation which only carne about by continual writing practice.

(Michael goes on to give further examples of how he has asked for direct input but not received it, eg. informal writing, and how to use the index of a Roget's Thesaurus).

My experience in open learning has shown me that it lacks enough in essential feedback and is also lacking in essential consistency if the tutors wish to avoid me losing my motivation then any lesson should run a course of consistency otherwise my drive will die a death. Is there not a need for greater flexibility from the tutors in open learning depending on the student's degree of confidence in his ability?

I think the idea of open learning is in reality a truly wonderful concept. All I am really trying to say is that the tutor should listen and value the students, their own views and experience as being relevant. I do not like it when personal teaching philosophies are literally enforced upon me. In fact I've taught myself to write even though my tutors have put me through unnecessary confusion, at home in my flat. Adequate support with consistency – my views listened to and respected and accepting them as mine. Humiliating and patronising feedback only serves to create needless gulfs. The tutors do not always know what's best for the student and they are not above failure themselves. Nor are they above learning from the students.

Wendy . . .

We feel that Michael's letter is very important in helping ABE tutors and learners look at some fundamental issues around adult learning.

Firstly: *what do we mean by 'self-directed learning and how should learners be supported in being self-directed?* Stephen Brookfield (see RaPAL 10 [*and reprinted earlier*]) pointed out that adult educators often believe 'that all you have to do as a teacher is show up, remove the artificial constraints of curriculum and content and institutional organisation from learners and they will gleefully bound off on these incredibly exciting voyages of self discovery and self directed learning'. In reality, he says, if adults start an educational experience without structure or form 'for the majority of people the immediate reaction is one of intense intimidation mixed with puzzlement stirred with anxiety'. He points out that people need to develop self directedness and even if you are very self directed in one aspect of your learning, you may still need a lot of teacher input in another. Michael's letter supports this view in many ways. He is clearly very self directed in many aspects of his learning, but would like to depend on a tutor in some areas (eg. punctuation).

Secondly, we need to really examine what we mean by *'responding to student needs'*. Do we mean students understanding of their own needs, or tutors understanding of what those needs are. Michael feels his tutors interpreted it as the second.

Thirdly, *how willing are tutors to treat students as experts in their own learning?* Are tutors willing to adapt their approach if the learner's learning still does not seem to fit with the prevailing philosophy of the centre he/she is in? This question is a subtle one as many adults in ABE only have the model of traditional education to work from and may see more experience based and self directed methods as 'not real teaching'. In time students may come to appreciate learning styles which ask them to become more self reliant. On the other hand, we all have different learning styles and we need the kind of teaching that responds to these.

Finally Michael says in another part of his letter that he feels bitter that he has been encouraged to do informal writing, autobiography story etc, but has been given little help with the more formal styles he might need in different settings. *How should the transition from personal, and story writing to more formal writing be managed?*

This is Michael's view of his own learning experience. His tutors would no doubt have other views and perspectives to add but they haven't had a chance to express those here. It should also be noted that Michael reports elsewhere that he and his tutors felt he progressed well in the early stages of using the centre when he was encouraged to write autobiographically. Nevertheless, Michael himself feels that he learned more by teaching himself in his flat after leaving open learning, and clearly feels angry, frustrated and confused by the teaching approaches he met with in the centre. We cannot deny that experience either.

I personally have felt worried before about examining critically in public, the student centred methods used in ABE, for fear of opening up opportunities for traditionalists to say – 'see, I told you so, we need to go back to good old fashioned tutor centred methods'. I hope we are far enough on now to enter a debate which accepts the principles of student autonomy, but looks at how we can develop our thinking and improve practice in this area. Reading Michael's account, I feel sad he has missed out on the experience of being in a group – this is what can be lost in the open learning approach. I feel being in a group might have made him less reliant on the particular teaching styles of his tutors and would have given him a great deal of support from others.

New writers research their own writing processes

Denise Roach and Peter Goode
RaPAL Bulletin No. 11, Spring 1990

In this article, Denise Roach and Peter Goode, both of whom began as learner writers, share their histories, their struggle and their insights with us. For Peter, we learn, writing is a new freedom: the vital link between his imagination and the outside world, his readers. It's the expression of his dreams, hopes and frustrations, as well as the demonstration of his abilities. His writing is taken from his recently published book of poems, 'The Moon on the Window'. Denise describes vividly, through the use of metaphors, how she had to learn to write in an academic language that was not her own, and about her struggle to find an identity and power within writing.

Both writers are actively directing their own development, using friends as sounding boards for ideas or as scribers or suppliers of the skills they are still developing. Both writers are actively trying to analyse how they write, as well as actually writing. Each very different account echoes the founding principles of RaPAL that learners, practitioners and researchers should work together on discovering more about shared areas of interest.

Peter Goode writes:

The Stallion
My shield is my past
My scabbard is my present

My future is my sword
My stallion is my inspiration.

Writing

I had set the reading up as Almighty God. The writing hadn't occurred to me. There was never, ever, any consideration of writing. It was never a priority at all. And yet, on the same hand, where the reading had almighty pitfalls, the only success I saw was to be able to read and this left me open to failure all the time. What's come out of it now is that I feel so calm about writing, although I can't actually write.

Writing, the sight of it, the way that I ride it, has become a very strong passion. I'm finding my own way of writing. My Gobbledygook. It's not ideal but it works for me. It's a private word system. I mean there's one word in ten that a reader can read. I don't know about commas and full stops. I don't want to set up any barriers with writing. I start with a thought and I finish with a thought. I keep my Gobbledygook on paper. It's my freedom. This is the beginning.

When this book's complete, that'll be my first attempt to read something in total that I'm aware of, but I don't know if I'll ever be able to read it.

Stallion

I'm not really sure why I was frightened. The nearest I've got is the stallion. It's a piece of poetry.

When I was about six, I was walking through this long corridor, very long, with glass so you could see into the classrooms. On one side of the blackboards was a drawing of a stallion, a very good one. Someone had actually created it. No one was in there and I went in and sat in admiration of its beauty. It was talking and walking with me. I used to write with my left hand at this time and I was drawing it and this pillock came in and started rubbing it off with a duster. It didn't exist for her. It was just pieces of lines to her. I wanted to stop her, so I threw the pen at her and it missed her eyes and stuck in the blackboard. You can draw your own conclusions what happened then. I don't know if this explains my fear of blackboards; or my fear of writing, but I was physically sick inside. It was too much.

The stallion came in talking. I'd written a poem with the stallion in it and hadn't connected the two at all until I had an interview with someone doing a thesis on life histories of working class people. I'd gone through something similar to her, bringing up children on our own and I was saying how horrific it is, but how much worse, to cut the umbilical cord when they're grown up and want to leave home. So we was relating to each other. She had a daughter and I had my children. *The Stallion,* I showed it to her and she related to it and asked me why I'd never finished it or thought of publishing. I hadn't worried about it because I'd just let it be born. 1 didn't know the connection until we started to talk about school. I thought I'd never see the stallion again and now I did, I thought it was more important to relate to it, smell, touch it, maybe even ride it. It was only then that I could actually let it go free because it had the same beauty. It still has.

Little Bits of Glass

I think I have a tendency to neglect the obvious in some respects. There's the wireless, the T.V. and subconsciously or consciously, I'm taking all this in. It may be just a fragment I'm interested in and that's when I peak. I may not think I've taken it in. All this is part of it. I don't like to reveal, probably because I'm a Magpie and I don't want anyone to take my shiny bits off me, because they're mine. I found them. They're only little bits of glass. It's just how they reflect at that moment. It's what facets that shine. If they shine in the colour that you're looking for, it is a discovery and when the sun goes in, it's just a piece of glass. That's always the way. It's always the poem of the moment that gets me going and after that it doesn't lose its lustre but it's no less no more important than the other ones.

It's not the discovery that's important. It's the moment of discovery. This is important to me. If I'm chipping away at a piece of granite, it's important that they don't take any of the pieces away until I've found the image. Equally, I find the pieces that I've chipped away are as strong as what I've ended up with. These pieces, I can rediscover. They happen not by design but by happening.

The discovery of The Moon on the Window is not only a personal discovery but the discovery of the progression of the knowledge of the written culture. Language across all continents. The first time that Plato's Cave was read to me was on the 17th of March, 1987 but the knowledge of the cave has been with me, since I was aware of me.

Who is this book by? Magpie

I chose Magpie, first time I wrote anything, because I am a Magpie. I listen to conversations, pieces of poetry, wireless programmes, and when they leave an impression inside me, either the jewel, sadness or the happiness, whatever it may be, I make it into my own vision. I don't know what I make it into; however I see it at the moment. I felt very comfortable working under Magpie. It was an advantage because people would discuss my poetry and not know it was mine. So there was an honesty about what they said. This was when we were selecting prose, poems, drawings and conversations and putting them all together in a magazine in an Adult Education Centre.

Style

There were two people, tutors, who'd say, "I knew it was yours because it had a certain style." I thought it would take me a long time to learn style; by this I mean about how a particular piece was written. And that's all I know about style. Coming aware of this fact, it seemed a challenge to write under my own name and acknowledge it by signing it. I don't know much about style but I still pick up now, what shimmers in between at the moment, what's attracting me.

Hooks and Triggers

Hooks to me are something I catch either with thought or subconsciously. Someone talking on a bus, seeing something in a shop window. Just last week, I did one of the dog howling. Or it can be just one word or the same as the poem that was a reflection of flowers against a wall. These are hooks. The triggers are the memory of the hooks. I can be walking on and something similar happens and the trigger will set the hook in motion and build a mental picture. That's when I want to write, which always spoils them up. I mean, I still write on the back of my hand. I haven't got a history of the written hand. If I'm by myself and I see something that reminds me of another happening, it fetches a mental picture. But if I'm with a friend or friends, it goes into the conversation and that's when .a poem gets lost, but I still carry the hook inside me.

What is this book about?

I think if this book is about anything, it's about people that try to succeed within the written word. Where every word or letter, piece of poetry. or prose is a challenge. In most education, it's failure or success but I would like it to be about trying again. Not to fail. Not to succeed. But to go on.

The Christening
You held me up to the Sun
That is your Day there!

You held me up to the Moon.
And said, That is your Night there!

I held myself up to Humanity.
And said, I am the Man there!

'*The Moon on the Window*' by Peter Goode is obtainable from Pecket Well College, 36, Gibbet Street, Halifax, HXI 5BA. 71 pages

Denise Roach writes . . .

Marathon

When I entered into academia it made me feel as if I was expected to run the London marathon whilst 10 stone overweight. As with any marathon you have to do the training before entering the race.

I had no formal qualifications but was accepted into university because of my experience gained via the 'university of life' (the real world, day to day living). Although I was pleased about being accepted into the 'higher echelons' of education – I found that gaining a place at university did not necessarily mean everything was going to be plain sailing.

Although my fresh start course (done on a part-time basis, one day a week for 2 years) did prepare me in some ways for the 'race' (university), it had only done so in the sense that it had given me the confidence and encouragement which confirmed that I was eligible to apply for a place in the race.

I hope that the following metaphors will make sense to students who are entering into study, and further, that tutors reading this piece of work may find some of my experience of relevance in their future teaching. Such metaphors were developed out of my experience of trying to accommodate myself within the academic community.

Academia has made me feel . . .
Like a puppet on strings,
Like an Alien from another planet,
Like a junk food addict in a vegetarian conference,
Like a working class person trying to talk 'posh' on the phone.

Being unable to identify with and express myself through academic language, I was left to use my own experiences of life to somehow make sense of my 'new world' (Lancaster University).

It was not until I had to explain to a friend (Roz) that my thoughts began to take shape. I tried explaining to Roz how I overcame same of my struggles in learning to write an undergraduate academic essay.

First I likened it to learning a new language. Secondly, I could not help but feel as if I had engaged myself in a detective agency. Strange parallels I know, but as anyone entering into study learns there are rules, conventions (taught and unspoken) that need to be followed and learnt.

The Skeleton Metaphor

I saw the bare essentials of an essay as being the bones of a skeleton. Being new to academic life I had first to find the bones. This was the beginning of my detective work.

I wandered around my new world feeling like an alien from another planet – after all, most undergraduates have their bones firmly in place before arriving at university.

So the hunt for the bones was on! Needless to say it was like trying to find a needle in a haystack or dare I say like trying to find a lecturer in a department!

After great pain and anguish I managed to find the bare bones (i.e. that an essay needed an introduction, middle and conclusion) and thus the skeleton began to take shape. I somehow worked out that the basic skeleton of an essay did not qualify me for a pass mark in my course work. Yes of course, the skeleton needed elaborating on before it could be accepted as an academic piece of work. In other words flesh needed to be put on the bones of the skeleton.

I began to feel myself getting lost again. I knew the flesh was specialised: it was grammar, spelling and sentence structure. Of course the thicker the flesh the better the skeleton is able to work and move. I was beginning to learn the new language, but the story was far from over.

The Clothes Metaphor

I began to tell Roz how different disciplines require different things from students – i.e. students are expected to wear different disguises for different disciplines. It was here that the clothes metaphor really took shape.

As different disciplines required you to use different disguises, thus my skeleton had to be dressed in. different clothing. Therefore as students we are expected to have a different writing identity within the many different academic disciplines.

It all seems clear writing about it now, but what an identity crisis I had!

Personally I have got a skeleton with not much flesh on it and few clothes to dress it up in. So what I have had to do, as one does when one has not got many clothes, is to resort to borrowing other peoples': academics from books and lectures, friends. After all using their clothes (i.e. their language) seemed to gain me access to the privileges of an academic student.

I still have a problem though and that is I want my own identity, I do not want to wear the clothes (the language) of others. So my present detective work is that I need to find the path to my own identity as a writer.

I do not want to be constrained, silenced and strangled by these tightly fitted, borrowed clothes.

I refuse to wear the 'off the peg' wholesale identity that academia has to offer. But yet I am presently restricted (coerced) into doing so, as my body needs clothing.

In this mixed bag of clothing that I wear, I am able to recognise my own clothes (that is my own words), just as I am able to recognise my own clothes in a jumble sale, or you are able to recognise your own words when you are arguing with a friend. The reason why I think that this is so, is because we/I have worked so hard to find those words.

Personal Power

A further issue that I am hoping to raise here, is whether you believe in your own personal power (which I would argue lies within us all). Personal power is what enables us to form our own identity as writers. A kind of disguised-free identity. Whether or not you do or do not seek/see to choose, to use that/your own 'personal' power, is going to be up to you. You may choose to play around with or go along with, the 'quick-yuck', wholesale disguise identity that the privileged powered academics' have to offer. The choice is yours. Here I merely want to draw your attention to the fact that nothing comes FREE. Everything has a price, and what is at stake here is your identity. If you choose the wholesale, off-the-peg identity, you have yourself to lose! For me I do not want to pay the price of a lost identity (written or unwritten). In many ways I hope to make you aware of what you are doing before you enter the Big Race.

There is no doubt in my mind, that I want my own identity as a writer. And although it may not be possible for me to obtain such an identity now whilst at university, I can see it being possible in the future. This is ironic in its self, as the higher echelons of education, is were one is supposed to develop ones thinking, which surely cannot be divorced from ones writing!

Personal power also comes at a price, but the price you pay to gain such power is paid back to you, through the amount of time you're prepared to put in. Surely you have nothing to loss, only personal power to gain by working for such times!.

The Academic Pentathlon

For me, personal power needs constantly working at, and I am constantly working at building new strategies to cope with my present academic work. Until the time comes when I do not have to wear these borrowed clothes, nor work at new strategies of how to keep up with my present work load; I can then put my time and energy into further developing my knowledge of how to perceive the world. But until then, I am left with my metaphors to get me through.

Here I share my strategies and ways of coping with academia . . . OR . . . here are my last few thoughts of how in my mind I see myself coping, until the days of empowerment. Below are some temporary supports that I use to resist having my personal identity eroded. I see myself as having to use aids to help me along in the Big Race (The academic pentathlon):

* Just as a person with a physical disability uses/needs aids to help them have a more freer and `abler' life, I use crutches to keep-up my speed whilst trying to run the race.
* When I feel myself drowning in the uncontrollable sea of knowledge, I grab for floats to keep my head above water.
* When my energy level is low, I liken myself to a puppet on strings. The weight of my body is being held up by the strings, and the strings are being worked by means that are out of my control, (on automatic pilot).

But (usually) on a day to day basis I am in control, and these last few words are only my way of coping. No doubt you have your own. If so why not share them and let others know. Start using your own personal power.

A5 Managing provision

Further RaPAL references about managing provision

McMahon, M. 'Research Report: Managing Adult Education', *RaPAL Bulletin* No. 6, 1988.

Clarke, J. 'Left Out in the Cold with Quality Standards For Basic Skills', *RaPAL Bulletin* No. 21, 1993.

Issues in management of literacy programs

Charles Lusthaus and Marie-Helene Adrien,
McGill University, Montreal, Canada
RaPAL Bulletin No. 10, Autumn 1989

This article is about management . . . Many of us are managers . . . because of the Education Reform Act, we are being asked to think about managing in very concrete "performance" terms. This encourages us to think of "output" and "throughput" and measurable quality, like exam results. Many management scientists would call this a naive way to look at managing. The danger is that we think about the product of managing, and neglect the process.

Organizations are constantly faced with the challenge of bringing together scarce resources to supply the products and services that people need or want. Managers are responsible for creating organizational settings in which these resources are used in the most effective and efficient way.

Although organizations active in the promotion of literacy are, of course, subject to the same necessity to make effective and efficient use of resources, their ability to do so has been limited by the very meagre resources available to them. In attempting to secure resources from national or local governments, literacy competes for attention with a range of other health and social programs. As a small, marginal program in the public sector, literacy in Canada and elsewhere has received few resources, so it is not surprising that the management of literacy programs has been virtually ignored. Now, however, with the increasing emphasis in Canada being placed on literacy, managerial questions are being raised: What is expected of managers of literacy programs? What are their roles and responsibilities? What tasks do they have to perform in order to provide effective and efficient services to the public? How do you build organizational capacity to provide sustainable literacy programs?

As literacy becomes a crucial national priority, managers of literacy programs will also face important questions of accountability. If such questions are not dealt with adequately, this will undermine the recent progress that has been made in developing literacy.

This paper presents a framework for thinking about the various functions that are required to manage literacy organizations. It suggests possible questions managers might ask themselves about what they do and how their organization is managed. The framework presented is meant to be heuristic in so far as we hope that it will stimulate reflection and thought regarding the management and organisation of literacy programs.

A Framework for Managing Literacy Programs

People organize to provide goods and services to the public. Organizations, in all their varied forms, are generally designed to meet the needs of their members and clientele. As you observe organisations in all their splendid forms, it appears that, regardless of their size or goals, organizations seem to need managers to carry out five generic functions. First is a strategic function. This deals primarily with managing the survival of the organization through understanding and acting upon the organization's relation with its environment. Second is the financial management function, which asks managers to be able to generate and control the money of the organization. Third is the marketing function. In this area the manager is asked to link the program, consumer needs and access, methods for service provision, and cost, in a rational fashion. Fourth is the personnel function which asks managers to deal with the needs of its employees in relation to the needs of the organization. Finally managers need to manage the program function. This is the organizational raison d'etre – the direct services it is providing to its clients. Unfortunately for literacy organizations, where our managerial size is quite small, these fundamental functions have to be fulfilled somehow and in some way. The creative question for literacy managers is: How can they be carried out?

Figure 1 provides a framework for the administrative functions associated with the management of literacy programs. Furthermore, it indicates that, for each of the five functions, managers need to plan; that is, to think about the future. Managers must implement, encourage, and require organizational members to do things that ensure the organization's success. Finally, managers must monitor the functions in order to ensure that the organization is "on track".

Clearly, it is a formidable challenge for members and managers in literacy organizations to carry out the various tasks implied by the fifteen cell grid. However, at the very least, literacy managers must begin to engage in the reflective act of reviewing the various tasks and functions and assessing their importance to the organization. Such reflection can produce new and improved ways of organizing so that we can provide better services. In Canada, our assessment is that if literacy organizations don't adquately come to grips with the managerial implications of this model then in a few years the gains made today might be lost.

To push our own reflection and those of literacy managers on these issues, the rest of this paper attempts to more fully clarify the managerial functions defined earlier and then provides a set of questions that can be used to stimulate reflection and dialogue regarding how to manager literacy organizations.

Figure 1: Framework for the administrative functions associated with the management of literacy programs

ADMINISTRATIVE FUNCTIONS

	Strategy Development	Financial Management	Marketing	Program Development	Personnel Management
Planning					
Implementing					
Monitoring					

Description of the Functions

I. Strategy Development

The strategic function of management deals with survival of the organization over the long term. It requires managers and other organizational members to be able to analyze various environmental forces – political, economic, social, demographic etc. and to link the future trends associated with these forces to the realities of the literacy organization. Managerial research in other spheres of activity has come to recognize that such "environmental analysis" is crucial to the development of the other strategic issues of literacy organizations, namely their mission, structure, and the basic character of the services that are provided.

A Canadian example of the need for good strategy development can be drawn from a recently announced program, called READ Canada. This is a national literacy program established outside of formal schooling. Its focus is on children and it is preventative. In developing its mission, it attempted to capture in just a few words the dreams of its founder. It describes its mission as follows "every child in Canada should have a story in their life every day outside of school". The program manager is charged with designing the organization's structure and activities in order to accomplish this mission.

The mission of the program is clear. How to organize and provide service is extremely complex. The program and its execution can be altered by pressure brought to bear by environmental forces outside the control of the manager. External factors, such as legislative and political decisions, immigration flow, demographic trends, and the socio-economic environment, can have an impact on the program and affect its chances of survival. For READ Canada to survive, the program manager must be able to plan, implement, and monitor its strategy for survival. In developing their strategy, administrators of literacy programs can ask themselves the following kinds of questions:

Do we have a clear mission that captures the essence of our program?

Do we adequately reflect upon and analyze the political social demographic economic environment?

Are we doing a good job of communicating our mission to the public?

Is our organizational structure congruent with our mission?

Do we adequately develop plans and take action based upon our concept of the future?

II. Financial Manangement

The Financial management function is concerned with the generation, classification and interpretation of financial information that will assist the manager to meet organization objectives. In this function the manager has a number of critical concerns. First, is to understand the types of possible funding avenues open to the organization. Historically, literacy program managers have paid very little attention to the development of revenue sources: Money usually came from one or two sources and that was that. Successful managers are more creative in the financial domain. Second, managers need to be able to generate financial information. Multi-year budgets, unit cost figures, and cash flow requirements are just a few of the myriad of data needs of managers. Thirdly, managers need a financial accounting system that will meet standards for financial control and, as well, be able to generate the management information needed. The level of complexity of the system should vary with the sophistication of the organization's structure. An effective accounting system should provide information for two main purposes:

- internal reporting to managers for use in strategic planning and controlling routine operations;
- external reporting to auditors and other outside parties. Conducted by adequate and qualified staff, the accounting system should generate information on budgeting, cash flow, current and future expenditures, and need for investment.

The following are typical questions that managers *of* literacy programs should ask themselves about the organization's financial management:

Are we able to generate the financial information needed to make effective requests for additional funds?

Is our financial management information timely?

Do we have an adequate financial control system which can compare our budget figures with actual expenditures?

Are our program activities adequately costed? Is there a way to adequately control program costs?

Can we justify the time required to collect financial data?

Do we spend enough time and energy in productive activities aimed at raising the revenues needed to make our program better?

III. Marketing Function

Marketing has not received much emphasis by those responsible for increasing literacy. And yet, literacy managers need marketing skills. In most not-for-profit organizations marketing skills are scoffed at and derided. People equate marketing with, trivial advertising. This is not the case. The marketing function requires the following question to be asked of managers of literacy: Are we able to provide the best available services to the most people at prices which are competitive? Managers of literacy programs must be able to get services to their target populations with the best delivery channel available. They must ensure that their programs are familiar to potential clients, meet current needs, available and accessible, do not duplicate other services, and are, using appropriate technology. The marketing function should be assessed periodically to see if the energy, time and money being expended are proportional to the results achieved.

Marketing efforts require a team approach since any member of the organization who deals with the public, whether by tutoring, teaching, fund raising, or lecturing, is a representative of the program and reflects negatively or positively upon its "image".

Despite this informal aspect to promotion, specific activities should be delegated to one or more individuals. The following issues should be considered in defining the marketing strategy:

Do I know who my clients are and where they are located?

How can I convince my potential clients that literacy skills are important and needed?

Are we using the most effective channels of distribution?

How can our potential users learn about our service?

What media will be most effective to reach our audience

Can we convince funders that the need and our costs are appropriately linked? Are there a reasonable number of potential service-users?

Are we offering the best possible service for the price?

IV. Program Development

Program development constitutes the "heart and soul" of most organizations. A program is a set of interrelated activities that leads to the fulfillment of the goals and mission of the organization. Typically, literacy organizations engage in a set of activities that define their program. Managers are expected to ensure that there are adequate plans, ways to implement, and a process for monitoring. These activities should lead to the fulfillment of particular objectives and be supportive of the goal and mission.

In developing literacy programs, managers should ensure that the program's objectives are realistic and accountable; all activities should be linked logically to these objectives. Attention should be directed to defining the needs of the target audience, to assessing the homogeneity of these needs and to adapting the program to this range of needs: Literacy programs have a life cycle (introduction, growth\maturity, saturation and decline) just as any program does. The stages in the cycle vary in length and depend on the context of the program as well as on the way in which strategy, finance, marketing and personnel are managed. Program managers must understand the current stage of the cycle for their program and act accordingly. Their focus should be on making the literacy program programmatically and strategically successful.

The following are specific questions that program managers should ask themselves concerning planning:

Are our objectives realistic, "do-able", accountable?

Do our activities make sense?

How do we make sure the activities are successful? What monitoring activities will we undertake?

Are we fulfilling our responsibilities to our funders?

V. Personnel Function

The key to any successful organization is its people: good management of human resources is vital to the achievement of organizational goals. Organizations active in promoting literacy have a complex task in that they use the services of volunteers and part-time workers and must operate on flexible schedules. Under these conditions, a good personnel system must facilitate a communication among all members of the organization thus leading to more effective programs. Managers of literacy programs should ensure that staff skills and personality match the organization's needs and goals. This concern should be reflected in the following personnel development activities:

Selection: deciding on an appropriate number of staff members, developing hiring procedures, as well as a strategy for attracting volunteers.

Training: designing a training program for all categories of staff members and deciding on retraining schedules.

Retention: delegating appropriately and ensuring a fair distribution of work load. Developing ways to increase communication and participation among employees. These issues should be resolved through clear job descriptions and the organization's organigram.

Evaluation: decisions should be taken on the frequency and the criteria for staff evaluation.

Renewal: as with any organization relying on volunteers, literacy programs can be faced with a high rate of staff turnover and should thus develop a system for staff renewal.

Rewards: all organizational members must feel that the services they are providing are appreciated. Clearly, salaries are the first level of reward. However, we know that in almost all literacy programs, organizational members rewards are built into the work itself. Therefore managers need to pay attention to the balance between intrinsic rewards – the work – and extrinsic rewards – the money.

The manager of literacy programs should enforce the accepted roles and responsibilities of all staff members on a daily basis. The manager should also review periodically the needs and goals of the literacy organization to anticipate changes in the personnel structure.

Managers might consider the following aspects of personnel development:

Do we have a staffing plan? That is, do we have a way to think about linking people to current and future activities?

Are staff roles and responsibilities clear and differentiated?

Do we provide adequate opportunities for communication, and participation to encourage all staff members to become involved in the literacy program?

Is supervision adequate?

Are enough time and resources allocated To staff development?

Do we have a reasonable approach to rewarding those who work for the organization?

Conclusion

Support to literacy organizations has recently become more "fashionable" to our society and our politicians. We could view the new-found popularity cynically and go on with our work substantially as before. On the other hand, it is possible to conceive that this an, opportunity that might bring considerable increased resources and much better access to services for those requiring them. This paper argues that if literacy organizations want to seize on this new opportunity, they must begin to approach their organization from a managerial point of view. A framework was provided along with a series of questions which, it is hoped, will guide a reflective intraorganizational dialogue. The process of dialogue and reflection will, we believe, lead towards improved managerial practices that are congruent with the program approach and its clientele.

'Management is not a one way street'

Julian Clissold

RaPAL Bulletin No. 11, Spring 1990

A response to 'Issues in Management of Literacy Programmes', Issue 10 [reprinted above] from Julian Clissold of Burnley College of Adult Education.

This article is about Management. It is also about Adult Literacy. Charles Lusthaus and Marie-Helen Adrien, writing in RaPAL argue that Literacy is a service, and that like other services, it requires management. Further, such Management should be efficient and effective. They imply that because literacy has been poorly resourced and marginal, insufficient attention has been paid to the Management of Literacy programmes. (They write about Canada but suggest that this is a much wider issue). In trying to correct this situation, Lusthaus and Adrien identify five key management functions:

Planning
Financial Control
Marketing
Programme Development
Personnel Management

In each of these areas, the writers suggest some questions that might help focus the attention of Managers in Adult Literacy Programmes.

On reading their article I have to confess to having been worried. I was worried about their view of literacy. There was often an unspoken assumption that literacy was a form of compensation, making good deficiencies in literacy students. I was worried about their view of Management. While it appeared to accord Management a functional and neutral role, it seemed at times to point towards practices that were manipulative. Finally I was worried about some of the absences from the article and in particular the absent student. Students appeared only fleetingly, often as clients, and always as if they had a common 'need'. In the remainder of this article I shall try to explain some of these worries. I shall also try to indicate a different set of issues that might be the focus for our concerns with respect to Managing Adult Literacy. Finally I shall try to show that many of these concerns are already at play within literacy even if they are not expressed in the conventional tones of Management-speak.

Lusthaus and Adrien begin their account with a simple assumption. Literacy is a service. That is, it is something provided by one group of people (tutors or volunteers) for another group of people (students). The student-tutor relationship is defined by the idea of service. It is a one-way relationship. Further it assumes that there are a set of skills and knowledges lacking in Adult Literacy students which the tutor can supply. Literacy itself is a product. It encompasses skills and knowledges that can be transported from person to person.

Lusthaus and Adrien do not present this as an explicit account. Nevertheless the core assumption of 'service' promotes such a view. It has various effects upon the sorts of questions they feel should concern Managers of Adult Literacy. Thus for instance, in discussing Marketing, they ask...

> *"Do I (the Manager J.C.) know who my clients are and where they are located?"*
>
> *"How can I convince my potential clients that literacy skills are important and needed?"*
>
> *"How can our potential users learn about our service?"*

These questions only make sense if you first understand the world of Literacy to be made up of service providers (tutors) and service receivers (students).

My own view of literacy is rather different. Literacy is not a set of neutral skills and knowledges. Nor is it a common set of skills and knowledges. Reading and writing do encompass skills and they do include knowledges but, they are also both personal and political acts. Doing either means making judgements, listening to judgements, making sense of the world, changing your ideas or those of others. There are too many commentators who have alerted us to the relationship between 'Communication' and 'understanding our worlds' to allow anyone to slip back into simple equations between literacy and skills/knowledges. (See for example "Writing and the Writer" F. Smith (1982) or the article by I. Winchester in Olson et al (ed): Literacy Language and Learning (1985)).

As a consequence when you support someone who wants to develop their litereacy, you are not simply transferring a skill. It includes an involvement at a personal and at a political level. When you help someone to read something you potentially help them to change their view of the world. When you help them to write, you potentially help them to change their world. Literacy is not a service in the sense used by Lusthaus and Adrien. It is not a one way street. "Students are not deficient in any straightforward sense, and tutors are not in possession of a product. All this has a direct bearing on how an Adult Literacy Scheme might be managed."

Lusthaus and Adrien try to provide a framework of functions that should guide the management of Adult Literacy. I want to make it clear at the onset that I find nothing exceptional in that list. However, it seems to me that there are other questions and concerns that should have a prior consideration. Critically it is important to ask "Who manages?" and "Why do they manage?"

For Lusthaus and Adrien it is managers who manage. That is not as obvious as it sounds. While it is not explicit, their model seems to identify a discrete group who have discrete functions in relation to tutors and students. Consider for instance their treatment of the 'Personnel Function'.

> *"Are staff roles and responsibilities clear and differentiated?"*
>
> *"Is supervision adequate?"*

The model is opaque, but is seems that management operates at a distance from those involved in teaching and learning. Managers organise a service that occurs at some distance from themselves.

The object of management ("Why manage?") is, similarly, defined within the terms of 'service'. The "mission" of an Adult Literacy programme becomes 'continuation of service'. ("Mission" is taken from Management studies in the U.S.A., for example T. Peters and R. Waterman (1982) "In Search of Excellence") But these are the terms of 'providers' not of the absent students. Managing Adult Literacy Programmes, for Lusthaus and Adrien, is given as a process removed from either tutors or students and which reflects the aspirations of neither group.

Is there an alternative? I believe there is, but it may not be in simple acts of faith or in establishing 'user groups' or 'student management committees'. Undoubtedly in some areas and in some literacy schemes such 'groups' successfully represent the aspirations of students at the level of scheme management. However, I would argue that management is not entirely - an operation that acts at that level. It also operates at the level of the group and at the level of the individual student-tutor relationship.

"Managing the Programme" for instance can be understood as a global process relating to the overall distribution of course provision. Equally on a day-to-day basis individuals or groups of students repeatedly negotiate their programmes 'with tutors. They 'manage' the direction, purposes and outcomes of their work.

It is possible to extend this analysis that can only be illustrated here. Nevertheless these are well documented examples of all manner of 'acts of collaborative management' that are carried on within schemes and that do relate to the objectives and aspirations of students (see for example the Special Development project sponsored by ALBSU into Special Activities, Project Co-ordinator Robert Merry, or the pages of *Write First Time*). They may not be about the total management of a literacy scheme, but they are certainly about 'managing'.

"Who manages?" is not then reducable to a simple statement. Many people 'manage' literacy provision. They may manage particular levels or regions of that provision; they may do so within all sorts of restraints and constraints. The process of management within Adult Literacy is a complex web of relations. Many of these may be uneven and unsatisfactory but very clearly management is not something simply invested in the roles and functions of 'Adult Literacy Managers'.

Similarly 'Why do they manage' cannot be reduced to a single statement related to service. There are as many reasons as there are motives for wishing to develop and extend communication. If those who have a formal responsibility to manage adult literacy, designated by job description and title, need a set of concerns or issues to focus their thoughts, then their starting-point should be "the who" and "the why" rather than any designation of management functions. Trying to answer these quite complex questions seems to me to provide a useful starting point to any attempt to develop management in Adult Literacy schemes.

A6 Curriculum content and process

Further RaPAL references about curriculum content and process

Gregory, G. 'Community Publishing and Writing Development', *RaPAL Bulletin* No. 4, 1987.

Upward, C. 'Simplified Spelling. Prospects and Perspectives', *RaPAL Bulletin* No. 5, 1988.

Collins, J. 'A Journey Must Begin with a Single Step', *RaPAL Bulletin* No. 37, 1998.

Roberts, G. and Prowse, J. 'Reporting Soaps', *RaPAL Bulletin* No. 38, 1999.

Frank, F. 'Students Writing Using IT', *RaPAL Bulletin* No. 40, 1999.

Good, M. 'On the Way to Online Pedagogy', *RaPAL Bulletin* No. 44, 2001.

Craig, B. 'Motivations For Literacy in the Information Age: Report on an Adult Literacy Internet-Based Home Training project in Multi-Ethnic Communities in New Zealand', *RaPAL Journal* No. 54, 2004.

Clearing away the debris: Learning and researching academic writing

Roz Ivanič and John Simpson

RaPAL Bulletin No. 6, Summer 1988

Introduction

Our goal is to blow away a lot of dusty old ingrained ideas about the teaching of writing. We are pushing back the boundaries which are imposed by formal learning and "acceptable English" to find the essential processes of writing what you mean. We are finding that it's all a question of WHO is the authority: traditionally, teachers have always been the authority, but when it comes to writing the only people who can possibly be the authority are the writers themselves. Being a writer means being in control, and being right. We are proposing this one fundamental principle for the teaching of writing: that teachers treat learner-writers as authorities, and help them find ways of writing what they mean.

We are a one-to-one tuition pair, arranged by the Adult Basic Education coordinator at Lancaster College of Adult Education. John has been a part-time student in A.B.E. for five years and is now taking stage B units in the Open College of the Northwest with a view to qualifying for higher education. Roz has in the past been a literacy tutor and organiser, and is now a volunteer tutor while working in the Department of Linguistics at Lancaster University. She is trying to find ways of integrating research into practice.

Learning First

We started with just learning. Although in our first year of working together John frequently said "Isn't there anything you want to do for your research?", Roz always answered "No, let's just work and see what happens." We worked out a set procedure for writing assignments and carried it through for each one.

This involved four stages:

(1) a plan,
(2) a 'stream of consciousness' draft,
(3) a hard graft draft to get the wording right, and
(4) a neat draft to hand in.

We quickly learnt that the very first draft – the 'stream of consciousness' – is a key element in the process of writing. It is both an essential step in the process and the version we always have to refer back to in order to retrieve John's intended meaning after several rewordings. John began to write first drafts without thinking about sentences, paragraphs or punctuation. We had lifted the restrictions associated with the academic; essay and turned it into a creative
act.

We also worked intensively on the 'hard graft draft', unravelling long sentences, rooting out repetition, and recognising academic waffle for what it is. This work resulted in a set of principles which John could use as the foundation for building essays in the future.

Research Later

Gradually research grew out of learning. At first Roz occasionally said: "Let me just write down what you said." To start with, John's insights about academic writing were random and spontaneous, but soon patterns began to emerge. We began to talk more, wrote down some tentative headings and started to make notes which related to our headings. Sometimes we tried tape recording our sessions, but it inhibited us both and neither of us had the time to listen through for the important bits, so we gave that up. At about this time, Roz started asking more and more questions like "Why do you think you did that?" or "What makes this bit difficult?" She had felt reluctant to ask too many questions before, because she was worried that they might waste time John really needed to spend working on his assignments. By the end of our first year of working together they had become an essential part of learning.

What We're Interested In

Here are some of the questions on our agenda. What causes you to repeat yourself? What is happening when you get tied up in a long sentence? What do you know about the style of academic essays? Which aspects of it do you try to 'copy' and why? We have found that all these problems are connected with a lack of self-confidence in the struggle for meaning. Repetition is the result of uncertainty over whether you've said what you mean – yet. Academic waffle gives the impression of 'getting it right' by the formal criteria John had been taught to apply to his writing.

What is the relationship between thinking, writing and communicating? No matter how complicated an idea is, we have found that it is possible to express it very simply so that everyone can understand it. The easier it is to understand something the better the writing. By pursuing simplicity we found we achieved a clarity which had not been in the writing before.

What use is the conventional advice on academic writing, and what harm can it do? We have found that laying down rules (for example, rules about how to construct a paragraph) actually has the negative effects of overloading, confusing and constraining the writer. What can you do to find out whether your writing will make sense to a reader? We have found that someone else reading your work to you shows you the exact sense and meaning that is coming out of your writing. Reading it yourself cannot show these gaps, because you know what should be there. What is the identity of the writer of an academic essay: who is the 'I'? What are the similarities and differences between academic writing and creative writing? We have found more similarities than we expected in that both are expressions of the writer's unique ideas.

The Value of Talking

We both realised together that discussing these things is actually an integral part of learning: John realised that by talking about his writing he was actually beginning to understand not only what was happening but why it was happening. As John took over control he was able not only to identify areas in his writing where things were going wrong but also to find solutions by asking the same sorts of questions as had arisen in our conversations. He felt he was improving more by discussing the nature and difficulties of academic writing than just by working on corrections of his own assignments.

Not Just Us

At the beginning of our second year of working together we began to wonder whether the picture we had built up of what's involved in academic writing was idiosyncratic, or whether it had any general validity. We feel it is really important to find out whether other people have similar feelings and experiences in relation to academic writing, and in turn whether the view we have developed is of use to others.

We believe that adults who are struggling cooperatively and consciously for the power of writing are likely to have a lot to contribute to research about writing development. Their insights can complement conclusions based on the experiments anti statistics of conventional research. Adults who have experienced years of misery and self-doubt have a strong commitment to saving future generations from suffering in the same way. One way they can help is by taking part in literacy research.

Recently we have talked to groups of students, tutors, organisers and researchers about our views on academic writing, and about this interplay between learning and research, trying to find other pairs who will compare their findings with ours. Writing this article for the RaPAL Bulletin together is another avenue in the same quest. We would like to contact other groups or pairs who will keep records of their observations on similar topics, so that we can eventually meet and compare notes. We believe the generalisations we draw from several adult literacy partnerships working on the same issues could make a major contribution to research in this field.

A lot of what we do will sound quite familiar to experienced tutors and students who feel in control of their own learning. Most student and tutor pairs or groups talk a lot about the feelings and difficulties involved in writing. This sort of 'language awareness' is being more and more widely 'recognised as an essential component of learning. What is different, what is RaPALish, is to recognise these insights as 'research findings', as 'knowledge'. Most of those discussions are disappearing into thin air, of use fleetingly to the participants, but not contributing to the general body of knowledge and understanding which can benefit the world. Our way of working suggests that the right questions for "clearing away the debris" and making progress with writing are the right questions for research.

P.S. Our next research topic is "Collaborative Writing"!

The significance of student writing

Jane Mace

RaPAL Bulletin No. 40, Winter 1999–2000

Jane is a founder member of RaPAL and author and researcher in adult literacy education. She currently works as Director of the Programme of Learning and Teaching in Higher Education at South Bank University, London.

Writing and classroom life

Writing is an act of making. Each time anyone puts words on paper or screen they are manufacturing something that was not there before. Adult literacy educators who have the pleasure of enabling adult learners to write are both witnesses to and agents of this manufacture – which, among other things, means convincing the learner to feel she is entitled to the word 'writer' at all. The writing might first be handwritten, worked on several times, with mistakes and corrections. In class, the learner works out with the help of the teacher the words she wants to convey her idea. At home, later, she may ponder these. Maybe two weeks later a deceptively short piece of writing, the product of many uncertainties, intervals and amendments, is as finished as it can be; and at some moment, by agreement with the learner, the tutor reads it aloud to others in the class.

At that moment, the writing, however short, has been published. Maybe six or eight people have sat there listening to it and they are its first readers. It is no longer a text muddled up with others inside the writer's head; it is theirs, too. The manufacturing work has resulted in something for others, as well as for the writer. Nobody who is involved in the work at this stage, however, sees this as a published work. Superficial things tell them it is not: such as the fact that it is handwritten; or that there is only one copy; or that it does not have a cover with a title on it. Yet the tutor's actions – first in asking the learner's permission to read it out and then reading it aloud to an audience, are the actions of an editor and publisher, as well of as a teacher.

The tutor takes home their writing and taps out copies on her word processor. Next day she copies the five pages of work on the photocopier and staples them together, with a cover sheet which says *Our Writing* – and, lower down, – *by the Tuesday Night Group*. This, too, is a student publication. The readers are familiar faces. The writers, this time, are several.

A week later, the tutor of the class next door, shows *Our Writing* to her class. They are impressed. She reads out one or two of the pieces. They talk. One says: that happened to me, what that woman wrote in that story. Another says: we could do a magazine like that, couldn't we? The following week, tutor (A) talks to tutor (B) and says: how about we do something together? Three months later, a forty-page magazine appears in the Centre. On the cover is the title: *The way we see it*; below, the words, *by ABE Students at Anystreet Centre, Placeham*. The tutors have written an introduction. The head of centre had agreed to a budget to print fifty copies which they give away free to authors or sell for 50p in the canteen.

You can continue the story for yourself. For now, I want to make a point, about the key difference between *Our Writing* and *The way we see it*. The difference lies not so much in the number of pages, nor the number of writers involved, nor even the number of copies printed, but in the fact that the writers will know the readers of the first publication, while they *will not know* the readers of the second. At this point the learners, who have just begun to consider they might qualify as 'writers', have to make another imaginative effort, and recognise that somewhere out there are other people who, despite the fact that they do not know them, might still be interested enough to be their readers.

So what is 'student publishing'?

In the history of UK adult basic education over the last thirty years, the phenomenon of nationally distributed published books of writing by adult learners was a relatively short-lived one. This was one of three main findings which emerged from a research study into 'the extent, rationale and effects of student publishing in adult literacy' which Rebecca O'Rourke carried out with me in 1991–2.[1] We knew from our own experience that the publications of the 1980s stimulated by *Write First Time*[2] and by the movement of community publishing had depended on a lot of voluntary labour. Much of the production work was a cottage industry, carried out by ABE tutors in their kitchens, discussed at editorial meetings held on Saturdays. The administrative and marketing work of selling and distributing books by mail order had often been done in the same way; the struggle often feeling an uphill one, trying to bring out into the world books kept under beds or in cupboards in the boxes they had arrived in from the printers. What Rebecca and I found in our study was that, in the face of the organisational and cultural changes in adult and further education which took place in the late 1980s, the energy and confidence to fund-raise for, produce and distribute such books had all but collapsed. People were no longer producing and selling printed books of writing by adult learners. An era had, apparently, ended.

The second finding from our research, however, was in stark contrast to this picture of doom and gloom. Nationally distributed books might have gone, but small-scale, home-made, short-print-run production of student writing on the lines I described above was, if anything, on the increase. ABE tutors up and down the country (and Rebecca travelled a lot, that year) were working regularly and often at the process of student writing. They saw it as an important educational process; students who participated felt empowered by it; one publication was inspiring another.

At the same time, our third discovery was that many ABE tutors did not see what they were doing as student publishing. There was a feeling that 'proper' student publishing was a special activity only undertaken by long-experienced experts in ABE who had access to extra time and finance and who knew how to produce printed books with a spine and an ISBN number. What seemed to need saying again with renewed energy – as I am saying here – is that the activity of *proper* student publishing is always, first and foremost, an educational one. It is about a *process of participation and learning*, with published products a by-product – whether these be photocopied sheets or printed magazines. In her contribution to the book which several of us wrote about community publishing in adult literacy education Stella Fitzpatrick enlarges on this idea. In this article, she stresses the *risk* entailed for adult learners to embark on this work, argues for the '*safe space* 'needed for them to feel confident to do so, and spells out the *attentive listening* required if learners are fully to develop a sense of their potential both as writers and readers (see reference in (1) below). Pause for a minute here and notice: these are, it seems to me, the functions of the skilled adult literacy educator, who knows that this is her job: to listen attentively to the experience and uncertainty of learners and to convey the kind of respect and rigour of attention which enables them to build their own sense of their potential.

On the desk beside me is a book called *Far and Near*. The editors are Rosemary Eggar and me. The publication date is 1979. The publisher was Cambridge House Literacy Scheme, in London where we both then worked. The publishing party for this book took place on the same day as the 'launch' of the first book I wrote myself[3]. Both publications are now out of print – which, looked at optimistically, means that those that were printed are out in circulation, in bookshelves elsewhere. The two publishing events happened in different parts of London; at the first, the publisher was launching two other books in the same series as mine, there were journalists present, interviews being fixed up, a lot of people I did not know who knew each other. At the second, most of the seventeen authors turned up, some bringing a brother or partner or children. We read aloud from some of the pieces in the book and authors picked up their copies of the book to take home.

I want to make two comments about this coincidence of publications. Although it did not have a spine, *Far and Near* did have an ISBN number and was typeset and printed; it was one of several books we published and; sold on the adult literacy circuit at the time, and were proud of it. But, as far as I am concerned, its significance lies in the educational value it had for both its writers end readers, in the thinking that went into the writing, the attention they learned to give themselves and each other, and in the learning about drafting, composing, editing and proof-reading which they gained from it. The second thing that occurs to me about the coincidence of these two publications is this. What the writers of *Far and Near* and I shared with each other was an experience which is at the heart of student publishing. It is true I had had articles which I had written in print before. But this time, the printed result was a book and, for me, the learning result – a result from all the process I had been through in writing – was that *printed books would never look quite the same again.* In a different way than I had never known before, I now knew something which literacy learners everywhere find so hard to believe: that behind every page of regular, linear, symmetrical typography lies confusion, uncertainty and numerous mistakes. It was a discovery of enormous liberation, for it meant that I knew at first hand that: however much their printed products appear coherent and complete, writers of books make mistakes and get stuck.

Looking back, wondering how it is that I have had such a long passion for student publishing in adult literacy, I suspect that the experiences of that day in 1979 might have had something to do with it. Over and over again I have seen this look of cautious revelation come over students in classrooms, reading writing from their peers in the group or elsewhere. It is not an easy outcome to measure, but it is unmistakable. Once a learner-writer has inside knowledge of published writing, it becomes possible for then to see other printed texts differently. It becomes possible for them to see that, if their own confusion lay behind the neat page of word-processed or typeset text, then some kinds of creative confusion might lie behind every page of print. It becomes possible for them to realise how much nearer to them the apparently distant, invisible and impossibly clever published writer actually is.

Notes

1. Rebecca and I produced a report out of the study, copies of which were distributed by Avanti Books, but which is now out of print. Both of us drew on the study for our contributions to the book I edited in 1995, *Literacy, language and community publishing: essays in adult education,* published by Multilingual Matters. This contains some of the best writing to be found on student publishing with pieces by Judy Wallis, Wendy Moss, Patricia Duffin, Roxy Harris and Stella Fitzpatrick (quoted above) – as well as Rebecca and myself; yet four years after publication this too is selling in relatively small numbers (at the time of writing this: just 450 paperback copies sold altogether in the UK and overseas).
2. *Write First Time* was published between 1975 and 1985. It was a newspaper-magazine of writing by adult literacy learners from across the country. Three issues were published a year, and at its peak, 6,000 copies were printed and distributed of each issue. The first of several residential weekends when students and tutors met together to work on ideas for writing and editing it was held in 1976. A full archive of all issues of the paper, minutes of editorial meetings, press coverage and correspondence is to be found in the Library, Ruskin College, Oxford, where visitors are welcome following contact with the librarian.
3. *Working with words: literacy beyond school* was published by Writers & Readers in 1979.

The title of this article is borrowed from the paper I gave to the RaPAL conference held in 1996, to be found in 'Living literacies: papers from the 1996 conference' edited by Stella Fitzpatrick and Jane Mace, published by Gatehouse Books, Hulme Adult Education Centre, Hulme Walk, Manchester M15 5FQ tel: 0161-226-7152

Exegesis Book Club: Adult basic education students write books for other students to read

Celia Drummond

RaPAL Bulletin No. 40, Winter 1999–2000

Celia Drummond is a literacy tutor at The City of Bristol College. She reports on the success of summer school activity, which was funded in part by the BT Reading Challenge.

To produce twenty new books for adult Beginner readers in under five weeks is quite a feat, but that is what a group of adult basic education students at the City of Bristol College managed to do this summer. As their tutor, I was carried along with the group's enthusiasm, determination and sheer hard work.

The project was made possible by an award £1000 from BT Reading Challenge, with matching funding from the college. We decided to run the programme as a Summer School activity delivered over two days a week for 5 weeks. It was offered free-of-charge.

Getting started

There was a core group of ten people who attended regularly, all basic education students. Also very involved with the group were a rota of four Personal Support Workers who gave much to the project. For me, every step was an experiment and I was very aware of my limited IT skills. (I was fortunate here to have the expertise of Ian Cumings and Teresa Thomas).

During the preparatory period, I became aware of how few books there were for first and second levels of reading. It seemed to me that it would be a good idea to concentrate on increasing the stock of books for this group. I also realised that before any stories could be written, students would need some underpinning knowledge. I found the Basic Skills Agency's publication, 'Making Reading Easier' very useful. It covers items such as 'Page Layout', 'Illustrations', Choice of Words' and 'Sentence Length', and the 'SMOG Readability Formula – Simplified'. These topics would have to be covered early in the course.

I started, however, by telling a story. I shared my experiences concerning a student who had borrowed her first book from a library. She wanted a 'proper book' – a book that looked like a book – not simply easy reading material. It wasn't all about learning to read; it was also about status and self esteem. The story was well received and set the scene. We felt that what we were doing was important.

During the first week, the students examined their own attitudes to reading, worked on some specialist knowledge and began to write stories of their own. We paid particular attention to using appropriate vocabulary. This meant concentrating upon the most commonly used words in the English language: social sight words and phrases, colours, parts of the body, words for numbers and names of people and places.

One activity that we enjoyed was appraising 77 books targeted at adult beginner readers. Working in pairs, the students listed pros and cons of the books before sharing their findings with the whole class and creating a list of 'desirable qualities'. This incorporated statements such as, 'We like books that 'remind us of ourselves', 'end on a hopeful note', 'have stories which are set in familiar backgrounds'. The books stayed in the room throughout the course so that whenever individuals had a few moments to space they could read them. We finished one busy day with a pleasant read round session – most chose to take part.

Having mulled over existing books, the students were clearer about what it was that they wanted to achieve.

Writing the Books

I took the group to a room where there was an enormous white board, stretching the whole length of one wall. By using the whiteboard, the first stories could be produced collaboratively. Each student was asked to name a setting for a story and three people suggested a theme park. I asked questions from which we derived a story line, such as 'Who went?' 'Why did they go?' 'What happened next?'.

Contributions came from everyone: some students gave their ideas shyly and hesitantly, others spoke with greater confidence. There was an air of good will that encouraged participation. Gradually we pieced the first story together and wrote it on the board. We discussed sequencing, spelling, punctuation, vocabulary and narrative structure. As I rubbed out, made additions and juxtapositions, I explained that this was a normal part of the writing process. I think that because I was responsible for the physical act of writing, students could be imaginative without the personal hamstrings of spelling, grammar and punctuation.

I was surprised by the quality of what we managed to produce. By the end of the second day, we had written about six stories, all of a reasonable standard. I would have been content to work on these six stories alone. However, most students turned up for the second week of the Book Club with stories that they had written at home. After editing, all of these were included in the series of books we produced. The fact that the stories were going to be published promoted a desire to 'get things right' and students welcomed feedback about punctuation etc.

Another of our story writing techniques was to table columns of names, colours, places, parts of the body and then tear up the lists so that each pair of students received several words from each list. They were then given a time limit to devise a story. This inspired what I regard as some of the best stories in the collection, such as 'Painting in the Bedroom', which appeals to me because it is such an original way of introducing beginner readers to 'colour' words.

When we had 20 stories on topics as diverse as outings, animals, crime, relationships, employment and education, I called a halt.

Illustrations

The group contained only one person who felt able to do illustrations. I was also able to contact an ex-student who had provided the artwork for books I had written four years previously.

With two artists at work, the problem of finding enough pictures was solved. The rest of the group and myself found it intriguing to observe the different styles of the two illustrators – one minimalist and one more detailed. They were a good balance.

The students decided they preferred books to have brightly coloured front covers. However, we had insufficient time so we followed the cover design of my previous publisher.

Battling with computers

With the stories and artwork more or less finalised, it was time for a serious IT attack and for most of the next two weeks we struggled to input our books on to the computer.

We began by typing the texts into Microsoft Publisher documents, deciding on page and sentence breaks and getting some idea of here the pictures were to go. As demanded by 'Publisher' we worked in multiples of four pages – eight, twelve etc. When we were ready we scanned in the pictures. In fact, to the end of the project, we were constantly reviewing and editing, words, pictures and layouts.

Throughout the production we considered ourselves a team, so that it was not important which book a student worked on. They would, for instance, design the cover or layout pages for a book another student had written. Students often consulted one other and worked in pairs.

Most of the group found the production period intense, demanding, sometimes frustrating and totally absorbing. Great creativity was shown in the use of pictures – clipping, shading, rotating etc. There was too much work for the students to handle alone during their two days a week in college, so Teresa and I moved things along bit by working 'out of hours'. However, we made sure that students were involved in very part of the process and could, genuinely, claim ownership of the project.

The prototype photocopied books were followed by a limited edition produced for us by the University of the West of England Press.

ISBNs and other distractions

Two hefty pieces of organisation we all undertook were the book launch and the acquisition of ISBN numbers. The latter was tedious, but rewarding; it made the publications 'proper books'. (See Resources and Contacts on page 33 [of *Rapal Bulletin No. 40*] for how to do this).

We become the Exegesis Book Club

Before the ISBNs could be issued, we had to find a name to publish under. This was fun. Late one afternoon, we had a group idea-storming session. We looked up 'reading' in a thesaurus, and one denotation was 'exegesis'. We found that 'exegesis' was to do with 'the interpretation or explanations of texts' and 'showing the way'. When we had mastered the pronunciation of 'exegesis' we liked the sound of it; we also liked the opportunity it gave us to be known as 'exegetes' : 'one who interprets or expounds'. Several students were excited by the fact that Inspector Morse, in that week's TV episode, used the word 'exegesis'.

Launch Day

From the first week, the students had started to think about aspects of the launch. Jobs were allocated to small groups. Periodically, the whole group came together to discuss progress, put forward ideas and consider ways of overcoming obstacles.

Collaboratively, we wrote a press release which was forwarded via the College's Marketing Department and appeared in several local newspapers.

The launch was held at the local W H Smith bookstore (thanks to the staff there) with food supplied by the College's caterers. Gill Impey, HTV weather reader came and mentioned us in two subsequent weather reports! At the launch, each student gave a short address in which they summarised their experiences in the Exegesis Book Club. The success of the launch was uplifting.

Overall the project was an enormous success. In their diaries and end-of-course reflections, participants focused upon the busyness and variety afforded by the project, the challenge and satisfaction of meeting deadlines and the reward of knowing that the work they produced will help others. They all wanted to continue their education.

We could have done with spreading the work over a longer period, and if we'd had more time, we might have devised worksheets to accompany the texts.

The books are being used with students and proving a hit. The beginner reader students like to know that other students wrote these books and that many of the stories are true. One group asked whether it might be arranged for them to meet the authors.

"It started off as a joke!": Women learning in a workplace Internet café

Julia Clarke and Sandra Southee

RaPAL Bulletin No. 44, Spring 2001

Sandra Southee is the full-time EDAP Co-ordinator at Ford's Dagenham Body Plant. Julia Clarke, now working at the Open University, was employed through the Trade Union Learning Fund to run the Growth Through Learning project during 1998–9.

In a recent government consultation document, "Skills for Life" (DfEE), questions were posed about the most effective ways to tackle "poor basic skills among adults". The RaPAL response to this document included a challenge to its focus on individual deficit, and an assertion that "the most constructive approach for promoting literacy is to support a combined programme of diverse opportunities for learning, together with research that explores the significance of literacy for participants in this programme".[1]

There are three points about such an approach that are illustrated in the following account of a workplace education programme. Firstly, although this was not set up as a basic skills programme, it offers an example of one of many possible educational contexts in which literacy skills can be developed through individual and social interests rather than on the basis of deficit or "needs". Secondly, adults may be motivated to join educational programmes for complex reasons, and their goals and purposes are likely to change as they engage with others in learning new skills and developing new interests. Thirdly, and perhaps most importantly, "progression routes" in adult education can run in different directions so that adults may wish, to enrol in a literacy or numeracy class *after* achieving success in other kinds of learning. There is nothing new in any of these assertions, but the current commitment of public funds to training and development in computer skills may provide us with further opportunities to demonstrate the importance of diversity in "basic skills" provision.

EDAP and 'Growth Through Learning'

The Employee Development and Assistance Programme (EDAP) is a joint programme which has been run by the Trade Unions and Ford for over ten years to provide Ford employees with educational guidance and sponsorship for learning opportunities. Each employee is entitled to an annual allowance for course fees, and at the Dagenham Body Plant there is a learning centre which also offers a range of courses tailored to meet the needs of people working a range of shift patterns. This includes an Internet Cafe, with open access and tutor support for those wishing to learn a variety of computer applications. Following an initial two hour induction programme with the tutor, employees can book time to browse the web, for example, or to engage in other learning activities using commercial software or to complete the series of assignments required to achieve a certificate in Computer Literacy & Information Technology (CLAIT). Other regular courses at the learning centre include Tai Chi, Photography, Music, Art, Bricklaying, Home Electrics, Washing Machine Repair and other workshop skills. In the same building, but now funded directly by Ford as part of the employer's job-related training programme, is a literacy and basic skills unit, Offline.

Many of us who have worked in adult literacy over a period of time are aware of the extent to which serendipity and chance encounters influence decisions about enrolling on, changing and withdrawing from courses (Clarke, 1989). Those of us who have worked in vocational or workplace education may also have come across adults who achieve high levels of practical competence or technical skills while developing ingenious strategies for avoiding literacy or numeracy tasks for which they lack either the confidence, ability or inclination. It is often assumed that adults who lack confidence in basic literacy skills should be encouraged to join a literacy programme as a 'first step' back into learning. While it is difficult to track the movement of adult students between courses run by different institutions in a community, the location of the EDAP learning centre and Offline on the same site has made it possible to collect evidence of interesting 'progression routes'. EDAP records provide clear statistical evidence showing that a number of employees taking up courses in practical skills, sport or even computer applications have later 'progressed' to literacy or numeracy tuition with Oflline (Southee, 1999).

From September 1998 to March 1999, the UK government's Trade Union Learning Fund supported an outreach project, 'Growth Through Learning' (GTL), based in the EDAP Centre at Ford's Dagenham Body Plant. The aim was 'to extend existing provision to the wider community, especially those taking the first step back into learning'.[2] The particular 'wider community' targeted by GTL included family members of Ford Body Plant employees, contractors working on the site (the majority of whom work in cleaning, maintenance and catering services), and employees in stores located in the neighbouring retail park. In effect this meant a focus on a predominantly female target group in a workplace where men comprise the overwhelming majority of the workforce. A variety of outreach and publicity strategies were used to encourage people from the three target groups to take up the opportunities on offer. Among those "first time returners" who responded, the most popular option was the Internet Cafe where the men in Ford's blue overalls, who previously comprised the main clientele, were now joined by women in their own clothes, or wearing green, orange or red overalls from ASDA, Homebase and the Wimpy Bar.

'Everything's computers now'

We arranged to interview nine of the women who had been coming in to EDAP's Internet Cafe on a more or less regular basis for several months and ask them about their experiences of learning and working; their feelings about themselves as workers/learners; their goals and aspirations; barriers and opportunities. Two of the women were family members of Ford employees, who had continued with their education after leaving school but had taken time out of full-time work when their children were young. The other seven had left school with few or no formal qualifications, one at the age of ten, three at fifteen and two at sixteen. We asked all the women whether they had learnt anything useful at school and most of them replied negatively, those who left at 15 saying that they learnt nothing until after they left school. Only the two Asian women mentioned the value of learning to read, write and do arithmetic at school. It seems that the other women took these skills for granted, although there was little call for what they saw as school learning in the jobs that most of them were doing.

Although this project was located in the workplace, when we asked why the women had chosen to learn about computers, several of them referred to home and family. For some of the women, there was already a computer at home, but, as Ghita said, 'I dared not touch it, my son says, Mum, not like that'. Yvonne said that she came along to the Internet Cafe because 'I've got to know about this 'cause my kids are doing it at school'. Similarly, Pat had decided to buy a computer because, when her grandchildren come round, '. . . it's nice to have something there for them to be interested in . . . and you get all the things for children don't you, History, Geography, Maths . . . which I think 'will benefit them as well'. Vicky said that '. . . all I wanted to learn for was basically for me... I don't know if I want to go anywhere job-wise with it'. But when asked what kinds of things she would do for herself, Vicky replied, '. . . well my husband's self-employed so I can do a lot of his work on it, his tax, whatever . . .'. This suggests that doing something 'for me' means doing something at home, which is clearly distinguished from something 'job-wise' even if it is really for her husband's work. Mina was a Health and Safety officer at work, so learning about computers would serve an immediate vocational purpose, as she was expected to write memos on a word processor and she didn't like asking for help with this. Yvonne also expressed the desire to switch from shelf-filling at night to a day job in the ASDA office now that her children were at school, and she believed that computer literacy would help her to secure such a job.

Another motivating factor was articulated in terms of a general observation that 'everything's computers now' and, as Pat went on, '. . . like you go to a doctor's surgery, you've got to be able to know how to use a computer. Everywhere you go now it's computerised'. Pat had expressed some ambivalence about joining the Internet Cafe, saying she would not have come on her own because she was afraid it might be like school and that she might be too old. Vicky, who had already decided to look for a computer course when the offer came from GTL, arranged to pick up Pat and Mary from their homes and drive them to the centre. These three women have been going along to the Internet Cafe together every week since their first visit. Mary had been even less confident than Pat about embarking on this course, saying 'it started off as a joke' and that she came to keep Pat company and thought, 'we'll go once, we won't come back'. But Mary did come back, despite her husband's scepticism, '. . . he said where are you going to go from here, which you can understand what he's saying, at your time of life, where are you going to go from here with it? 'But' she added, 'it's for satisfaction now, more than anything, I think. To prove the point that you can do it you know'. Having decided to 'prove a point', Mary echoed Pat's reference to the doctor's surgery, suggesting that this group of women had rehearsed the arguments that would justify their decision to learn about computers. 'I'd like my own firm actually and a computer would come in handy, you know, in catering, or anything . . . or if you go to the doctor's surgery now isn't it, what do you need? A computer. So, yeah, I think eventually it will be all, wherever you go . . .'.

'Very rarely do you use spelling'

Mary's comment that she only came along 'for a joke' suggests a defense against the risk of damage to an apparently fragile perception of herself as a learner. Mary, like most of the others, went on to achieve a CLAIT qualification, despite saying she had not liked the idea of an exam. She reflected, '. . . It's a funny world . . . you don't want to but then you suddenly want to don't you'. After a short spell in an office, Mary had worked as a sewing machinist until she got married, '. . . and then of course when the children were little I went into catering 'cause the hours seem to fit with the children doesn't it . . .'. So for Mary, learning about computers had helped her to regain other skills she had acquired at school but rarely used since,

> *"It livens up your brain again . . . You get a bit dead as you get older don't you? . . . and it's made me think again. My spelling's atrocious and it's made my spelling a lot better so that has pleased me actually 'cause, I mean, it was pretty good when you leave school but you don't use it much do you, you might write an odd letter, very rarely do you use spelling."*

The statistical surveys that come up with figures like 7 million adults in Britain having "poor basic skills" are derived from surveys in which people like Mary are asked at random to spell words like *necessary* or *sincerely*. It is hardly surprising that so many people get it wrong when so many jobs, particularly those that are generally available to women who want to balance paid work with "being there" for dependants, are those in which, as Mary says, you rarely use spelling. The EDAP Internet Cafe is just one small example of the kind of learning opportunity which may have "started off as a joke", but then "livens up your brain again . . ." and provides an incidental remedy for "poor basic skills among adults" along the way.

References

Clarke, Julia (1989) *This is a Lifetime Thing: outcomes for literacy students in Hackney*, London: ALFA (Access to Learning for Adults).

DfEE (Department for Education and Employment) Skills for Life: the national strategy for improving adult literacy and numeracy skills, Nottingham: DfEE Publications.

Southee, Sandra (1999) *Team Work* in 'Adults Learning', Vol. 11 No. 1 pp 26–27

Notes

1. Details of the national strategy for literacy and numeracy can be found on: http:www.dfee.gov.uk//readwriteplus

   RaPAL responses to this and other government consultation exercises can be found on the RaPAL web site at: http:www.ling.lancs.ac.uk/groups/literacy/rapal/RaPAL.htm
2. We used phrases like "first steps back into learning" or "first time returners" in this project as shorthand terms to denote people who had not previously participates in organised adult education provision since leaving school. This does not mean that these adults haven't continued learning all kinds of things in many other ways throughout their lives.

Hawkwood rhythms

Kate Tomlinson
RaPAL Journal No. 49, Autumn 2002

Kate is an experienced practitioner who currently works as a Basic Skills Tutor at Stroud College, Gloucestershire

---

Gloucestershire was part of the Pathfinder project trialling some of the activities of the "new world" of post-Moser Adult Basic Education. Because in the "old world" I had run several residential writing events for students, I was asked if I would undertake the organisation and delivery of a residential event for the Pathfinder project. There were various constraints such as the need to map to the Core Curriculum, to deliver twenty hours of "direct teaching", and to assess before and after to "measure progress". One of my chief concerns was to preserve, within the new frameworks, the enthusiasm and confidence-building ethos of the old workshops, and the idea of writing as a process as well as product.

Pathfinder and the Core Curriculum

One of my colleagues, Julie Brailey, agreed to do the project with me. We didn't want the Core Curriculum to become a straitjacket, particularly as we felt that the main benefits of the experience could not simply be seen in terms of the acquisition of disparate skills and involved outcomes which were very hard to measure. The aim, as laid down by Pathfinder, was primarily to look at the effects of an intensive form of Basic Skills provision on the learning of the adult participants; and to measure any differences in learning between this group and a control group attending sessions of two or four hours a week on a regular basis.

The residential writing event (focusing around the topic of *describing a place*), was held at Hawkwood College, a private Steiner based college, in a rural setting outside Stroud. It is the venue for a wide variety of courses, often around such subjects as personal development and alternative philosophies. It also hosts art and music workshops.

Once it had been decided that the residential should be a writing event, and that the venue lent itself particularly to developing descriptive writing skills, the Core Curriculum at text, sentence and word focus was used both as a frame and a reference for the activities. These were planned to build on each other and lead the learners through the writing process until the final afternoon, when students drafted a short description using skills and materials they had been working on, during the weekend.

What we did

We introduced each section working with the group as a whole, but encouraged interactivity by asking the volunteers present to write on the flip chart and keep a continuous record of their thoughts and feelings at each stage. This provided a subjective record of what they considered they were getting from the experience.

We tried to make activities as varied as possible. Learners used laptops for an activity on recognising and writing sentences, moving round the room to add adjectives or descriptive phrases to what had been typed by the person before. We spent some time discussing poetry and pictures on the theme of peace and collected words by wandering in the garden and exploring it using all the senses.

Learning styles and spelling

It was important to use students' varied learning styles in our programme. We thought about spelling by looking at syllables and how they correspond with the beats in music. We looked at visual patterns in the words they had collected to describe Hawkwood, splitting them up into syllables, root words and letter patterns. I had read a brief page in "The Spelling Pack" suggesting the use of music to help spelling.

> *"The right brain responds to emotion, art, music and patterns and grasps the whole picture quite quickly. The left brain tends to take in information in stages in a logical sequence. To maximise memory and learning, we need to try and involve both sides of the brain fully; and stimulate all of it by the learning environment we create.* So, learning to spell to music might be a good idea!"
> ("The Spelling Pack")

I recognised that this was probably based on an over-simplified view of brain function but felt that, even so, we could explore this in relation to spelling. With this in mind, we asked the local samba band to help us.

Samba, as most people know, is a dance from Brazil. The Stroud School of Samba, creates the rhythms of the dance through use of percussion instruments: huge drums such as the Primera and Secunda, small ones like Caixas, Tams and Agogo Bells. Starting with our own names and then using phrases that students had collected, such as "tranquillity and quiet" and "'beautiful environment'; we counted and clapped the syllables, and then played them on the drums.

Evaluation

At the end of the two day writing workshop, all of us, students and tutors alike, felt that this was a good way of delivering a learning experience for those who can find the time and means to attend. It may be a solution for people with very irregular shift patterns who find coming to either weekly sessions or short courses a problem.

Participants at the events evaluated them positively; they liked:

- the opportunity to focus in an uninterrupted way on a particular topic
- being able to complete work started without any gaps and keep a sense of continuity
- learning in a closely knit peer group
- learning in a quiet place without the family or other distractions
- learning using a variety of methods and approaches
- all focusing on the same skills and processes with learners at a similar level, using
- whole group teaching for long periods of time.

The comments from each group were very similar.

As a requirement of Pathfinder, we had devised pre- and post-course assessments using multiple choice questions (in the manner of the National Tests) targeting the Core Curriculum skills, which we were teaching. We thought that though multiple choice questions provided some objectivity and ease and speed of marking, they were not a good way of assessing writing skills. Most of the students I have talked to who have done the National Tests have commented that it was relatively easy to get the right answer when selecting spellings or looking at wrongly punctuated sentences. They achieved scores they knew did not reflect their ability in free writing. We also found that they did not reflect the wide skills they demonstrated in the course of the activities we had observed and discussed. Confidence, for example, was a predictable but not easily measurable outcome. However, together with the students' evaluations, gathered and recorded at frequent intervals during the two days, the overall assessments made us feel that a great deal of learning was taking place.

Nevertheless, we concluded that only by making residential events an accepted part of provision can their true value be assessed. Funding, assessment, and logistics must continue to be explored. The funding was relatively high for the numbers attending and this will always be an issue, especially if the value of this kind of learning experience is not fully recognised. Such recognition is vital given the outcomes in terms of experiencing learning as fun and in the development of learner motivation.

Releasing potential in the dyslexic writer (using voice recognition software)

Ellen Morgan

RaPAL Bulletin No. 27, Summer 1995

Ellen Morgan is a Senior Lecturer and Co-ordinator of the Dyslexia Support Service at the University of North London. In 1993–4 she received funding from the Higher Education Funding Council in England (HEFCE) to run a one-year project training graduate "mentors" to teach undergraduate dyslexic students. Part of the project included exploring the value of specialised technology including a Voice Recognition System for severely dyslexic adults.

All writers face similar challenges. To achieve effective written communication, they must organise ideas, carefully select words and then crystallise their arguments. My experience teaching dyslexic adults in basic, further and higher education has revealed a common thread in relation to developing writing skills. Whatever the nature of the *content* the writer wants to convey, dyslexic writers face interference in *transcribing* their thoughts.

This article explores ways in which writers with limited technical skills can be liberated to focus on their composition skills by using technology (computer or tape recorder) or the human "scribe" (amanuensis). The degree to which limitations in technical skills inhibit composition skills varies along a continuum of mild interference to major impediment. The means of intervention must be determined by the nature of *the student's* difficulty. Before recommending alternate transcription possibilities, tutors should evaluate potential obstacles created by the possible need to acquire new skills.

Writing Challenges for the Dyslexic Adult

Frank Smith argues that the transcription and composition skills involved in writing are conflicting tasks which compete for the writer's attention. He suggests that this competition can be the source of writing difficulties[1]. Analogies such as that of the lord of the manor dictating to his scribe, or the business executive dictating to her secretary, illustrate cases in which the creator of the ideas is separate from the executor of the words. However, most people do not have the luxury of dividing these tasks; undeniably, students are expected to be able to accomplish both roles simultaneously. This is problematic for many writers, but particularly so for the dyslexic.

Encouraging the Composer

According to Nicolson and Fawcett, "dyslexic reading and writing performance appears more effortful, more prone to error, and more easily disrupted than normal performance"[2]. They propose an "automatization deficit" hypothesis which suggests that dyslexics are inhibited in reading success by a failure to automatize the sub-skills necessary for fluency. This hypothesis could serve equally to explain the classic dyslexic complaint of "knowing what I want to say, but being unable to put it down on paper". The *composition* is not generally the problem; rather the *transcription skills* which non-dyslexic writers take for granted (eg spelling, punctuation, paragraphing, handwriting) constantly interfere with the thought process; these are the very skills the dyslexic has been unable to automatize.

The battles between composing and transcribing must be resolved if dyslexic students are to be judged on an even footing with their non-dyslexic peers. The teaching challenge is to establish avenues for dyslexic students to gain competence in communicating their ideas, without worrying about the transcription issues. There are many options for separating these tasks; current trends suggest there will be increasing opportunities in the near future for machines to perform the mechanistic tasks involved in writing. The road to liberation for written expression is rapidly widening.

Granting "Permission" to Concentrate on Composition

I was astonished recently by the relief expressed by a dyslexic student when I suggested she write what she wanted without worrying about her spelling errors. Sarah had explained that her simple sentences, immature vocabulary and short paragraphs were the result of trying to get down the bones of her argument with minimal distractions from her thoughts. Reliance on a dictionary to access unfamiliar spellings interfered with her ability to "hold" her ideas. The embarrassment of submitting work that would be judged by weak spelling was worse than the worry of producing written work that did not reflect her level of conceptual thinking. My suggestion that she concentrate *only* on her thoughts and put them down *in any form possible* (including pictorial symbols) had a liberating effect on her. She appeared simply to need *permission to* compose without worrying about transcription issues.

Students such as Sarah, whose mild to moderate dyslexic difficulties affect their transcription skills, can be helped to separate the two aspects of writing. The transcription skills can be developed by working directly from a piece of student-generated writing and concentrating on spelling, handwriting, punctuation or paragraphing, according to the student's wishes. Students and tutors can engage in valuable dialogue about the importance of drafting and the need to proof-read for different purposes (i.e. content vs. transcription).

The Role of Technology

Students in Higher Education generally have little time to devote to improving transcription skills. For students with severe difficulties, the amount of intensive tuition and practice time necessary to impact significantly on the writing skills is simply unaffordable. Most of the students' energy must be expended on reading and producing the relevant *content* required for success on the course. These are the students who require full-time secretaries (or the equivalent) to ensure that they are being judged fairly with their non-dyslexic peers. In these cases, I would argue that the use of technology and/or human "scribes" (amanuenses) can overcome the serious disadvantage that the severe automatization deficit creates.

Voice Recognition Systems are still in their early infancy, but are rapidly developing in sophistication. The technology enables the user to train a computer to produce correctly spelled words in response to an individual's voice. This may seem like a panacea; indeed it may very well become just that for many dyslexic users. However, the current state of the art is not sufficiently advanced to warrant recommendation for all. In cases where students' difficulties severely impede their ability to write, alternate strategies are essential.

John was a severely dyslexic student who relied on a tape recorder to replace the pen. He recorded all lectures, listened to the tapes and took "oral notes" on a second dictaphone. He then hired an audio-typist to produce his notes on disk, which he could manipulate on the computer. Similarly, he dictated all essays on tape for subsequent audio-typing. This expensive and time consuming strategy enabled him to complete the first two years of his degree. In his final year, he took advantage of the Dragon-Dictate Voice Recognition System. For John, this technology represented the independence previously existing only in his dreams. Now, his spoken word could generate instant visual feedback.

John's explanation of his problems and reactions to this new freedom are described on the next page, in both his handwritten and computer dictated versions.

First Attempt

See Figure 1.

Second Attempt using a Word-Processor

> I am a dislicie stundt doing a BSc dedreg I have expence diffulcties in wighting all my life. It can be vary frustuating trying

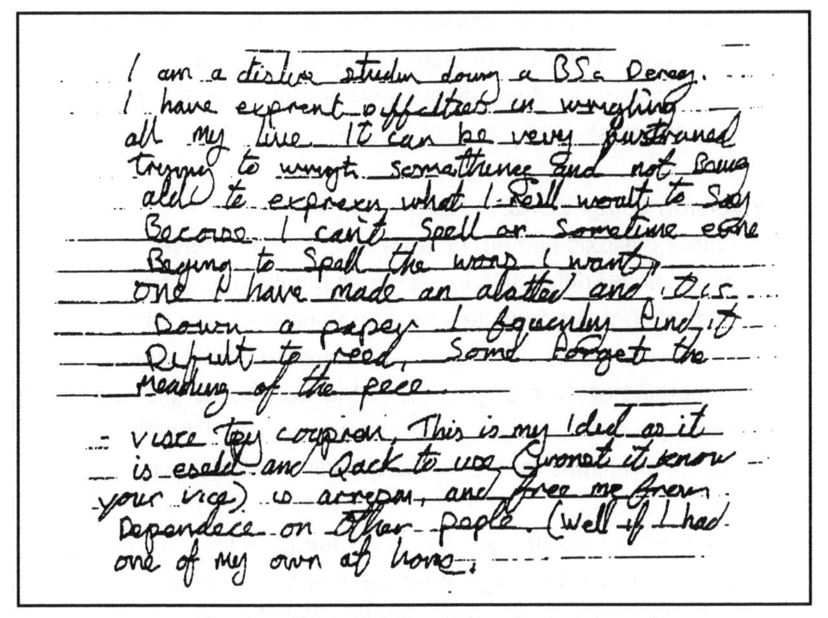

I am a dislex studen doing a BSc Deray.
I have exprent offcltres in wrighting
all my live. It can be very frustraned
trying to wright somethinc and not Bawe
able te exprexe what I Rell woalt to Say
Becouse I can't Spell or Sometime esne
Beying to Spell the worp I want,
one I have made an alatted and tis
Down a paper I fqucenly find it
Difult to reed, Somd Forget the
meaning of the pece

- vioce tpy compion. This is my Ided as it
is esald and Qack to use (wonst it know
your vice) is arrepsu, and free me from
Dependace on other peple. (well if I had
one of my own at home.

Figure 1: First Attempt

to wright something and not being able to express what I relly want to say becese I can't spell or sometime even begine to spell the words I want. wonrs I have made an atemed and it is down on paper I find if diffulit to reed, sometimes forgetting the miening of the pece.

– Vice tpy computer, this in my ideal as it is esze and quck to use (once if know your vice) it is arront, and frees me from dependsie on other peple. (well if I had one of my own at home that is)

Third Attempt after Using Spellcheck

I am a dyslexic student doing a BSc degree I have expence difficulties is wighting all my life. It can be vary frustrating trying to wright something and not being able to express what I relly want to say because I can't spell or sometime even begin to spell the words I want. Wonrs I have made an atemed and it is down on paper I find if difficult to reed. sometimes forgetting the miening of the peace.

– Vice tpy computer. this in my deal as it is esze and quick to use (once if know your vice) it is arront and frees me from dependence on other people. (well if I had one of my own at home that is)

Final Version Using VOICETYPE (unaided)

I am a dyslexic student doing a BSc degree. I have experienced difficulties in writing all my life. It can be very frustrating trying to write something and not being able to express what I really want to say because I can't spell something or even begin to spell the words I want. Once I have made an attempt and it is down on paper I find it difficult to read, sometimes forgetting the meaning of the piece.

– Voice type computer – this is my ideal as it is easy and quick to use (once it knows your voice). It is accurate, and frees me from dependence on other people. (Well if I had one of my own at home that is).

Overcoming Obstacles to Using Voice Recognition

The main advantage of this system is the writer's emancipation from worrying about how to spell words. Once the computer is trained to recognise the user's particular pronunciation, it will always spell words correctly. However, if the user doesn't ensure that the initial computer selection is the correct match for the voiced word, the computer will always respond incorrectly to a particular utterance. For example, if the author says "dyslexia;" the computer may offer "dinosaur" as its guess, along with other words that might or might not approximate the correct word. The user must choose the correct option, or spell the word into the computer (either orally or by keyboard) to "train" the computer for future recognition. This obviously poses a real obstacle for the dyslexic, who is using the system *because* of spelling difficulties.

There are two possible solutions to this "Catch-22" situation: the user can read text into the computer, correcting incorrect guesses by referring to the printed version. The more words that are trained into the computer's memory, the better the computer's guessing mechanism becomes, and the fewer errors it will make. Additionally, it is important for someone other than the dyslexic writer to check for errors and supply correct spellings when necessary. This may be a laborious process initially, but the degree of support needed becomes increasingly less as the user's lexicon increases in the memory store.

One ironic disadvantage to the Dragon-Dictate system is that it actually requires the writer to concentrate on each spoken word to ensure the correct word is entered. Thus, there needs to be considerable concentration on the transcription which obviously detracts from the composition. A new Speech Recognition System produced by IBM addresses this problem. The IBM Voice Type Dictation System adjusts the words on the screen by employing contextual guessing. Thus, the computer will figure out the spelling of homophones once the remainder of the sentence is dictated. Moreover, this system has the additional advantage of recording the speaker's voice, allowing the user to speak into the microphone without looking at the screen and to edit or correct the text after completing the dictation. This offers both a multi-sensory input and the opportunity to postpone the transcription tasks until after composing is complete.

The Use of An Amanuensis: Whose writing is it?

Dyslexic difficulties are often exacerbated under stress conditions (eg examinations) and therefore adults who lack fully automatized skills will be disadvantaged in a testing situation which taxes short term memory and which places high demands on reading, writing and sequencing skills. Many adult dyslexics feel more confident in their oral skills than in their writing skills. Indeed, it is often the observed discrepancy between oral contributions in seminars and performance in written tasks that alerts lecturers to identify dyslexic students. Employing the skills of an amanuensis removes the transcription function from the writer and allows for fuller concentration to be placed on composition.

Who can benefit?

Some students who cannot successfully resolve the competition between transcription and composition rely on oral strengths by developing dictation skills. This ensures that their thoughts are recorded, either by taping or by using a scribe. However, not all students are comfortable with composing orally, a skill which is essential for effective use of an amanuensis. Additional challenges that must be considered for maximising the benefits of using a scribe include proper timing (making sure the dictation speed conforms to the scribe's writing speed), ensuring that the scribe has clearly legible handwriting, and mastering the use of any specialist vocabulary.

The Amanuensis – a Borrowed Hand

The advantages of having an amanuensis can best be described in the words of a student. Sue explains:

> *"Hearing what I was saying by speaking out loud acted as a memory prompter. Verbalising my words allowed me a wider choice and use of vocabulary. This in itself allowed me to remember what I had revised. I was able to draw on my strategies which the Dyslexia Unit has shown me, like thinking and talking out loud and hearing what I was saying. It felt like I was in control of my thoughts; I was aware I was not jumping from one idea to another. Using an amanuensis took away the pressure of losing too much time spent on my confused spelling and sentence structure. The amanuensis temporarily became an extension of me. She represented my hand memory. There are no words to describe the confidence this gave to me as a dyslexic student."*

Thoughts for the Future

The use of amanuenses and/or developments in new technology can be liberating for severe dyslexics. The freedom to concentrate on ideas without the worry of how to get them down on paper empowers the writer and places her on an equal footing with her peers. Perhaps this can best be expressed in the words of Sue, whose summary on the experience of using an amanuensis was:

> *"I can only describe the way I felt as being free to express my thoughts in such a way that I wonder if this is what it is like to write as a non-dyslexic student."*

I think we owe all severe dyslexics the opportunity to experience this sense of exhilaration and to express their true potential.

Acknowledgement

My own observations and analysis of the benefit of intervention in enhancing the writing process have been greatly enriched by the feedback received from both the mentors and the students themselves.

References

1. Frank Smith, *Writing and the Writer*, Heinemann Educational Books, 1982, p.21.
2. R.I. Nicolson and A.J. Fawcett, *Automaticity: A new framework for dyslexia research?*, Cognition, Volume 35, pp.159–182.

Some challenges posed by reading research

Ann Finlay
RaPAL Bulletin No. 23, Spring 1994

Ann Finlay is currently employed by Loughborough College teaching basic skills. She is also carrying out research as part of her Ph.D. studies at the University of Nottingham

There has been growing debate concerning the ethics of research; particularly with regard to the relationships and rights between researched and researchers. As an ABE tutor aware of my obligations towards students, and as a research student facing the demands of academe, I have struggled to reconcile what at times seem like conflicting demands.

As a starting point I would like to briefly review the different approaches to literacy made by different 'experts'. Du Vivier (1992), in a neat classification, divides society's response to the need for literacy provision into five areas: schooling, training, counselling, learner centred, and social action (see table).

Each of these five responses is answering the needs of a different aspect of the person engaged in literacy studies. For example, *schooling* centres on the cognitive functioning of the student as defined by cognitive psychology. For those unfamiliar with cognitive psychology/cognitive research, this approach to learning concentrates on how language, memory, perception and intelligence operate, without considering emotional, social, physical or political factors, almost as if the human mind were a computer and nothing else. The providers of *counselling* aim to remedy the damages of

Mode of Response	Focus	Definition of Illiteracy	Presumed Causes	Response	Goals	Means of Assessment
Schooling	Cognitive Domain	Lack of knowledge of correct forms and usage	Dyslixia. Missing school through truancy or circumstantial factors	Remedial instruction with individual attention	Improvement in assessable standards	Norm-referenced testing
Training	Practical Application of Literacy Skills	Inability to apply skills in real-life situations	School fails to teach the skills needed in modern society	Practive and application of component skills in simulated exercises	'Functional' literacy and enhanced employability	Criterion-referenced tests, checklists, profiling
Counselling	Affective Domain	Damaged identity through social, cultural and educational deprevation	Corporal punishment, ridicule and stigmatisation of failure in school	Listening, reassurance and self-expression through writing	Development of self-esteem and Integrated identity	Learner's own impressions. Attempting what had previously been avoided.
Learner Centred	Felt Needs of Learner	Multi-dimensional depending on the learner's experience and intentions	Combination of some or all of the reasons above	Facilitation – tutor adapts their response to meet their student's needs	Full participation in society. Access to mainstream adult education	Self-referenced. Joint negotiation of goals and assessment of progress.
Social Action	Political Potential of Literacy	Marginalisation through the differential distribution of knowledge	School recreates social inequality and reinforces the notion of a dominant culture	Consciousness – raising through situational analysis linked to skills development	Empowerment of group or community to take action for social change	Increased student involvement. New sense of autonomy and power for participants.

social and educational deprivation. The *learner-centred* approach (the traditional Adult Literacy response) concentrates on the felt needs of the learner, and so on.

Political correctness intervenes here, as some approaches use terminology which is currently acceptable to students and tutors in ABE, and some use language which is unacceptable. The language used reflects, of course, the attitudes of those offering provision (or research into such provision) to those receiving it. And however carefully we try, I doubt that it is possible to use language which is not loaded in some way or other. I suggest that in ABE today there is concentration in research on those approaches which use language in a more politically correct way than on those which do not.

A close look at some of the words chosen and used shows why some approaches tend to be rejected and others embraced. For example, it is normal in cognitive research to label people participating in research 'subjects', to speak of 'treatment groups'. These words and others in cognitive psychology overlap with the language in medical literature, which has been shown by Kress (1985) to be highly suspect in the ways in which it uses language to maintain power over, and distance from, the rest of the community. It is not surprising that possible cognitive solutions to literacy tuition needs should be ignored, when the language used in the cognitive approach can be demeaning and disempowering to those who are researched. With the use of the label 'dyslexia', in particular, medical terminology is damaging with its assumptions of neurological malfunction and disease. Furthermore, in schools the cognitive approach has tended to dominate the literacy scene, at least as far as research goes. (This is not meant as a criticism of school teachers who have a difficult enough life without criticism from the likes of me!)

Understandably, ABE research and practice have chosen different routes. The work of Freire (1972), with its concern for peoples' rights, has been eagerly taken to heart. 'Empowerment' has become a buzz word. The role of literacy in the community is being explored. The classic work on the success of literacy provision by Charnley and Jones (1986) includes peoples' feelings about their literacy abilities as legitimate evidence of progress. Finally, functional literacy, with its emphasis on the actual literacy priorities of students, as judged by students themselves (Hayes and Valentine 1989), and the workplace requirements of employers, have added to the roundness of the adult literacy picture. Four of Du Vivier's five categories are respectable areas for the ABE researcher to investigate. This is one of ABE's major achievements, that it has changed the face of literacy by taking a wider and more humane view than reading tuition sometimes took formerly.

But, however valuable they are, these approaches are not able to solve those problems which relate to cognitive difficulties. Four out of five dimensions are being addressed, but the fifth dimension is largely ignored in ABE because, in the past, research into children's reading problems was carried out and verbalised in pejorative terms, which were particularly hurtful to adults. Also, educational researchers treated reading as if the cognitive dimension took precedence over the other four. These past mistakes do not mean that cognitive research is wrong but that its place, its language and its methods need careful thought. How can solutions to cognitive problems be found if we ignore them because we have not got politically correct past practice to guide us? I believe that it is in the interests of us all, as well as being ethically desirable, to take into account the whole person. This means considering all five of Du Vivier's categories, even though the schooling/ cognitive approach has been traditionally researched in ways many of us in ABE find repugnant.

At the moment my own research into the strategies used by non-fluent adult readers is attempting to use some aspects of cognitive psychology without abandoning traditional ABE philosophy. My interest originated in my failure to find adequate explanations in the educational literature for some ABE students' reading difficulties. For example, why does a reader who appears perfectly normal in every other respect, and in spite of years of tuition from a variety of people, still have difficulty in reading as fluently as her or his peers?

Dyslexia is not always an adequate explanation and none of Du Vivier's non-schooling 'modes of response' gave me the complete answer. What, exactly, are such readers doing to try and work out what a word is? When such a reader meets a word s/he cannot read accurately, Freire's work is of no practical help to me, invaluable though it is in other contexts. Words which cannot be easily decoded by the reader may make a text incomprehensible. And it may well be a text which is required reading for that person. For example, a student I know who is working for a City and Guilds car maintenance qualification cannot gain his diploma without reading and understanding certain car manuals. If I knew more about what he did when he encountered words he could not read accurately I might be able to give him more effective tuition. It is this kind of question which has led me to look at cognitive research in my search for solutions.

One of the tools of cognitive psychology is a technique called *protocol analysis*. It has been used, among other things, to research how people solve technical problems such as working out answers to numerical calculations. The participants are invited to explain their thoughts out loud as they work out a solution to the problem, and then the resulting transcripts of their explanations are analysed in depth. By treating reading as a technical problem solving activity, and by using protocol analysis to find out more about it, I expect to gain more insight into the various cognitive mechanisms involved when readers are faced with words they cannot read with ease. (Incidentally, this technique, when applied to reading, is not to be confused with the better known *miscue analysis*.) From there, I hope to move on to seeing how the insights gained may be applied in teaching situations.

As an ABE tutor I look for ways in which the participants may gain credit for taking part in the research. For example, those participants working towards gaining Wordpower Certificates are able to count their participation in the research towards achieving some of the oral units. Finally, when writing up the research, I have to try and avoid using the traditional language of cognitive research and replace it with alternatives acceptable in ABE, which means having the confidence that sticking to convictions will not penalise me in an examiner's eyes.

In conclusion, one of the challenges facing researchers in ABE today is to extend the areas covered by research in order to take into account all aspects of literacy tuition but to do so in new ways which are reflected in the language used. I have spent many hours trying to trace other cognitive research into ABE students' reading difficulties and, with one or two rare exceptions, I have found virtually no British research in this field. We need the courage to break new ground, without necessarily having precedents to guide us, and the confidence and creativity to write for academic audiences, using language compatible with our teaching ethics. To finish hopefully – I believe these demands can, and are, being reconciled in some instances, and those of us who are ABE practitioners involved in research are in an advantageous position to help bring this about.

References

Charnley AH and Jones HA (1986) The Concept of Success in Adult Literacy. ALBSU.

Du Vivier E. (1992) Learning To Be Literate. Dublin Literacy Scheme.

Freire P. (1972) Pedagogy Of The Oppressed. Penguin Books.

Hayes ER and Valentine T. (1989) *The functional literacy needs of low-literate adult basic education students.* Adult Education Quarterly 40 (1) p.1–14.

Kress G. (1985) *Socio-linguistic development and the mature language user: Different voices for different occasions.* In Language & Learning: An Interactional Perspective edited by G. Wells and J. Nicholls.

Writing for change

Jane Pinner, Val Watkinson and Sue Bergin

RaPAL Bulletin No. 7, Autumn 1988

Bolton Royd Adult Education Centre, Bradford

Traditionally writing in ABE has been used as a diagnostic tool for the teacher – to find out what the student does not know. This is what we, as teachers, have been taught to do. Therefore it is safe. This automatically reinforces the prescribed teacher/student role. The student does something wrong and the teacher knows what to do about it.

However, if the focus of attention for students and tutors in ABE changes, if it is no longer skill diagnosis, grammar, spelling rules, punctuation and so on – but becomes instead the use of 'writing' as a medium for self-expression and experimentation, 'getting down on paper' the ideas/loves/hates/experiences which we all have – we are then in a completely different situation – everyone is a learner and everyone a tutor. Roles are balanced. All experiences and writings are valid.

Writing becomes self-diagnosis or therapy. It can be cathartic, enjoyable and rewarding. Students do not have to learn the rules as a precondition to being able to write. The role of the tutor has changed. It becomes now to encourage and assist students to find suitable means of expression – a voice. This may not necessarily be written – but can be spoken, scribed, sung, play-acted, or whatever!

Accentuate the Positive, Eliminate the Negative

We were aware of how ABE schemes around the country had extended the notion of 'writing' with much success – for example in London, Brighton, Bristol and Manchester, and wondered how this would work in Bradford.

From the vaguest sense of all this, five years ago, we began to experiment with various ways of working with ABE students – using the students' own experiences as the tool.

With financial support from Yorkshire Arts Association, we employed Ian McMillan, a working-class poet from Barnsley as a Writer in Residence. Ian worked with ABE tutor/student groups, not only in institutional educational settings but also in community centres and unemployment drop-in centres. This was an attempt to break down the traditional barriers between "published writer" and "awed audience".

The lively way that Ian approached groups and the exciting methods he employed, stimulated ABE students to produce writing of their own – some for the first time ever. For many students and tutors, the result was a complete release from the established, hidebound notion of writing as a private and isolated struggle. Writing could be fun – a joint activity. It did not even need to be committed to paper. Words could be sung, spoken, transcribed from a cassette player or even acted. It was all valid.

To build on the initial success and keenness generated by the residency, we began to plan other projects. Residential writing week-ends were integral to this development. They provided opportunities for adults from basic education groups in Bradford to spend a weekend away, immersed in an atmosphere of writing, learning, fun and a sharing of ideas and work. After 4 years the popularity of weekends, facilitated by local writers including Kitty Fitzgerald and Jan Maloney, has still not waned.

Perhaps the most exciting and encouraging features of these residential events have been:-

- the demystification of writing as a craft
- the sensitive learning atmosphere generated during the weekend

- the diversity and high quality of individual and group writing produced
- the impetus for women's writing groups to emerge within the ABE scheme
- an increased openness to taking risks, sharing life experiences, thoughts and writings
- the obvious fun and excitement, derived from spending a weekend away from families and other responsibilities

and probably most of all:-

- increased student participation in the planning and organisation of residential weekends and consequent writing projects.

Student participation and a collaborative style of working has enabled us to become more responsive, as tutors, to student demands or goals. The formation of the West Yorkshire Association of Community Writing Groups (known colloquially as the 'Wigwags') – a student group, keen to promote ABE student writing and community-based writing projects, was a direct result of this way of working.

In 1987, the Wigwags, with our support, organised a community writing fair, held in a community arts venue, in central Bradford. This all-day event combined:

- free workshops, including play-writing, short stories, poetry writing and performances, dance, editing, publishing and marketing
- stalls run by local community-based writing groups
- evening entertainment: bands, clog-dancing, folk music and poetry readings
- cafe, buffet and bar
- signing and creche facilites
- disabled access
- volunteer tutor support for ABE students attending the fair.

The aims underpinning this community writing fair were:

- to create a focal point for ABE writing students/groups in this area;
- to act as a forum for those interested in learning, experimenting with words, exchanging ideas and information;
- to provide an environment where ABE students were able to meet and work with other individuals and groups;
- to destigmatize Adult Basic Education by heightening its profile in Bradford;
- to change the emphasis of ABE provision from 'closet literacy' to 'a celebration of student writing'.

Other one-day writing events have been organised, which have provided continuity of contact for ABE students and the opportunity to test out other forms of self-expression, through writing. One such event involved working jointly with Yorkshire Arts Circus (a community publishing group from Castleford) and students, to organise a project around the theme of music. Another writing event, appropriately named 'People's Patter', offered students, tutors and members of writing groups a diversity of workshops. Always, the overriding aim of the projects remains to provide a productive yet enjoyable day.

It is important to state that whilst residential and one day courses are perceived to be an integral part of the learning experience for students in higher education – for ABE students, as the 'poor relations', such experiences are all too often viewed as non-essential. In such a climate of opinion we are faced, as tutors and students, with the perennial and time-consuming problem of fundraising. To ensure that writing weekends and other projects work, we need to fund, at the very least, workshop fees to guest poets/writers, creche facilities and student travel expenses.

Finance for ABE is currently very limited and literacy provision appears to be moving away from the liberating notion of student control of learning processes and curriculum towards a more restrictive approach, in which students are required to work on tailor-made vocational packages. Demand is no longer student determined but labour market led. This is happening against a background of the Training Commission demonstrating increased interest in Adult Basic Education. The educational rationale for ABE is taking second place to meeting the demands of private industry for a better trained workforce.

Indeed, given that the educational model of provision may give way to the training model, it is hardly surprising that substantive funding for writing projects is rarely incorporated into ABE budgets. At best, nominal funding is allocated. Writing projects are not generally seen to be directly related to enhanced employment prospects. We believe that this hypothesis could be successfully disproved – but that must be the subject of a further article!

Despite continued funding constraints, we are strongly committed to writing developments and continue to consolidate and promote writing projects. Certainly there are observable changes taking place for ABE students participating in writing events – changes not only in self-confidence and self-image, but in attitudes to education.

Currently we are working with students on the following writing projects:

- a second ABE Community Writing Fair
- a residential Writing Weekend in the Yorkshire Dales
- a series of regular projects to include playwriting, poetry sessions, drama workshops and readings
- a student/tutor-run Publishing and Writing Committee
- the establishment of a Community Literature Worker post for West Yorkshire, funded jointly by Bradford and Ilkley Community College, Yorkshire Arts Association, Bradford Literacy Group and local industry.

For us the philosophy directing our work remains:

- a belief in the power and importance of self expression;
- a desire to expand the boundaries of literacy provision, making it more responsive and relevant to students' lives;
- a recognition of the need to create opportunities whereby ABE students and tutors can meet to develop autonomy, self-

confidence and skills, within supportive and enjoyable learning environments;

- a commitment to promoting a dynamic student/tutor partnership in all aspects of our work;

and, above all,

- a conviction that writing must be for change – in students' and in tutors' self-perceptions, in the development of Adult Basic Education provision, and in the everyday lives of students and tutors within ABE.

However, rather than just take our word for it, let us leave you with some comments from students and tutors who have been converted to working in this way.

- *'However nervous you may be, you find yourself eager to participate . . . It's surprising how soon you become one of the group.' (Joan – a student on a writing weekend)*
- *'This was a day for gaining confidence in that even the most nervous performed.' (Gill – ABE sessional tutor)*
- *'You learn a lot out of it – and I like meeting people. We have a real time here.' (Alison, ABE student, Women's Writing Event)*
- *'A day concentrating not on themselves, problems, families, etc. – but on words and music.' (Josie – Bradford Writers' Circle, workshop facilitator)*
- *'There were books, pens and you name it – it was there for our needs.' (Paul, student on Writing Day)*
- *'Folk begin to have fun without thinking about it. Energy was injected and the result was spontaneous energy 'and creativity from the students.' (Loma, workshop facilitator, Sing a Poem)*

Two workshops on critical literacy practice

Margaret Jessop, Gillian Lawrence and Kathy Pitt
RaPAL Bulletin No. 35, Spring 1998

Margaret Jessop (Senior Community Education Worker) and Gillian Lawrence (Community Education Worker) both work for the City of Edinburgh Council. Kathy Pitt is a part-time research student at Lancaster University.

This article reports on the work of two conference workshops that incorporate different approaches to critical literacy practice. Kathy Pitt explains that Critical Language Awareness (CLA) is the pedagogical application of research being carried out by linguists, social theorists, political scientists etc. This research, sometimes called critical discourse analysis, links a close look at the language of institutions with social theories of how power works in societies today. These issues are of interest because they offer greater understandings about how this power works, and therefore more possibilities to change or resist these practices, and work towards a less unequal society. Kathy links this resistance to the definition of democratic freedom as "the freedom to answer back" (given in a talk in 1997 by Professor Mick Dillon of Lancaster University).

In their workshop, Margaret Jessop and Gillian Lawrence reported on their recent Photographing Literacy course in which they sought to apply ideas from Critical Discourse Analysis (CDA) in ABE programmes. The editors of this Bulletin have tried, with the authors, to merge the two articles submitted: despite their differences in terminology and approach, both explore key areas of critical thinking and practice.

Margaret and Gillian have proposed the term critical literacy practice to incorporate these approaches to literacy. They explain that critical literacy practice in ABE not only applies the insights from these areas of critical research but also enables adult literacy students to carry out their own research and reflection in this area.

Photographing Literacy

In the 'Photographing Literacy' course we sought to apply critical literacy practice in ABE programmes. Following Fairclough (1989, 1992) we worked through three different stages of reading texts with students and explored how these were framed by power relations in our society. In the first stage the main features of the texts were identified and described and in the second stage the social relations between writers explored. During the final stage the interactions were explained within their wider social context. Our intention was that learners could move from a reflective stage to further action using their literacy skills, if they chose to challenge or resist dominant messages in our society.

Without exception students at first considered the only valid reading material to be books which would be read as an isolated activity indoors. During the course, however, the group was able to move away from this 'school model' as different texts and writers' intentions were considered in detail. Students became aware of the importance of reading between the lines when analysing their photographs of text within their local community.

During the second session of the programme (following input from a photography tutor) students were provided with cameras and asked to photograph examples of literacy from public places around them. The group had already expressed an interest in women's issues: the subjects photographed were many and varied, including adverts on billboards and buses, graffiti on the local Job Centre and the wording in shop window displays. The photos were categorised thematically and used afterwards in group discussion and decoding exercises. These exercises helped students unravel the complex ways in which their selected texts and images were indeed framed by power relations in our society.

At the end of the programme the group collectively wrote an introduction and commentaries to their work which were mounted on exhibition boards with photographs and displayed locally.

Decoding exercises

During the course itself (and in the Conference Workshops) we found that by using a framework of questions on selected images and texts, hidden messages, values and beliefs emerged for group members. This framework was used throughout the course and applied to the photographs taken by students around their neighbourhoods.

For us as presenters of both workshops 'critical' questions, to ask about language we hear or read, involve considering:

- What are you looking at? Who produced it and who is intended to read it? What views of the world are assumed?
- What is the message? What are the values and beliefs that go with it? What is made to seem "natural"?
- How is the message communicated? Why this language and/or choice of grammar and vocabulary? What has been left out? Consider both images and language.
- Whose interests do these language choices serve? Why was this piece produced?
- What can this piece of writing tell us about our world? Who is powerful and makes decisions in our society?

The workshops looked at a number of examples. This first is taken from the photographing Literacy course, which recorded graffiti on the outside of a Job Centre (see Figure 1).

Student readers considered how the writer uses an order: "smash", wanting claimants to stand up and protest against government snoopers investigating benefit fraud by following people, photographing them and questioning neighbours etc. One reading of this is that claimants feel degraded: it is an invasion of privacy to be spied on. The government has targeted the less well off in society. This graffiti is one person's way of speaking out against this in three words.

Figure 1: Smash Snooper Scum

Kathy's conference workshop analysed a local newspaper article entitled "Beggars Beware":

Police ready for crackdown on tramps

BEGGARS BEWARE

A city centre crackdown is on the way to stern the rising tide of vagrancy in Lancaster. The police are all set to launch a major initiative to control the growing problem. The move comes as the number of tramps and beggars seen on the streets of the historic city has increased alarmingly throughout the summer. And with it the complaints to local officers of abusive begging and drunken behaviour. In recent months attempts by the police to curb the problem have made little impact. Inspector David Langley of Lancaster police said: "The problem is an on-going one and it has its peaks and troughs. It has become apparent, however, in recent weeks that there has been an increase in the number of vagrants in Lancaster.

Problems

"It is a policing and a social problem and we are, over the next few weeks, going to intensify our patrols in relationship to the city centre. We hope to stamp out the problem altogether,"

Vagrancy was at the centre of discussions at a recent Chamber of Commerce meeting when local businessmen pointed out the problems they were facing because of the vagrants. A spokesperson for the Chamber of Commerce said: "Over the last two years we have seen the problem increase. We have appoached the local authority several times and simply been given the usual reassurances that the police will move them on. What is concerning our members is the impression they give to people visiting the town, particularly as we are trying to promote tourism in the area. It is a difficult problem because, of course, we have every sympathy for the problems that these people have."

Understand

Superintendent Geoff Armstrong has issued a warning to the public to be wary of bogus beggars. He said: "Many of the people claiming to be homeless are not. We know that they have homes. We understand that people are kind-hearted but we ask they they do not encourage the beggars."

Lancaster Citizen; date unknown

As is often the case in this kind of writing, we found it interesting to look closely at the subjects and objects of the action in the text. We saw how textual representation of the actors fell into three main categories:

- "The beggars" represented as tramps, vagrants, them, bogus.
- Inanimate, abstract "actors" represented as vagrancy, the growing problem, the complaints.
- Those involved with "the problem" in the article represented as the police, Inspector David Langley, local businessmen, a spokesperson of the Chamber of Commerce, we.

Clearly, the wording for the three groups is different: the group of people seen as creating a problem are represented either as negative and impersonal groups, or as "they"; the other. Pronouns are interesting; if there is a "we", then who is in this group, and who are the "they" outside it? In the inanimate grouping, the "beggars' 'are made more abstract and impersonal by being called "the problem". This wording asks the reader to take it for granted that the presence of these people constitutes a social problem. Inanimate subjects can also hide unknown actors (e.g. "the complaints" – who complained?). Individuals are only named in the police and business group – the group that holds authority. Different patterns of naming the groups reflect the choices made in the article for whose voice is heard; the voices of the police and businessmen are there, those of the "beggars" or the fourth small group of "the public", are absent.

Critical practices

In CLA, reading involves looking at what is absent from the writing, as well as what is present, and how it is worded. Metaphors, actors, vocabulary choices, absences, these are some of the features of language that are looked at in CLA in order to see how writers position the people or places they write about. There are no "correct" answers; there can be many "readings" of one piece of writing, but an analysis of words and grammar can often provide evidence when a reader feels that only some perspectives have been represented.

In our workshop discussions, what was emphasised was the need to avoid imposing particular readings, or the reading of particular pieces of writing. One of the values (for tutors as well as students) in looking closely at language and image choices, in looking behind the words at how and why they were produced and what assumptions are implicit, is that such analysis provides the reader with "the freedom to answer back" and the tools to do so publicly in written form. It does not magically provide power to the powerless, or change material conditions, but knowing more about how language works can encourage other ways of looking at the world, a refusal to accept certain positioning and the beginnings of an argument for change.

The Photographing Literacy student group summed up their learning from this approach:

> *"We are surrounded – whether it be at home, in the street, at work etc – by writing and images. How many of us stop and see beyond the*

words that are written? As a group, we did stop and read beyond the words and images. We found there are countless, silent messages in public places. Some messages are hidden – behind a double meaning – others are very clear, to the point. Not only do they try to influence our choice but also seek to set their standards and social values upon us."

A Photographing Literacy ABE tutor pack is now available.

References

Fairclough, N. (1989) *Language and power.* Longman.

Fairclough, N. (ed.) (1992) *Critical Language Awareness.* Longman.

Janks, H. (ed.) (1993). *Language and Position and Language and Identity.* Critical Language Awareness Series. Hodder and Stoughton.

Lancaster Literacy Research Group (Centre for Language in Social Life). *Photographing Literacy Practices*, Working Paper No. 41, in *Changing English* Vol. 1, No. 1 . Published by Institute of Education. London (1994).

Lothian Council (1993) *A poverty profile of Lothian Region*, Social Policy Sub-Committee Report, Dec 1993. Macrae, C. (1997) *Struggling with authority: texts, power and the curriculum* RaPal Bulletin 32. 1997. pp.9–14.

Shake, rattle and write

Mary Wolfe
RaPAL Bulletin, No. 35, Spring 1998

Mary Wolfe is a tutor at the YMCA George Williams College, London.

I have recently become interested in exploring graffiti as an example of public literacy. This article draws upon both a conference workshop and my own current work with undergraduate students of informal and community education. I see graffiti as transgressive writing in that it is generally illegal but more significantly in that it transgresses those conventions of stability, authority and orthography associated with more mainstream literacy practices. Like the graffiti artists whose work underpins this piece, I have transgressed the time regulations framing the workshop and included more recent texts. Graffiti, like yesterday's newspapers, constantly rewrites itself.

I tutor a course for informal and community educators, on a programme which includes a professional qualification in youth and community work. The programme area I work in relates particularly to the area of literacy practices and education. I don't 'teach literacy' any more although I have done so in the past. I spent the early part of my working life as a tutor pleading that literacy or basic skills understandings and practices could and should inform wider areas of formal educational practice. I now find myself arguing for an appreciation of the impact of literacy practices upon the work of informal educators. What interests me, and the reason why the programme includes such a course, is how to work as informal educators within the intensely and intensively written environment which we all inhabit. In the conference workshop, I sought to explore with RaPAL colleagues how an understanding of informal education and its literacy practices can enrich all of our work with students.

Informal education theory frequently recalls Dewey's focus on the significance of the environment in which we work:

> *"We never educate directly but indirectly by means of the environment. Whether we permit chance environments to do the work, or whether-we design environments for the purpose makes a great difference. And any environment is a chance environment so far as its educative influence is concerned unless it has been deliberately regulated with reference to its educative effect" (Dewey 1966: 19).*

I want to focus upon this link between the environment and education and a Freirean view of reading the world whereby:

> *". . . by perceiving the relationships among objects and their reasons for being, the cognizant subject derives an understanding of the objects, the facts and the world. Reading context and reading text: the one implies the other." (Freire 1996: 182)*

Learners as much as educators are social subjects: we read and make sense of the world around us. I also want to stress that we inscribe that world with meaning. By that, I mean that we bring abilities and motivations as meaning makers to our environment; we 'transform' (Kress 1997) our surroundings by writing or inscribing them with meaning. The texts of graffiti constitute a constant rereading and re-writing of our context or environment: the writing on the wall makes that wall newly available as an ideological challenge, as a declaration of love, as an establishment of identity. The message realises a particular significance in a particular place and form: text in context.

The presentation included some slides of the written environment in which I live or work. My focus is on exploring the essentially oppositional nature of graffiti, which while working against the grain, remains highly regulated and conventional. The question which I want to address is, I think, one of the most useful questions I remember to ask myself: what is going on here?

Written authority

I started off by looking at the public notices which inscribe regulations, particularly of time and place: the bus timetable, the electricity sub-station marked danger and the road sign for FLEET ROAD. Graffiti is used to subvert the authority of these texts by imposing, palimpsest like, an overwriting. Black felt pen demands to know of the bus timetable "So why don't it come then?" The electricity sub-station is first of all decorated with the printed orthodoxy of an SWP poster, illegally fly-posted, which itself is then overtaken by handwritten scrawl. FLEET ROAD becomes EDEFT ROAD, the sign now designed to confuse the outsider and amuse those of us in the know. The texts pile up, one upon the other, subverting written authority at its very point of delivery and presuming for itself that immediacy of response otherwise only allowed to the e-mail community.

Thinking of the reader . . .

It seems to me that this is supremely dialogical writing which challenges the traditional anonymity of the social regulator, normally hidden behind timetables, street signs, safety notices and the supply of energy or of revolutions. Of course, all writing is based upon a motivation to communicate. In graffiti, writers constantly recognise their audience, either through a general address:

or through specific messages:

To Fish and Danny Why did you run away?

"OI, BAZ YOU WANT TO START SOMETHING?"

In all these cases literacy is made available as a means of social exchange, as writers demonstrate their awareness of their reader but reject the,traditionally private nature of their message in favour of public writing. In another example, workshop participants noted the politeness with which the writer addressed the reader directly in order to issue a reminder of the rules of place and appropriacy:

"Brick wall is
4 guests"

The personal made public

There is a subway crossing an inner city motorway near where I work which gives access to a geographically distinct neighbourhood. This entrance way records the events and lives of local people. It is a noticeboard like the personal column in The Times and, like The Times; has established its readership and its conventions to address many shared preoccupations, if not about marriages or deaths at least about who does, or does not, fancy who . . .

Thus the wall becomes a noticeboard to provide a necessary alternative to other, and differently valued, public declarations.

Writing and identity

Signing one's name remains a very public statement of an individual's identity, and was of course, a key early determinant of literacy levels. The author or artist signs high culture and designer sportswear. Similarly in graffiti we see signing – or tagging – everywhere. Often this is extraordinarily artistic as the reality of illegality becomes in itself an art form.

Sharks

There is an odd tension between the assertion of identity and the need for secrecy. "Shark's" careful calligraphy mirrors the development of spoken argot: a written equivalent of a code only known by the initiates. It mimics but debunks high culture since much of the value of tagging lies in the signature being instantly recognisable – but only to the initiates. It must remain unrecognised by, if not hidden from, outsiders. This is writing as counter culture; in a prohibited public space, it is the personal made public.

Spelling and punctuation

The group at the workshop noted the strictness of graffiti's regulations concerning spelling and orthography. "TiKB", a Turkish group, can sign its tag from top to bottom, from left to right, large or small but the use of the lower case vowel among the upper case consonants is observed as stringently as that rogue a in RaPAL. On the side of a former concert hall, the message that:

"Ken Dodd's dad's dog's dead"

is punctuated with all the correctness of a Wordpower worksheet. This last phrase was then borrowed as the name of a briefly successful North London hip-hop band – or maybe was tagged later by fans. The origin is of little significance in the dynamic exchange between text and context.

Writing and stability

Graffiti mounts a head-on challenge to Goody's infamous claim that:

> *"writing, and more especially alphabetic literacy, made it possible to scrutinise discourse in a different way by giving oral communication a semi-permanent form; this scrutiny favoured the increase in scope of critical activity [. . . and . . .] the potentiality for cumulative knowledge"(Goody 1997: 37)*

The ephemeral nature of graffiti with its overwritings, its immediate responses, its openness to collaboration and to amendment, establishes a street culture which prizes the ephemeral at the expense of the implied permanence of high culture. I am not sure whether Clare did L Terry 4 ever because the wall has been wiped clean by the local council. Shark's tag has been overtaken by a tribute in intricate calligraphy:

> ***"So long Motha Teresa and Diana Princess of Wails".***

The sense of temporality is clear as is the critical challenge in the demands "What are you staring at?" (the essence of the questions suggested by Critical Language Awareness?) or (and I first saw this on September 16, 1997) "Who killed Diana?"

Creative writing

Much graffiti realises a meeting point between pictorial art and writing with its elaborate calligraphy and careful attention to colour and form.

At times, the texts offer an almost surrealist pleasure in art for art's sake, in writing for the sheer pleasure and fun of it. My own favourite, now wiped away but re-recorded here, remains the cautionary:

"Never trust a dog with orange eyebrows".

To draw this together, my interest lies in the way in which this culture reflects but undermines mainstream culture in much the same way as spoken language has transformed the significance of epithets such as wicked or bad etc. I would like to suggest that not all these graffiti artists enjoy or attend or succeed in engaging with school's literacy practices. And yet here is an obvious enthusiasm to write according to intricate and complex conventions. Certainly we are seeing here the making and taking of culture through literacy practices. It echoes Morrison's description of narrative in that it is 'radical, creating us at the very moment it is being created' (1994: 27). An art form which, unlike high art, is free, public, ephemeral and collaborative. Graffiti reveals writers seemingly at ease in understanding, transgressing and transforming the often apparently stifling conventions of the literacy practices of formal education. The writers demonstrate their energetic ability and willingness to rewrite their world in marked contrast, it seems, to the often slow and painstaking reading and writing of our classrooms.

References

Dewey, (1966) *Democracy and Education: An Introduction to the philosophy of education*. New York: Free Press. First published 1916.

Freire, P (1996) *Letters to Cristina, reflections on my life and work.* London: Routledge.

Goody, J. (1977) *The Domestication of the Savage Mind.* Cambridge: CUP

Hodges, N. (1988) *Literacy and Graffiti: Alternative Reading and Writing in Milton Keynes*. CLAC Occasional Paper Number 16. Milton Keynes: Open University School of Education

Kress, G. (1997) *Before Writing, rethinking the paths to literacy*. London: Routledge.

Morrison, T. (1994) *The Nobel Lecture in Literature, 1993*. London: Chatto & Windus

Who says so? . . . Who? Can their views be challenged?

Gaye Houghton

RaPAL Bulletin No. 28/29, Autumn 1995/Spring 1996

Gaye Houghton is a former ABE co-ordinator who now works part-time for CENTRA in Chorley as an external verifier for the Wordpower and Numberpower schemes. She combines this with her part-time MA studies at Lancaster University.

Since starting an MA in Educational Studies, relating ideas to experiences in adult basic education has been an extremely positive development for me. Within the overall experience, however, some aspects have been contentious. My strong yet mixed reaction to the assessment process and also to the factors which influence my choice of a suitable academic writing style appear to be at the root of my 'difficulties'. Although I write about 'difficulties' I am referring to things which are very stimulating. They are challenges which encourage my academic writing much more than they discourage it.

Assessment is a process which, when positive, gives confidence to students and their tutors about the students' ability to complete the MA. Sometimes, however, the fact that different styles of academic writing exist has raised queries about which style to adopt when producing writing for assessment. The 'styles' I am talking about are (1) the traditional, formal and impersonal academic writing usually written in the third person and (2) the less formal but equally valid academic writing which allows the writer to use the first person, 'I'. Although I am quite capable of producing both styles I have recently started to use the latter because it encourages me, personally, to be more 'honest'. This does not mean that formal writers are dishonest; it just means I feel writing in the first person has elicited more genuine responses from me as an individual.

When the idea of using the first person was first mooted I felt completely at ease with the principles of academic democracy, honesty and writer identity that I, myself, saw in it. Putting the principles into practice, however, was another matter. I had no previous experience of producing academic writing using the first person. I was reminded of my past ABE students who had talked about and practised the writing they needed for their everyday purposes. When they had to write in 'real' situations, however, they sometimes lacked the experience and confidence. Now I was in the same position. It seemed strange but exciting to encounter these 'challenges' at this point in time. They did not exist when I first began to write academic essays.

For my third assignment I decided to evaluate academic writing practices. The first section (which is quoted below) concerned the social context in which academic writing is produced. The second was about my personal development as a student writer of academic essays. (Please note the impersonal style of writing in section 1 and the use of 'I' in section 2.)

The Social Context for Academic Writing [section 1]

Recent research into academic writing has highlighted the issue that academic writing is a social practice. This, says Brodkey (1987, p5) has led to studies which examine the circumstances under which people in the academic community read and write. She describes the academic community as consisting of "readers who write and writers who read." Whilst engaging in these acts academics (and students) acquire a distinct set of reading and writing conventions which they absorb from the academic communities in which they operate. The mastery of these conventions is considered crucial if students and academics are to succeed in their chosen field. Brodkey (p25) points out that many academics produce articles designed for academic journals which are read by small, select, educated audiences who wish to keep abreast of issues in their specialist areas. The articles are not common reading for a general, popular audience. In fact, the process of producing an academic article for a particular academic journal to be published by an academic press can, and does, create an exclusivity of language and readers which many academics aspire to in order to achieve 'success'. In this sense, if an academic article were to succeed with a wider, popular audience it might be regarded as a failure by some members of the academic community.

An academic hierarchy then, with the academic journal at the top, appears to be the norm in most academic communities. Researchers such as Benson, Gurney, Harrison and Rimmershaw (1994 pp 64–65) have identified the two main conventions of academic writing as 1) the adoption of an impersonal, formal language wherein 2) all ideas and views have to be 'substantiated' by academic references. They point out that these well-established conventions can sometimes cause conflict within academic communities, ie. ideas become contorted rather than communicated. Language is depersonalised and students' individuality is removed from the text. Consequently, many students feel they have nothing of value to say. Or, if they do have independent ideas, they are unable to express them because the ideas cannot be 'verified' by academic reference. Some students feel they have valid professional and personal experiences which are worth referring to.

Today, there are moves in the academic world to 'democratise' the research and writing processes. Hamilton, Ivanič and Barton (1992) have criticised the mainly 'top down' approach of academic research and writing. They argue that the ideas of student writers and the people they might interview for their research are often just as valid as the ideas of professional academic writers.

One fundamental way of challenging the conventions of academic writing already mentioned is to put the 'I' back into academic writing which immediately confers personal and intellectual ownership onto the actual writer. Ivanič and Simpson (1992 p 142) advocate that the identification of 'I' is important for meaning and credibility in academic writing. Students need to establish the identity of ". . . the people behind the texts they read, establish an identity for themselves as readers and writers, and recognise the relationship between themselves and the tutors who will read their work."

Ivanič and Simpson take the view that all academic writers, whether student, tutor or professional, become more 'visible' if they use 'I'. Therefore their texts are easier to read and understand. Using 'I' stops people from 'sitting on the fence' or hiding behind the ideas of others. It is important to note, however, that this does not advocate ignoring the ideas of others. Rather, it advocates not allowing the ideas of others to take precedence over the sometimes valid, individual, personal ideas of the unestablished writer.

The following extract is from the second section of the same essay and deals with my personal development as a student writer. It shows the writing 'contortions' I went through just because I did not know the tutor's preferred style.

My Personal Development as a Student Writer [section 2]

Most of this section was written before section 1. I decided, however, to place this section in second place. There were two reasons for doing this. Firstly, although I am sure in my own mind about the validity of using the first person and use it in this section, I am not sure whether the tutors who will assess the assignment will feel the same. This leads naturally to the second reason: In case my tutors want me to write in an impersonal style, I thought it prudent to write the first section in such a way (without 'I') in order to 'prove' to them I was capable of producing more formal writing. For effect, this had to be done at the beginning of the assignment although it is my intention to use 'I' throughout from now on.

These actions, of course, raise issues about the academic hierarchy, tutor and student power (or lack of it) and the academic validity of my personal experience. I hope to become more confident in expressing the latter as the assignment progresses.

Many years ago my first degree tutors encouraged me to use a traditional, impersonal academic style in my writing which had to be 'objective' and 'value-free'. The ideas, theories and generalisations of accepted authors and 'experts' were to be cited, quoted, criticised and celebrated but not my own (a mere student). Personal ideas and theories from me were regarded as too 'subjective'. For this reason the use of the personal pronoun 'I' was not acceptable.

Reflecting back on my first academic writing experiences I am surprised at how emotional they were. Some reactions were positive: I remember the relief when an essay was finished, the satisfaction when the writing was beginning to 'flow' easily, the joy when the essay passed the tutor's scrutiny and the elation when it received a good mark. These positive reactions meant that I had mastered, to an adequate level, the traditional academic approach and was 'winning'. In short I had become academically 'literate'. I had gained what Ivanič and Roach (1991 p l) call "privilege power". They compare this with "personal power" which means writing for personal purposes using the writing conventions only when it suits. With "personal power" the writer has the confidence to flout the conventions and not let them rule.

Trying to achieve "privilege power" can be disempowering in some cases. The positive reactions I had when I felt I was "winning" with my academic writing were sometimes counteracted by negative ones such as the wearying struggle to adopt a traditional, academic style of writing which meant using language different to my own everyday language, the frustration of trying to write long, complicated sentences which sounded contrived and unnatural, the misery of thinking I had no ideas of my own, the anger caused by not being able to say what I wanted to say and, sometimes, the sheer boredom of it all. These negative feelings came when I was fed-up and 'losing'.

When I arrived at Lancaster University for my MA course all the memories about academic writing, both positive and negative, began to flood back. After a short while, however, I was able to confront and cope better with the negative side of academic writing. Tutors stressed that course members' ideas and the ideas of participants in their research were considered just as valid as the ideas of well-known established authors and practitioners. Not only was the 'I' in academic writing considered valid but it was also imperative. My turn to develop as a writer had arrived. In the past I had tried to develop ABE students by encouraging them to "put themselves" into their writing. Often, in order to do so, they had to confront old, established; sometimes negative writing (or non writing) habits and attitudes. It was necessary to 'unlearn' in order to learn afresh. Now, it was my turn to confront my own writing habits.

I began to feel that I had buried some of my intellectual and personal writing identity because of the writing conventions I had adopted over the years. Now, however, I felt I was clambering out of the 'academic 'rubble' of the past. It has been an 'up and down' climb over the last few months. I know, however, that the use of the first person has provided a handy rock (me) on which to rest my research and writing. This 'rock' has to be 'dragged' to the top of the 'rubble' heap and 'shored-up' before it takes all the weight. (I certainly need more practice and confidence in writing in the first person). When this has been achieved, hopefully, my place on the summit of my personal academic 'rubble heap' will be secure.

Being able to impose my own authority on my research and writing was a liberating process. I was able to read, think and write about developments in ABE from a genuine, meaningful perspective – my own! In my early writing days I would have thought "I should be so bold" as to think my opinions and views were worth anything. However, after many years involvement with ABE I was made to feel I had something to offer, ie. knowledge, experience and understanding. Now my ideas were as important as those of anyone else . . . or were they?

Soon after I started attending my second MA module I began to 'stumble' in the 'rubble'. Some of my negative feelings about academic writing began to trip me up. There were so many tutors presenting the module I felt unsure which academic writing style would be generally preferred. I wanted to write in the first person but I found myself wondering if the tutor-markers would accept it. For security I reverted back to the old methods of academic writing which had been instilled into me years ago. I did not write in the first person and tried to use an impersonal 'objective' style of writing. At the same time I tried to include my own ideas. Consequently, the writing process was much more difficult. Looking back, this was part of a developmental stage I had to go through.

Writing the third assignment (the one quoted here) was a much more positive experience than writing the second (where I 'stalled' on using the first person). Even so, I was obviously still wary about the tutor's response which is why I used both formal and informal styles. Somewhere I seem to have grasped the 'notion' that some tutors (and students) feel that using the first person is easy(?) and therefore not as academically rigorous as more formal academic writing. This could be the influence of my own traditional writing background but there does seem to be a prevailing university culture which supports my 'notion'. In fact, for me, using the first person for academic writing is just as rigorous as writing impersonally. It demands from me an honesty which can be refreshing but sometimes quite painful. It is not easy to sit here as an MA student and write about 'challenges' to my writing for other people to read about. Halfway through this piece I nearly decided to abandon the whole thing in order to keep up the 'front' of an efficient, 'laid back' academic writer. I was worried in case my comments would 'enfeeble' me in some way in the eyes of others. However, I have spoken to tutors and fellow MA students and some find the issues surrounding the use of the first person just as challenging as I do, which is reassuring.

Dilemmas about writing MA Assignments

When I completed my fourth assignment (written in the first person) my wariness concerning the above 'notion' was well-founded. A tutor (not the actual marker) 'suggested' that I change my style to a more formal approach. As the assignment was to be assessed it was difficult to ignore the 'suggestion'. The tutor who made it did not 'demand' the change. As a colleague of the marker she knew the latter's preferences and simply wanted me to obtain the best mark possible. When I remarked that some people advocated writing in the first person she said that they were "very different". This was not said in any critical sense. She merely wanted to indicate that different opinions on 'style' existed. All her comments were made with my best interests at heart. At that point in time, however, I gave up the hope that all tutors would accept both valid academic styles no matter what their personal preference. I tried to contact the tutor responsible for the marking to find out for myself what she would accept but failed to reach her. In the end it was I alone who decided to change my style. At first I was perplexed with my self-imposed(?) predicament because the link between assessment and writing style had become an issue that was beginning to frustrate me. Once the actual rewrite had started, however, I found the task intriguing. Playing around with words and writing styles really interests me.

Conclusions

Writing in the first person was not a valid style twenty years ago when I studied for my first degree. Many tutors (and students) also started studying and writing at the same time. This is bound to influence the way they think about and produce academic writing today. Perhaps there are some tutors and students who have not had the opportunity to look at the issues surrounding the use of the first person yet? Others probably have and find the subject as challenging as I do. Some may openly reject the use of the first person and find their 'honesty' unaffected by adopting a more impersonal writing style. Others are probably not bothered by it one way or another, and just carry on writing in the style they are used to; usually the traditional formal style because it has been around the longest.

Two questions now arise which concern both students and tutors. Can students become proactive writers and write in the style they prefer and not worry about how this might affect their tutors assessment of them? Or, must students become reactive writers and choose a writing style that suits their tutors? Writer integrity wants me to pick the first option but common sense with regards to the assessment issue forces me to consider the second. What do YOU think?

References:

Benson, N., Gurney, S., Harrison, J. & Rimmershaw R (1994). The Place of Academic Writing in Whole Life Writing in Hamilton, M., Barton, D. &

Ivanič, R. (Eds.) (1994). Worlds of Literacy. Multilingual Matters: Clevedon.

Brodkey, L. (1987). Academic Writing as Social Practice. Temple University Press.

Hamilton, M., Ivanič, R. & Barton, D. (1992). Knowing where we are: Participatory Research and Adult Literacy in Hautecoeur, J. P. (ed) Current Research in Literacy: Literacy Strategies in the Community Movement. Unesco Institute of Education: Hamburg.

Ivanič, R. & Roach, D. (1991). Academic Writing: Power and Disguise, Centre for Language and Social Life, Department of Linguistics, Lancaster University. Working Paper Series No. 19.

Ivanič & Simpson (1992). Who's Who in Academic Writing, in Fowler, N. (Ed) (1992). Critical Language Awareness. Longman.

Student status and the question of choice in academic writing

Theresa Lillis and Melanie Ramsey

RaPAL Bulletin No. 32, Spring 1997

Theresa Lillis is a Ph.D. student based at the Learning and Teaching Institute at Sheffield Hallam University, where, for the past three years, she has been carrying out research with a group of mature women students into their experience of writing at university. Melanie Ramsey began her full time studies in Sheffield, and is currently a second year undergraduate student in Psychosocial studies at the University of East London.

Introduction

In the Autumn 1995/Spring 1996 edition of RaPAL, Gaye Houghton invited discussion on the following two questions:

> *"Can students become proactive writers and write in the style they prefer and not worry about how this might affect their tutors' assessment of them? Or must students become reactive and choose a style that suits their tutors?"* (p.14)

We would like to contribute to a discussion of these questions by drawing on our experience of talking together about academic writing over the past two years.

The main focus of our talk has been Melanie's writing for her undergraduate coursework, but we have also talked about Theresa's writing for her PhD. The issue of choice in our academic writing has been a recurring theme in our discussions. We both realise that because we are bound to the institution's practices, primarily through assessment, the possibility of choice is limited. However, we also think that the possibility of choice differs according to our different student statuses: as an undergraduate and as a postgraduate research student. This is reflected in the extent to which we each feel we have been able to act on understandings generated from our talk.

In the following two sections we attempt to explain this by focusing on some specific examples from our talk and our writing.

Melanie's experience: talking to learn about the conventions

Learning about the conventions

> *I've never experienced talking to anyone about my essays before so I find it very interesting and I appreciate . . . maybe other people in the class don't (laughter). I do, cause I can see the benefits of it. Nobody's ever sat down and talked to me about my essays. They've just said 'hard to fathom at times'. (laughter)*
>
> *(Melanie in a taped discussion with Theresa)*

I (Melanie) was pleased to be able to talk to someone about my academic writing because this was not something I had done before. By talking about specific bits of my writing, I was in a better position to work out why tutors might criticise it. Through talking, I could understand why some things in my writing might be considered 'wrong'. For example, Theresa and I discussed my use of *however* in the following essay:

Extract from my essay:

> In fact all types of West Indian Creoles should be viewed along a continuum. *However* (our italics) there is a large number of Creole speakers, but no one speaker uses a creolised speech to the same extent.

Extract from my discussion with Theresa about *however*:

M *I thought however meant, well, another change of thought.*

T It does but it means a change in direction of thought.

M *I thought it meant the same direction . . . oh a completely different idea?*

T It's like for example I might say 'I like shopping. However, I'm not going today'.

M *Whereas I've been saying, 'I like going shopping. However, I'm going to buy some'. (laughs)*

'However' is not a word that I would normally use when I talk, so perhaps it's[1] not surprising that I would use it incorrectly in my writing. I obviously understood a part of the meaning of 'however' but it was only by talking it through that I could work out a more exact meaning of it.

Our talk also helped me to make sense of more specifically academic conventions, for example conventions relating to *referencing and plagiarism*. In my first piece of writing, I was trying to outline what Bernstein meant by 'elaborated' and 'restricted' codes.

I found it difficult to write about these ideas without using his words. We talked about how I might be accused of plagiarism if I wrote too closely to his text, and how in academic writing you were expected to show where ideas were from by referencing. We also talked about why it was important to separate out the voices in writing. The ideas of separating voices was useful to me. The following is an extract from our discussion, where I talk about this:

> *And I find pleasure in it now you know . . . because I like to be able to distinguish between what I'm saying and what's being said by someone else. When it comes together, it's not good at all, cause you're not making yourself clear. You ought to let the person who's reading the essay know what you are; what you're saying. I think that's very important that. Separating the voices is one of the best . . . I think it is one of the best things I have done. I never really thought about the voices before, separating them. I didn't realise I was making them converge, making them link in. I didnt realise it.*

So through our talk, I feel that I have learnt more about standard conventions and I feel positive about that. I want to be able to communicate my ideas in a way that will be understood by those who are reading my writing. And because they follow, and expect me to follow, certain conventions I have to do so too. But . . .

Whose conventions are these?

I also feel really frustrated at having to write in a way 'they' want me to write, especially when I can see no good reason for it. For example, most tutors will not allow the use of 'I'. When Theresa told me I could use 'I' in my writing, it felt uncomfortable at first, but I did try using it and liked it.

The following is an extract from my essay on Bernstein's 'restricted' and 'elaborated' codes:

> *I am aware of the certain expectations that it (the academic context) imposes upon me, and I am compelled to comply with them . . . Therefore my use of the elaborated code is context tied, because it is put into practice only in situations concerning education.*

I think it's really important to be *in* your writing and to let the reader know who you are and what you think. The 'I' is really powerful because it makes you feel a part of it. When you're not allowed to use 'I', you feel like an outsider. If we as student-writers, are not allowed to use 'I', we're kept outside our writing.

Not allowing the use of 'I' is part of the whole emphasis in academic writing on formality. It comes into writing in lots of ways, but one particular aspect which frustrates me is the insistence on full written forms, for example, *I am* rather than *I'm*, *cannot* rather than *can't*. Everybody knows what *I'm* means, so it's not a question of clarity. I think it's a question of exclusion. It seems to me that the only reason for using full forms is to separate the writer from the reader, building a boundary between them that sets the writer of academic texts above and apart from many people. Why do we have to do this? I don't want to do that in my writing.

I feel strongly about these last two conventions: I'd like to be able to use 'I', and I'd like to be able to use contracted forms such as 'I'm'. But I don't feel that I, as an undergraduate student, can choose to write in the way that I want. Tutors assess my work and are therefore the ones who can fail me. Although at times I imagine the possibility of challenging some convention, at this stage in my life as a student, in my position as an undergraduate in the institutions, I feel I must conform to the conventions.

Theresa's experience: talking to explore the conventions and to make choices in her writing

Talking with Melanie about writing has helped me make choices about what I want to say in my writing in specific ways. I will focus on three instances below:

Talking about 'voices in the text'

Talking about 'voices' in Melanie's text, which she mentions above, helped me to focus more clearly on significant voices in my own texts. I consider more carefully which writers I am drawing on, the extent of their significance for what I am trying to say, and importantly, which aspects of their ideas and work I want to identify myself with. It links in with my understanding of Bakhtin's notion of the dialogic nature of language and the ways in which all utterances, spoken and written, resonate with the voices of others (see Bakhtin,

1981, and Wertsch, 1991). Our contribution as meaning makers – when we talk and write about anything – lies not so much in our individual and 'original' utterances, but in our own particular orchestration of these voices. I used this notion explicitly in my writing of my transfer document from MPhil to PhD. This document is a report of the progress of research to date which many postgraduate students have to complete.

Extract from Theresa's transfer document:

> 4. CONTEXTUALISING THE STUDY: SIGNFICANT VOICES IN MY APPROACH TO LANGUAGE AND LEARNING (see Bakhtin 1981 and Wertsch 1991 for exploration of the notion of 'voices').
>
> *The partiality of knowledge*
> In this section I want to point to writers whose work, or aspects of work, have become significant for me in this study . . .

By talking with Melanie about voices in the texts, I came to see the notion of orchestration of voices as a possibility for *real* choice in *actual instances* of my writing, rather than a more distant, theoretical concept which had little impact on my practice.

The use of 'I'

As can be seen from the extract above, I have used the first person 'I' in my academic writing. Although more widely accepted now than in the past, the use of 'I' in academic writing still raises hackles, as I discovered when my Director of Studies criticised its use in my transfer document. Melanie's clear articulation of the importance of using 'I' as a way of being part of the text has encouraged me to reconsider how I want to position myself in my writing. This involves taking greater responsibility for my meaning making and questioning aspects of some of my previous academic writing, for example my MA dissertation where I rarely used 'I'. Taking responsibility for how I want to write, rather than how dominant practices dictate I should write, has involved me thinking more carefully about where and how

I might want to use 'I' and has enabled me to argue my case with my Director of Studies. For example, in the excerpt above, I argued that the use of 'I' both reflected and reinforced my emphasis on the partial nature of knowledge.

Moving beyond a 'timid professionalism'

As researchers and workers in higher education, I think it is easy to be drawn into the dominant ways of making meaning in academia. Through concerns – unacknowledged perhaps most of the time – about what we *should* say rather than what we want to say, we constrain our meaning making. Sara Ruddick, a philosopher, exploring her own fear about whether, as an academic, she should be asking certain questions, calls this 'timid professionalism' (in Belenky, Clinchy, Goldberger and Tarule, 1986:96).

A small example of me moving away from such 'timid professionalism' is the extract below. I was trying to write about the way in which academic writing conventions regulate what students can mean, based on discussions with the students in my research project around their writing for course assignments. The wordings marked in bold are the ones I found most difficult to say; it took me some weeks to decide that, given that I thought it was important to talk about ideology, I, as writer-director of voices, should do so.

Extract from one of Theresa's drafts on the nature of academic writing conventions:

> **We need** to look more closely at this notion of appropriateness and I would suggest that in order to understand how it affects what students can mean **it is useful** to draw on Fairclough's work on language and ideology.

The most difficult decision for me was to express authority through the categorical statement 'we need' and 'it is useful', even though I was clearly lessening such authority by 'I would suggest'. Previous drafts included 'it seems useful' and 'it may be useful'. I'm not suggesting that we should always use categorical statements, but that we should use them when we feel it is important to do so.

Conclusion

In conclusion we'd like to make the following points:

- It is not easy to write what you want and how you want in Higher Education. As participants, even if we disagree, our choices are constrained by what have come to be seen as the acceptable, rather than the dominant, practices of the institution. In our experience, undergraduates are in a less powerful position, and are thus more constrained than postgraduates at PhD level.
- Talking about writing has helped both of us in our writing in different ways. Theresa feels she has been encouraged through talk to take more active control over how she wants to write. Melanie feels that talking has helped her to sort out what the conventions are, and to be more aware of how she is writing. Although at times she imagines a future where she might challenge some convention and write how she wants, for the moment she feels this is not possible.
- We think it is important to recognise that whilst there are dominant conventions, some of which may be enforced in overt ways (*don't use 'I'*), there are also differences and divisions within institutions, between tutors, departments and faculties, where student-writers may be able to find small spaces to say what they want to say, regardless of their status. We have tried here to give an idea of the type of spaces we would like, and we would be really interested to hear what other participants in HE – students, tutors and researchers – would like to do in their academic writing.

Notes

1. Melanie is consciously using a contracted form here and elsewhere because she wants to move away from formal academic language.

References

Bakhtin, M. (1981) *The dialogic imagination, four essays by M.Bakhtin*. M Holquist, traps.' C.Emerson and M. Holquist. University of Texas Press.

Belenky, M. Clinchy, B. Goldberger, N, and Tarule, J. (1986), *Women's ways of knowing*. New York: Basic Books.

Wertsch, J. (1991) *Voices of the mind. A sociocultural approach to mediated action.* London: Harvester Wheatsheaf.

A7 Assessment and accreditation

Further RaPAL references about assessment and accreditation

Thompson, P. 'Wordpower for Who?', *RaPAL Bulletin* No. 11, 1990.

Morphy, L. 'More About Wordpower', *RaPAL Bulletin* No. 12, 1990.

Herrington, M. 'Work in Progress – OCN Accreditation for ABE Students', *RaPAL Bulletin* No. 20, 1993.

Pieces of paper: A survey of student attitudes towards accreditation in adult literacy and English classes in Southwark

Tanya Whitty

RaPAL Bulletin No. 21, Summer 1993

Tanya Whitty is the Co-ordinator for English and Literacy in the Faculty of Adult Education, Southwark College.

Background

I have been an adult literacy worker since 1974. In those days, I talked about goals in terms of students' personal fulfilment and group empowerment; I worked with students to explore their experience, using it to develop their reading and writing skills, and publishing the end product in books of student writing. In 1982, I left adult literacy work, and when I returned in 1989, I discovered that things were no longer the same: we had moved into the era of the big A for Accreditation. I felt both suspicious and challenged, but climbed aboard the Wordpower bandwagon, determined not to be left behind when the certificates were being handed out.

To my surprise, I found great benefits for both myself, as a tutor-organiser, and for students who opted for the new accredited courses. Never before had I been able to demonstrate successfully to students the need for careful record keeping, well organised files, for drafting and re-drafting. I found a clear syllabus comforting: it gave a common framework for tutors, while allowing them to use their own materials and ideas. While Wordpower seemed to downgrade the importance of creative talking and writing, it provided a strong motivator and emphasised areas of literacy which may not have received much attention in the past.

Three years after I had first started courses in City and Guilds Wordpower and Pitman's English for Office Skills, it seemed time to ask students what they thought about accreditation.

Questionnaire and Respondents

We designed a questionnaire, asking six questions. The sample of 47 students was about a fifth of all students currently in classes. They came from ten different classes on four sites and included students from the five different courses on offer, at a range of different levels from Level 1 (Basic Reading and Writing) to GCSE English Language.

Characteristics

	Male:	50%	**Female:**	50%	
	Day:	60%	**Evening:**	40%	
	Working	36%	**Unemployed**	64%	
Age:	**Under 20**	**20-29**	**30-39**	**40-49**	**Over 50**
	2%	34%	30%	21%	13%
Classes:					
Level 1	**Level 2**	**Writing for Bi-Lingual Students**		**English for Office**	**GCSE**
15%	40%	7%		19%	19%
30% attended more than one class a week					

Method

Before asking students to answer the questions, it was explained to them by their tutor that the purpose was to find out about their attitudes towards accreditation and certification; to find out whether the priorities of the Government, the Adult Literacy and Basic Skills Unit (ALBSU) and the Funding Council were also priorities for students. They were then asked to fill in the questionnaire individually, with help from the tutor with reading and spelling, where necessary. The responses were from individuals, given without prior group discussion.

Responses

Q1 What was your main reason for deciding to come to an English class?

To improve job prospects	55%
To gain access to Further Education/training	17%
To communicate better	15%
To feel more confident	13%

Although this was an open ended question, most responses mentioned one of the above. 17% of students included a reference to wanting a qualification in their response, but no one gave this as their sole reason for attending.

Other responses included:

to write a book
to stand on my own feet with reading, writing and spelling
to understand things more clearly and improve my state of mind
to feel confident in writing without worrying about spelling
to prove to myself I can do it

Q2 Have you already got any certificates of any kind?

A surprising 43% of students already held a certificate; 4 students had certificates from their countries of origin. Others included:

C & G:	Maths, Plumbing, Joinery and Construction, Electronics, Wordpower
RSA:	Computing, Typing, Wordprocessing, Book-keeping
GCSE:	Mathematics
PPA:	Creche worker
B.TEC	Business Studies with Computers
CSE	Various

Q3 Are you currently working for a Certificate in English?

55% were currently working towards either City and Guilds Wordpower, Pitman's English or GCSE.

Q4 Would you like to work for an English Certificate in the future?

79% said they would like to work for a certificate:

GCSE	40%
Wordpower	20%
ESOL/LIT	9% (all)
Pitman's	10%

The importance of GCSE as a qualification in students' eyes came out clearly. There were three GCSE students who wanted to go on to take Wordpower courses. This shows the confusion over the value and ranking of different types of accreditation.

21% of students stated that they did not want to work towards a certificate.

Q5 Do you think certificates are important?

91.5% of students answered that they did think certificates were important.

For employment: 49%

For personal achievement, proof of learning for yourself, confidence building 30%

Other responses:

- a second chance
- important for mothers to show example to children
- encouragement to work hard
- to be proud of

8.5% of students did not think certificates were important.

Responses included:

- they are not important if you can read and write properly
- not to me but to employers
- most jobs do not want certificates, just skills

Q6 Different people think different things are important.

How true are these statements for you?

	Level of Importance	1. Most	2. Very	3. Fairly	4. Not
a	I want to feel more independent about reading and/or writing	47%			
b	I want to feel more confident about filling in forms and applying for jobs	32%			
c	I want to gain a Certificate				
d	I want to communicate better with people	30%			
e	I want to gain the skills to go further with my education and training	32%			
f	Other	0%			

Note: students were asked to distinguish the statements a – d by ticking boxes 1–4 on the above table.

Results:

i) The percentage breakdown of students ticking MOST IMPORTANT is shown on the table.
ii) The percentage breakdown of students ticking I WANT TO GAIN A CERTIFICATE is shown below:

Level of Importance	1. Most	2. Very	3. Fairly	4. Not
	19%	38%	24%	19%

The responses to this question echo those of Question l: all other factors were seen as more important than gaining a certificate by most students, and a minority did not think they were important at all. However, most students ranked them as very or fairly important and 19% saw gaining a certificate as the most, or one of the most, important reasons for coming to classes.

Results and Conclusions

From the results of this survey it is clear that most (almost 80%) of students want to gain a certificate in English and an even higher proportion (91%), thought that certificates were important for employment purposes. ESOL students in particular, seemed keen to gain a certificate (although the sample was very small).

However, a sizeable minority (20%) said that certificates were not important to them, and not a single student stated that gaining a certificate was the sole reason for attending the class. Gaining independence and confidence in their communications skills was seen as more important by 80% of students. However, many students saw gaining a certificate as a means of becoming more confident.

The fact that a surprisingly high percentage of students already held a range of certificates and yet were also unemployed, may have meant that they were realistic about the usefulness of certificates as passports to jobs. The results show that although 50% felt that potential employers were looking for people with certificates, 30% said that they wanted a certificate to prove something to themselves rather than for employment purposes.

Only one GCSE student said he was attending the class primarily to gain the certificate, showing that although this was the certificate that the highest percentage were hoping to gain in the future, it was not the piece of paper alone that was important to the students.

It is interesting that so many people who had gained craft certificates which must have involved them in substantial reading and writing activities, should feel so unconfident about their communications skills that they subsequently have chosen to join English classes.

The results seem to indicate that most students welcome the opportunity to gain accreditation in English. This is not, however, the reason they came to classes in the first place. A certificate is seen as an extra benefit, which may be valued by employers or colleges, but which is also, importantly a proof of personal achievement.

Students showed a lack of clarity over the relative levels of different certificates. Most students had heard of GCSE and wanted to aim for it at sometime in the future.

Implications for planning a programme of English classes

The survey indicates that it is important to keep a balance between accredited courses, which inevitably have a prescribed syllabus, and the more open-ended classes where students have the opportunity to explore their strengths and weaknesses and to gain confidence in their skills, particularly when they first start a class.

It is clear that many students enjoy the challenge of accredited courses and the opportunity it gives them to prove they "can do it". Although many students (and tutors) feel uncertain of the marketability of Wordpower and Pitman's Certificates, students value them for their own sake, and as mileposts towards GCSE.

Despite the increasing pressure to accredit every course, it is important to recognise that for a minority of students, gaining a

certificate is an irrelevance. Others lack both the skills and confidence and need time to explore their learning without the pressure of assignments or exams. At all levels, there needs to be a choice for students – a choice which is clear both in terms of lateral alternatives and progression routes.

Wordpower and the publishing of student writing

Rebecca O'Rourke, Josie Pearse, Jill Ross and Adele Tinman
RaPAL Bulletin No. 19, Autumn 1992

Rebecca O'Rourke was a research officer at Goldsmiths' College. Josie Pearse is an adult education tutor in Brent. Jill Ross is the organiser at Leyton Open Learning Centre. Adele Tinman is a part-time tutor at Hammersmith Reading Centre.

The Background

In September 1991 Goldsmiths' College began a one year, Leverhulme Trust funded research project into student publishing in adult literacy. The brief was to:

- survey the extent and nature of such work over the last 15 years;
- seek the views of students and tutors on how best to develop student publishing within the ABE curriculum;
- consider the kinds of learning it enables.

During extensive visits and through a questionnaire survey of Britain and Ireland many issues emerged and were debated. (1)

One key issue was the way in which student publishing and, to some extent, writing development work was changing under the pressure to accredit adult basic education, especially since the introduction of Wordpower at the end of the 1980s. The research encountered a range of opinion about this:

> *"Wordpower is a real killer for this kind of work. It's all tasks, tasks, tasks. Writing development is too often seen as a kind of optional extra." (Tutor, Swansea)*

> *"We almost used student writing as a bench-mark, a way of saying you've achieved something. Now with accreditation, different goals and progress emerge earlier, and in dialogue." (Organiser, Norwich)*

The majority view was that accreditation was here to stay. There was a strong sense of polarisation between old and new styles of work: the one characterised as a commitment to long stay students, group work, student publishing and creative writing; the other as Open Learning, individual work, accreditation and rapid progression. These polarisations, whether real or imagined, are in practice both inaccurate and damaging.

During research visits, many tutors wanted examples of how to successfully combine student publishing and writing development work with Wordpower. There were no examples of completed work but two groups, one based in Hammersmith Reading Centre, the other at Leytonstone, were producing magazines and planning to accredit this work through Wordpower. Tutors from both schemes came to a meeting convened by the research project which had also been advertised via word of mouth and Wordpower training events (2). There were three meetings which provided a forum to discuss the links between student publishing, Wordpower and Numberpower and began exploring models for student publishing within their frameworks (3).

Introduction

Writing this article is a chance to widen the conversation. We have not solved the problem of integrating student publishing into Wordpower, but by sharing some of our observations and experiences we hope to initiate debate about how to carry this important aspect of work forward in a changed context of delivering adult basic education. We begin by setting out some of the issues we discussed and then give two fuller examples of work done by Josie Pearse and Adele Tinman.

We think some of the problems stem from a perception of Wordpower which identifies it with a functional approach to literacy, emphasising competences and mechanical skills, rather than an approach which includes personal development and takes a more rounded view of the students' interests and needs. We have two observations to make about this. Firstly, the need for understanding and insight should underlie the first approach, but this needs to be made explicit, especially for new tutors. We have seen instances where the zeal with which Wordpower had been promoted has been read as a criticism of all past practice and led some inexperienced tutors to reject work which focuses on personal development.

Secondly, the need for students to demonstrate critical insight is required for successful attainment at Stages 2 and 3 but not mentioned at Stage 1 or Foundation. It is unclear, and perhaps needs discussion, whether this is an oversight or an implied, deliberate omission -reflecting the view that cognitive skills depend on the acquisition of functional and mechanical skills. Josie Pearse's example drawn from work in Wormwood Scrubs illustrates this issue well.

Further problems arise when activities are contrived to fit the accreditation, rather than fitting the assessment to the work. 'Off with the old, on with the new' thinking exacerbates this. Adele Tinman's account of work at Hammersmith Reading Centre shows that skills and competences are perfectly capable of being assessed through existing activities. Student publishing, in its widest sense, is an obvious example of a set of activities that require and develop a wide range of communication skills.

What is at stake?

There is a dilemma for those who want to include student publishing but are wary about doing so. Does the Wordpower framework really limit creativity, writing development and publishing or not? Although the scheme is aimed at broad contexts, the range statements of the scheme show mainly functional examples. The ALBSU published pack of worked examples states:

> *"The pack should not be seen as a course or syllabus which learners*

> *follow. Instead it gives examples of the type of materials that can be produced and the ways they can be used. Materials should always be produced to meet the individual needs and situations of the learners."* (Cox et al: 3)

Despite this, the cumulative effect of examples drawn from workplace or domestic situations emphasising practical and functional skills biases tutors towards that area of work.

The role of trainers and assessors

The range of statements of the scheme and the support and training materials developed for Wordpower include very few creative examples. This is likely to affect tutors differently, but with the same result: that creative choices are not often made. Tutors new to ABE are more likely to follow the Wordpower path as set out for them. Experienced tutors who are new to Wordpower may adopt a more conservative approach because they are understandably unwilling to risk their students' work failing to meet assessment criteria because a regional assessor disagrees with their choice of work. Perhaps too, the emphasis on functional work is partly a reaction to Wordpower itself.

The scheme can be daunting and when tutors are feeling overwhelmed, they are less likely to experiment with it. Perhaps more creativity will emerge as people get more comfortable with the framework. Currently, though the role of the assessors and the trainers is crucial in developing that permission to stretch and test the framework by experimenting with creative ways of working.

> *"The pressure was on last year. This year people delivering it are more relaxed. We were constantly second-guessing what they wanted."* (Tutor, Liverpool)

> *"The Wordpower trainer emphasised that the concepts and strategies could be got at through all sorts of work. She kept saying that people should not feel constrained by the suggested examples."* (Tutor, HMP Preston)

The argument

This is perhaps a good point to look at the creative versus functional issue which so often informs the positions people take on Wordpower. Advocates on each side have been stereotyped: the one as failing to deliver useful skills in a cost-effective manner, the other as de-humanising learning in the service of the market. The association of Wordpower with a functional emphasis has led to the assumption that it must be delivered in a functional way. It is partly this that accounts for the hostility towards Wordpower from those who see education as more open-ended.

Undoubtedly, the current political and economic climate, and the ideological atmosphere it generates, favours a model of education which promises a useful, literate, socially engineered workforce. Three things need to be said about this:

1. The claims made for purely functional and vocational education sound less and less credible as the recession resolutely refuses to bottom out and unemployment continues to grow.
2. Just as words such as 'freedom' and 'choice' have been distorted so that their appearance now alerts us to the absence of the thing itself, it is important to refute the claim that curriculum development and assessment were not taken seriously until Wordpower happened along.
3. Even if you have a restricted end in view, there is no reason to believe that this can only – or will best – be achieved by restricted means. A variety of approaches works best for people new or returning to learning, as it does for tutors.

A case in point

The curriculum vitae. What could be more functionally orientated? Isn't it something every student should have when they leave a work-orientated course? Set it against, as many will, the life story. Autobiography: the quintessential form of student writing. We argue that before students can attempt the CV they must possess the narrative of their own lives. To achieve this, they must have the opportunity to reflect upon their experience. For many students, autobiographical writing performs just such a role. Some will want to take it further as writing, others will not; some will want to use it to develop a CV, other will not. It is impossible to predict beforehand who will take which option, just as it is to legislate the period of time it will take for a person to move from one place to the other.

Making a case

When we began collecting examples of how student publishing could be used in the Wordpower framework, we encountered the negative polarity again. People assumed we wanted a wholesale swop:

> *"I don't see that it's necessary to make the whole of a Wordpower course revolve around publishing." (Tutor, London)*

Neither do we. We wanted an end to the idea that you choose between one or the other: accreditation or publishing. However, as many tutors pointed out, both activities demand extra, unpaid work, so perhaps it does become harder to choose publishing if publishing means a substantial book or magazine. It is also important to remember that students cannot be forced into any learning activity, publishing included. However, the research project as a whole revealed broader definitions which shift the emphasis from the product: (a 32-page magazine or a bound book with a spine etc) to the process – the preparation and presentation of work for others to read. This change enables two things to happen which allow publishing to become the crucible of learning about reading and writing, and communicating with others. These are:

1. The length of the published piece can vary from a single sheet upwards, which increases access often limited by considerations of time and cost.
2. The emphasis on the relationship between the writer and reader, who you are writing for, develops critical reading:

> questioning your own and others writing. Working on a smaller scale means there is more likelihood of this happening than has sometimes been the case.

We must check nostalgic renderings of the glorious old days of student publishing. It often was wonderful, but it often was not. Tutor led, skimped editorial work and rushed or drawn-out production often meant fewer learning gains than there could have been.

Scrutinising what potential there is to bring student publishing into Wordpower provides an opportunity to look at the quality of publishing as well as the quality of Wordpower.

Ideas for WP accreditation in student publishing

See Table 1 opposite.

Two examples

1. A reason to use the computers: Hammersmith Reading Centre (by Adele Tinman)

A computer course, designed to give basic instruction about computers, developed into a publishing project. The idea of a magazine arose because women had produced pieces of work on the New Direction course and in the Wordpower group which they wanted to present to others. The course was funded for two terms.

Many adults report a lack of confidence when they attempt to write their thoughts down on paper. Using keyboards often reduces this lack of confidence, by enabling students to produce clear and attractive work in a printed format which gives encouragement to many who have a low opinion of themselves as writers. Students therefore develop critical judgements linked to the reading and writing process in a practical way, which in turn links in with Wordpower criteria.

Table 1: Publishing Task	Wordpower Unit/Element			
	Foundation	Stage 1	Stage 2	Stage 3
Writing a poem story	U4/E2	U4AE2	U3/E4	U3/E2
Reading a piece of writing	U1/E1	U1/E1		
Critical reading of same			U1/E3	
Group discussion			U4/E4	U5/E3
Writing my evidence of same	U4/E2	U4/E2	U3/E4	U3/E2
Taking minutes at group meetings		U7AE1		
Producing minutes	U4/E1	U4/E1	U3/E2+3	U3/E1
Reporting back: to groups		U6/E2	U4/El or 4	U5/E1 or 3
to 1	U5/E1			
Starting up the computer	U1/E2	U1/E2		
Follow menu instructions	U1/E2	U1/E2		
Showing another student how to proceed	U5/E1	U5/E1 + 4	U4/E1	U5/E1
Getting other students tutors to attend meetings/reading evening:				
on the telephone				
face-to-face	U5/E1	U5AE1		
visiting groups		U6/E2	U5/E1 or 4	U5 E1 or 3
by letter		U4/E1	U3/E2 + 3	U3/E1
by publicity			U3/E2 + 3	
giving directions and use of map	U2/E2	U2/E2		
reading evening:	U5/E3	U6/E1		
welcoming people		U5/E3 U5/E4		
"M.C." role		U6/E2	U4/E1	U5/E1

This chart is a trawl of ideas relating to the experience of the writers and is by no means comprehensive.

The knowledge that they can produce a successful piece of work which will be read by others is one of the most beneficial parts in the student publishing process. Sharing of experience, and understanding that experience in its personal and social dimensions, is often an important element in producing writing. The group dynamics in play as these delicate processes happen offer many opportunities for developing – and accrediting – communications skills. Publishing is a focus for the development of mechanical or functional skills: punctuation and proofreading, the recognition of sentence structure, paragraphing; which many students in ABE classes have difficulty understanding.

The magazine was used to raise funds for the student committee. During the reading evening one person had the responsibility for selling the magazine and approximately £45 was collected. This will be used to fund future activities, for example writing weekends or group visits. Student publishing can, therefore, be seen as an on-going project, involving skills as marketing a product.

Linking into Wordpower

The publishing process was linked to Wordpower in the following ways:

1. The practical skills were used to assess students in giving and following verbal and written instructions. For example: showing another student how to turn on the computer, put in a disk; the use of various function keys and printing out their work.
2. The planning of the magazine, from its initial stages through to the celebratory reading evening, provided much scope to assess communication skills. For example: selecting writing, illustrations, cover design etc; attending meetings and small discussion groups on the progress of the magazine; arranging times of meetings, telephoning or writing letters to get other students to attend; giving geographical directions, using A-Z; reporting back to other students who could not attend

meetings; taking notes of what was decided.
3. The 'creative' aspect of students writing down their own stories for inclusion in the magazine provided opportunities to assess students' writing development. Some students, whose poor levels of concentration prohibit their writing in groups, found it far easier to produce their work on a computer.

The increasing use of computers in everyday life, especially in our children's lives, means that the use of computer techniques and 'know how' is a fundamental basis for progression by students in ABE classes. Success, however, depends on how often students are able to use the computers. Editing does become easier and more graphic when using computers, but students do need to be familiar with the keyboard and function keys.

2. Real debates over language: HMP Wormwood Scrubs (by Josie Pearse)

A group of Stage 2 student editors were working on a magazine of life's writing at Wormwood Scrubs prison. The group was in the evening and extra to their usual Wordpower classes, in fact it was set up by me deliberately to get away from what I saw as Wordpower's restrictions. The editors met regularly and contributors dropped in to the discussions frequently or occasionally.

Four different issues were raised during a discussion over changes to a poem. These were:

1. The value of real situations in the accreditation process
2. That Wordpower need not focus on workplace examples.
3. The accreditation of group activities, including oral work.
4. The assumption, at the lower levels of Wordpower, that cognitive skills depend on the acquisition of functional skills.

Michael had written a poem in non-standard English. The student doing the typing wanted to change it to standard, but he consulted the rest of us first. The discussion that resulted really pleased me because I had been trying to flesh out the issue all year. Some students had benefitted from our occasional discussions but others were still sceptical about whether a piece of writing was valuable if it was non-standard.

The poet himself was a Foundation level student but, as he was also a well respected mucician and songwriter, the group was not able to write him off as just a poor speller.

The discussion at first centred on the spelling. But it was not just the spelling, the grammar was not standard either. This was a teaching point in itself. What was grammar? What would happen if we changed 'Dom ah go drown inna it"? It sounded completely different. Grammar, they decided, had a lot to do with rhythm and sounding good.

We then realised there were standard English parts in the poem, especially at the beginning, as if Michael had begun in a 'writing' frame of mind and then, as one student, Calvin, wrote in reflective writing on the discussion:

> *". . . about the seventh line he starts to feel very strongly about what he is writing, which brings out the West Indian in him."*

On closer study we also realised that, as far as spelling went, he had reverted near the end, writing 'them' instead of Vern' as earlier. At this point we decided we would have to ask the author in.

Michael did not want anything changed at all:

> *"No way. That's my poem, man. I wrote it like that."*

A heated discussion ensued. In the end Michael agreed that the progression from standard to non-standard was what he had written and so he compromised on the later spellings for the sake of consistency. But none of the grammar was to be changed and by the end of the discussion even the typist was wanting to emphasize the non-standard.

To me, they learnt more about choices in writing than I had been able to teach them with designed materials over the previous few months. It was later when I was thinking about Wordpower for the daytime group that I realised this discussion could fit the Stage 2 element which requires the student to identify intention and meaning in a text. The examples given in the range include fiction and biography, but even if they had not I would have been prepared to put the case to my outside assessor.

My outside assessor was also the regional assessor, responsible for training other assessors and very helpful. She was surprised I was checking, as she was well aware of the exigencies of the prison environment, especially with lifers, where the vocational element in Wordpower has little meaning to people who will not be in the world of work for 15 years or more. (Eventually we did develop work for accreditation around understanding complaints procedure and knowing your rights.) Being forced to downplay the functional helped us to realise that Wordpower can be delivered using non-vocational situations and materials.

In writing up the assessment for Unit 1 Element 3 (Critical Reading) I faced the problem of evidence. Could each member of the group be individually accredited for the group discussion? I felt they could, but to make sure I asked for a short piece of reflective writing on the process that led up to our final editorial decision. They agreed with this but not all the students did it. I only accredited those who did.

I wrote up a detailed account of the process on the assessment sheet. The assessor always wanted as much on the sheets as possible, as feedback for the students. It is important for them to understand that a writer has choices. This is an integral part of functional skills. Because students are often sceptical about the value of discussion it is important to demonstrate that this understanding of choices only comes through discussion between the writer and others. Wordpower is valuable in offering accreditation, and therefore status, to this activity.

However, the experience also showed up the anomalies between the levels. Why can't these two fundamental learning issues be accredited at the lower levels? Michael, being a Foundation level student, had no accreditation out of the discussion. His tutor credited him with the writing of the poem, but his higher skills could only be accredited at higher levels. Foundation level assumes that he will not feel comfortable in a group discussion. As a practising artist, he was certainly used to interpreting song and poetry but he had to be at Foundation level because standard English writing was new to him.

There is no doubt that Michael is an exceptionally gifted person, but he is not the only one. The question has to be asked: Do these assumptions reinforce the low self-esteem of students?

Final thoughts

These examples give a glimpse of what is possible and what needs more debate and discussion, in particular working on this article we became very aware that all the debate so far about Wordpower is tutor talk. Nobody is asking students what their experience and opinion of Wordpower is (4). We hope this debate will happen within existing training for the improved delivery of Wordpower. Separate training events in student publishing and Wordpower run the risk of reinforcing the notion that student publishing represents the opposition – the voice from the margins – rather than its acceptance as a valid mainstream component of the ABE curriculum.

Phyllis Thompson, writing just after Wordpower had been introduced (5) was cautiously optimistic about holding on to ABE's life enhancing aspects. So are we. Like her, we believe strongly that for this to happen there must be debate and discussion from everyone involved in Wordpower. We offer this article in that spirit.

Notes and references

1. A full account is given in *Versions and Variety: A report into student publishing in adult literacy education* by Rebecca O'Rourke with Jane Mace, distributed by Avanti Books.
2. The meetings were attended by Linda Duhig, Pat Hulin, Alison Jeynes, Pat Marden and the authors of the article.
3. We did not discuss numeracy, student publishing and Numberpower at our meetings. However, the research project did find tutors were enthusiastic about the possibility Numberpower opened up in this respect.
4. We hope to go on and explore this. Please get in touch if you want to get involved too.
5. Thompson, P. (1990) Wordpower for Who? RaPAL No. 11 Spring 1990.

For whose benefit is accreditation? An investigation into perceptions of basic skills accreditation amongst tutors, students and employers

Jill Kibble, Leeds Open Learning Centre

RaPAL Bulletin No. 22, Autumn 1993

Accreditation is now the means of funding and justifying the existence of much adult education work. As practitioners are aware, outcomes in ABE/ESOL are often subjective and can be problematic to quantify for funders. This can provide an excuse for lack of planning rigour and progression routes but has frequently been a strength of our work in developing uniquely flexible and responsive individual programmes. For managers accreditation is only a means of producing tighter vocationally related programmes with clearly defined outcomes and increased participation.

Contemplation of these various and perplexing dichotomies that are inherent if accrediting all ABE/ESOL work, led to this investigation into perceptions of accreditation. I carried out a small survey which encompassed tutors and students in a variety of vocational and non-vocational groups offering different accreditation and tutors from a range of provision attending a study day at Lancaster University. One of the workshop groups produced the following picture – see Figure 1.

Clearly, students in provision do find accreditation motivating and

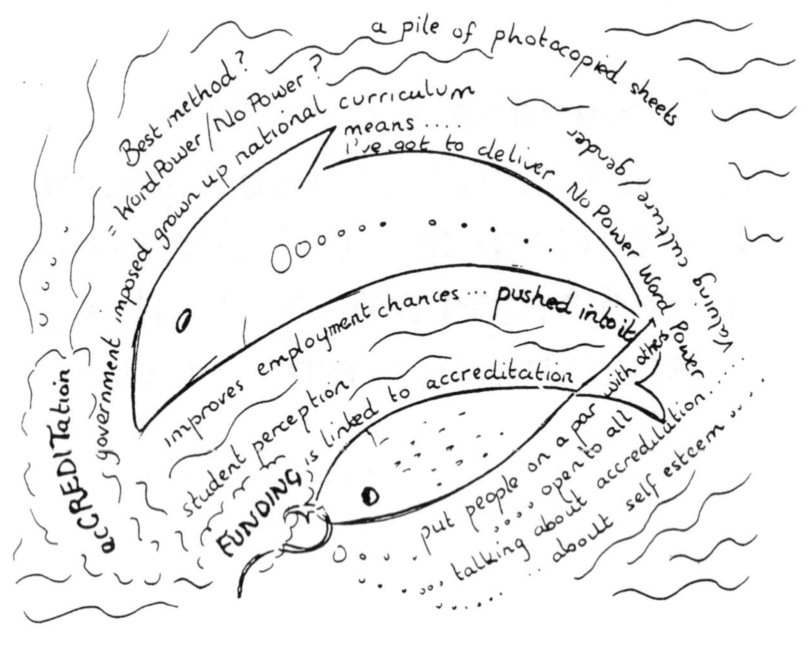

Figure 1.

tutors are putting in considerable effort to ensure that the diversity of students needs are being met. Interestingly, many students wanted to take an 'exam' and considered that these were the only 'real' recognition of achievement. Administrative systems involved in competence based accreditation present a challenge to students and tutors alike, particularly at foundation level, and can involve excessive amounts of time. Generally there was a greater degree of satisfaction with courses, such as those accredited by Open College, offering more flexibility and ownership. However, concern on validity and recognition of such courses was borne out by the survey of ten major Leeds employers. Most are unaware of, or do not recognise, certificates not of NVQ level 2 or above or general education GCSEs. Since most students hoped that accreditation would improve their employment prospects this is *a* cause for concern and indicates a need for careful counselling and advising. Despite a considerable publicity drive, Wordpower was not recognised outside of college personnel departments. In summary the main points arising from this survey are:

- the importance of adopting ABE/ESOL accreditation that is flexible to student needs and not merely vocationally oriented
- the need to develop administrative and record keeping systems which retain the benefits of increased organisation and regular review but are not extensively time consuming
- the need to use accreditation which challenges students and provides opportunities for new, relevant learning. This may well involve offering a range of accreditation within one group
- the need to educate and inform employers and Adult Education funders in the range of work undertaken in ABE/ESOL and the variety of positive outcomes our students achieve
- the value of 'first step in' provision in terms of recruitment
- accreditation is improving student and tutor organisation and rigour in planning (also inventiveness in producing relevant situations in which to demonstrate competence)
- accreditation is motivating some students and improving retention and attendance patterns, particularly amongst younger students
- genuine concern that accreditation is only being used for funding purposes, particularly when the system is imposed and removes traditional flexibility and sensitivity to need

Talk this way: An exchange of views on the assessment of speaking and listening

John Arnett

RaPAL Bulletin No. 41, Spring 2000

John Arnett (lecturer in Basic Skills, College of North West London) asks:

Is it me, or is there something more than a little dubious about grading people, especially when they are adults, on their ability to use their own, native language? Assessment of speaking and listening has featured in GCSE level exams for years, and currently accounts for 20% of the overall marks. It is my impression that teachers of English, probably because they know only too well that the whole procedure is suspect, have tended to pass everyone in speaking and listening and concentrate on reading and writing which, after all, is what is tested in the exam. Mindful of this, and of annual complaints about falling standards when the results come out, the exam boards, propelled by SCAA, have tried to enforce a more rigorous approach to the assessment of speaking and listening.

My objection to the formal assessment of spoken English, as currently enshrined in GCSE syllabuses, is twofold. Firstly, it has an assumption of objectivity and "scientific" rigour that cannot be sustained. Secondly, it undoubtedly favours certain modes of expression over others, so that speakers of non-standard and regional varieties of English – working class people in the main – are penalised. At its worst it seems to me that students are being rewarded for having the right sort of personality and class credentials. Adult students such as those taking the SEG Mature syllabus in colleges are especially likely to find this patronising and nonsensical, and are not afraid of saying so (perhaps this could be accredited?).

Michael Stubbs (1986), lists a number of further difficulties –

> *"the logical contradiction involved in trying to test informal language in a formal test; the lack of consensus about what constitutes effective language – some linguistic styles may define social group membership: our lack of knowledge about language development after the age of about 5 . . . (1986:150)."*

Furthermore, with speaking and listening, there is no hard evidence upon which the teacher can make constructive comments – unless everything is video filmed (and there is never time) it is in its very nature ephemeral.

Speech is at the same time more personal and more public than writing. You can use long words and fancy turns of phrase in your essay without anyone but your teacher knowing about it, and without inviting the scorn of your peers. Your speaking voice is an expression of your own identity in a much more fundamental way than your writing could be said to be use of a dialect is an expression of solidarity. The fact, if it is a fact, that middle class students do better in speaking and listening assessments, is hardly revelatory – they do better throughout the education system and in exams generally. What I find objectionable is, firstly, the inbuilt bias in GCSE speaking and listening assessment methodology which seems to disadvantage a large proportion of candidates; secondly that so many teachers (myself included) have, for too long, gone along with this uncritically, and thirdly, that the concept of "oracy" that much of the current orthodoxy is based upon is, as Michael Stubbs argues,

> *"not sophisticated or explicit enough to support the teaching practices or testing techniques which are based on it (1986:143)."*

Of course it would be splendid if everyone, in all walks of adult life, listened attentively and spoke articulately and appropriately and coherently at all times. How many people do though? Assessment of speaking and listening ability, particularly as it applies to adult, native speakers, is artificial, discriminatory and nonsensical. Until we can do it with the same degree of objectivity with which we assess reading and writing we shouldn't do it at all.

A reply from Ellayne Fowler

Ellayne is a lecturer in Adult Basic Education at Trowbridge College.

No, it isn't just you John. For a long time I've questioned the inclusion in City and Guilds Wordpower of the Unit 'Talk to one other person' at Entry and Level I. This accreditation may be more familiar to many of us than GCSE. Similar arguments hold however. I have long found it difficult to justify to perfectly competent native speakers that they have to present evidence of holding a conversation with someone they already know. Where is the evidence that difficulty in expressing yourself on paper is matched by a difficulty in spoken English, which to many students is implicit in the inclusion of these Units in Wordpower? I suspect that the entry level requirement was written with the idea that this was a qualification particularly aimed at Adults with Learning Difficulties. Does this explain some of the underpinning knowledge and understanding, such as "speaking loudly or quietly influences audiences and can affect communication"? Was it also aimed at the ESOL learner? Does that explain the necessity for knowing the "conventions of word order"? Perhaps in aiming at too wide a market you end up pleasing no-one. There is more suitable accreditation available for ESOL and competent native speakers, with or without learning difficulties, can be offended by the simplicity of the tasks.

The problems in this area intensify when you reach Level One. Again the accreditation refers to "talking to one other person". The step up appears in the underpinning knowledge to relate mostly to "range of spoken language styles according to purpose, topic, audience"; "summarising" and the use of "visual aids". It is unclear whether you need to be able to demonstrate, or simply recognise "the range of spoken styles".

Taking up John's point on Standard English, I think it would be easy to conjure up pressure towards the Standard within the underpinning knowledge – the inclusion, for example, of "pronunciation and intonation can affect the communication process" and "tense is important". I think though that this underlines the need for accreditation boards to write a rationale for these standards. What was the audience/context expected to be for this accreditation? Is there any basis in existing linguistic research on spoken language? (I suspect not). In the end, it's a matter of interpretation. I do not think a native speaker has any problems with tense in speech unless they are inhibited by a language disability. Another tutor might interpret the same clause to mean that a local dialectal use of tense should be discouraged. How meaningful is accreditation that can allow such a wide range of interpretation? How rigorous can the verification of oral elements be, unless many are physically recorded and then verified?

From Level 1 we leap to talking to groups of people at Level 2, where many of the tasks are reminiscent of former Communication Skills certificates. At least here, students can build self-esteem and confidence through group based discussions and presentations.

In the end the same questions remain. Why are we asked to accredit oral communication in native speakers and how do we make it meaningful? Are we being pushed towards encouraging an oral Standard English? I'm not sure I can answer those questions, but I do think we need to continue asking them. I think accreditation boards should produce rationales for what they ask. I think those standards should be informed by current linguistic research on spoken English, rather than politicians and employers' wish lists.

On a personal note, I continue to encourage students to take part in group discussions and to act as audiences to presentations. I accredit their work for the lower units and elements, because they are performing at a higher level. I continue to eavesdrop and also write up conversations I'm involved in. I do it because students want to achieve City and Guilds qualifications. Is that good enough?

John concludes . . .

I agree entirely with the points made by Ellayne Fowler in relation to Wordpower, which I have felt similarly unhappy with. The very designation 'Talk with one other person' somehow brings home the absurdity of the whole procedure even more forcibly than with GCSE. It is worth noting that the draft proposal for a new accreditation framework for basic skills, post-Moser, gave equal weighting to speaking and listening as to reading and writing, and needs to be challenged for the same reasons.

Reference

Stubbs, Michael (1986) *Educational Linguistics*, Oxford: Blackwell.

B1 Numeracy

Further RaPAL references about numeracy

Colwell, D. 'ALM, A New International Research Forum on Adults Learning Maths', *RaPAL Bulletin* No. 23, 1994.

Tomlin, A. *et al.*, 'Maths – All Our Ideas Came into One', *RaPAL Bulletin* No. 38, 1999.

Cooper, S. 'Review of ALM6 Conference Proceedings', *RaPAL Bulletin* No. 44, 2001.

What is this thing called numeracy?

Joy Joseph

RaPAL Bulletin No. 13, Autumn 1990

Joy Joseph, Tutor Organiser for Adult Basic Education and Numeracy Group Tutor, South East Bristol Adult Education Centre, and ABE Development Worker (Numeracy), Avon County Further Education Development Centre.

Contributing a few entries recently to the RaPAL bibliography, I was somewhat staggered to find that numeracy appears to be outnumbered by literacy in the contents list in a ratio of roughly 1:20. As a veteran of the campaign to see numeracy recognised in its own right and given equal status with literacy in training programmes, I was moved to reflect on what it is about numeracy that has brought about this enormous difference in weighting.

My reflections led me into a number of areas of confusion around the definition and nature of numeracy, which shed some light on the problem, though not enough, and which I feel might be worth sharing as a way of setting the numeracy ball rolling for RaPAL. So here goes.

"Public" Definitions

The main clue to the imbalance of literacy/numeracy literature is in dates: literacy has a long headstart. The Oxford English Dictionary (OED) reports that the first recorded use of the term *literacy* was in **1883**. In contrast to this, the terms *numerate* and *numeracy* were coined "to represent a 'mirror image of literacy'" as late as **1959** in the Crowther Report[1]. The terms are defined at some length in Paragraph 401, and summarized (para. 419) as: *not only the ability to reason quantitatively but also some understanding of scientific method and some acquaintance with the achievement of science.*

The OED follows this lead, defining *numerate* as *"acquainted with the basic principles of mathematics and science"* (although the Concise Oxford Dictionary (1990 Edition) omits any reference to science, and defines innumerate as having no knowledge of or feeling for mathematical operations).

Collins dictionary is more modest in its definition: *able to use numbers, especially in arithmetical operations.*

A somewhat different angle on numeracy is given by Le Roux[2], who defines it as *the ability of mathematicians and non-mathematicians to communicate with each other.*

The best discussion of the meaning of numeracy is probably to be found in the Cockcroft Report[3] (1982), which summarises Crowther, and states that none of the submissions received by the committee confirm that view, suggesting a shift in meaning over the 20 year period. The Cockcroft view of numeracy as essentially a coping skill, an "at-homeness" with everyday numbers, coupled with an appreciation of information presented in mathematical form, is probably the view most widely accepted today by interested people.

Popular Definitions

Another problem of definition that numeracy tutors face is that of the popular conception (or misconception) of numeracy. To begin with, many people (especially those in greatest need of help) simply do not know the word. So publicity material has to use something else, and this adds to the difficulty of trying to separate basic and functional numeracy tuition from basic maths, or even sometimes from not-so-basic maths.

Another popular misconception goes even deeper than this: the view that numeracy (or whatever you choose to call it) is about "learning how to do it". This is the legacy of the rote-learning approach to number work, and is very deeply ingrained in large numbers of the population. Bringing about the change in attitude that enables students to forget (temporarily) about methods, go for an understanding of concepts and develop an interest in numbers, is probably the most crucial, as well as the most difficult part of the numeracy tutor's job. I'm sure most practitioners could quote examples of students who, despite all one's efforts, cling grimly on to the memory buoy, until at last the sheer weight of half remembered facts and methods drags them under.

"Professional" Definitions

As if the difficulties engendered by confusions in public and popular definitions were not enough, we are also up against divided opinion on what we are about from within the Adult Basic Education service. The reason for this is in part at least historical. Most of the early help in numeracy was given by literacy practitioners in response to immediate demand, and it has been largely literacy practitioners who have come to hold positions of responsibility and influence in ABE in general. As a result, much of the "received opinion" on numeracy has been influenced by the literacy philosophy of basing work on students' expressed needs, and taking a solely "functional" approach. While the theory sounds convincing, and has some merit in numeracy, experience suggests that it is a simplistic view which imposes severe limitations on the progress that students can make. In addition, it is an approach that is well-nigh impossible to apply in a mixed ability group, whose functional needs and interests will be extremely varied, but who may well share a motivation to work on number *concepts* that have (different) applications for all of them.

Now as soon as we begin to move away from the idea of solely functional numeracy, towards an understanding of number concepts in general, we begin to stray over the border into Maths, which does indeed have a vast and ever expanding literature. I suspect that many numeracy tutors spend most of their working hours on the fence between numeracy and maths, leaning one way to use life skills to illumine pure number work; leaning the other way to link life topics into major concepts, thus enabling the transfer of learning to take place. It is always a delicate balance.

A Personal Definition

For me, balance in a different sense is an important aspect of being numerate. I see this as the possession of sufficient understanding of the world (or language) of numbers not to be floored by the numbers that are flung at one in one's own life. And perhaps the chief difference from literacy is that it is essentially a **mental state** which can exist quite independently of print and paper.

This is perhaps why so many people with literacy difficulties are quite confident about handling money, time, and the everyday calculations they need in their work. One does not have to be able to read and write in order to understand basic number concepts or think logically – if recording is necessary as a memory aid, there are other ways of going about it (cave painting was probably one).

I'd like to conclude with a favourite tag (adapted from Sewell[3]):

> *Maths is what I don't understand;*
> *What I do understand is just common sense.*

My job, as a numeracy practitioner, is turning maths into common sense.

Notes

1. 15 to 18: A Report of the Central Advisory Council for Education (England) (The Crowther Report) HMSO 1959
2. Withnall, Osborn & Charnley: Review of Existing Research in Adult and Continuing Education – Vol VII Numeracy/Maths for Adults NIACE 1984 (My notes on this contain the reference to Le Roux in International Journal of Mathematical Education in Science and Technology Vol 10 No. 3 1979; also to Sewell: Maths in Everyday Life, Reading A.E. Service, 1978)
3. Mathematics Counts: – Report of the Committee of Enquiry into the Teaching of Mathematics in Schools (The Cockcroft Report) HMSO 1982

Numeracy in parent literacy sessions

Jean Milloy

RaPAL Bulletin No. 24, Summer 1994

Jean Milloy is a literacy and numeracy tutor and has been working on the Walk in Numeracy project for the last four years.

Walk in Numeracy is a project of Hammersmith and Fulham Council for Racial Equality and is funded by the London Borough of Hammersmith and Fulham. *Walk In Numeracy*, sister project of Hammersmith Reading Centre, provided numeracy input to four of the Parent Literacy courses set up by Viv Bird and Kate Pahl (see 'Parent literacy in a community setting', No. 24, pp. 6–15). The numeracy sessions arose out of, and attempted to respond to, issues raised by participants, and covered a wide range of topics.

The Issues

Most of the parents involved in the sessions had primary-age children and their concerns covered the following areas:

- what maths is taught and how is it taught in primary school?
- what is a child expected to know at a particular age (i.e. as specified in the National Curriculum)?
- how can we help our children with maths?

One parent raised the specific question of base numbers, which her daughter had been working on.

Basically, there were two large questions behind these concerns: "what is National Curriculum maths?" and "how can our children learn maths when they never seem to 'do' maths (in the sense of working through pages of sums)?", the latter being a question often raised by numeracy students with school-age children.

Responding to the issues

Jenny Creighton and Jean Milloy, who taught these sessions, tried to incorporate these concerns into a 2 hour session. Although we did not know the students, they had worked together previously in parent literacy sessions, so we were able to anticipate a certain amount of group confidence. At the same time, we needed to take account of their feelings about maths, and to expect some anxiety about the subject and their own learning experiences. We therefore used the idea of a *maths diary* as an icebreaker at the beginning of three of the sessions, asking the question, "what maths have you done today?" to begin a discussion about everyday maths and individuals' own maths skills. Most people respond that they have done *no* maths; the tutor then elicits information about using money, using time, estimating, preparing food and so on – all activities requiring the use and application of number and mathematical skills. The point of such an exercise is to examine both the range of students' own numerical skills and to what extent these reflect experiential learning ("I never weigh quantities – I just know how much to put in"; "I glance in my purse and know if I've got enough"). The exercise also points up the belief of many people that if they can successfully carry out a numerical operation then it can't be "real" maths because maths is hard and they were never any good at it.

In one session this belief led into a very useful discussion about mathematical thinking and its place in the maths curriculum. We looked at ideas around *ordering, sorting* and *place value* (which helped in understanding some of the activities undertaken in the early primary years), *comparison* (bigger than/smaller than), and *position* (next to, under/over, up/down and so on). In turn, this, with some tutor direction, moved us into a consideration of *problem-solving* and *decision-making* and the whole issue of *investigations,* the most important aspect of which is perhaps being able to explain what you are doing and why – a long way for most people from the conventional notions of maths being about *learning rules* and *getting sums right.* The significance of such discussion lies in the fact that you are actually doing the work as you talk; we followed it up with some maths puzzles which, pleasingly, provoked spirited arguments about the best way to solve them rather than disputes over the results gained.

Initial discussions like this not only broaden people's sense of approaches to maths teaching and learning at primary level, and provide a safe environment in which to articulate thoughts and feelings about maths, but also introduce the whole question of the organisation of, and thinking behind, National Curriculum maths, which puts emphasis on using and applying maths and problem-solving. At the same time, the ability to carry out mathematical operations is clearly indicated, and this raised the question of *methods* and the use of *calculators.* One group produced at least three methods of doing subtraction, and we looked at the difficulties that arise between parents, teachers and children when different methods are offered. Similarly, learning the multiplication tables always raises hackles: another group agreed to differ on the most useful method, but all were intrigued to learn that the associative law (6 x 7 is the same as 7 x 6) means you don't actually have to learn the whole lot. As tutors, we feel that it can often be helpful for parents to have some idea of the language of maths: expressions such as *place value, number bonds, sharing* (rather than *division)* not only enable people to understand what teachers talk about but also shift thinking from the functional to the conceptual.

All the discussions mentioned above were interspersed with brief examples of maths teaching and learning. For example, the maths diaries raised the topic of percentages, and the group members themselves provided a number of ways of working these out. It is important for tutors to maintain a highly flexible approach in these sessions, which are difficult to structure because of their open nature.

Teaching specific topics

One group member had asked to do work on base numbers, and this provided an opportunity to carry out an investigation into number bases and place value. Students were given coloured counters and grids marked off in different number bases, and were asked to produce the same number and to think about how they would describe it. They worked in groups of two, and the exercise and discussion took up most of the two hour session. This was a difficult and not entirely successful exercise, partly because the students did not have a model (for example, base 10) to work from. However, it did provoke an interesting discussion about both the process and its results, and questioned the apparently fixed nature of mathematics. The students were encouraged to think more about how they were working than to worry about the end result, and this led into questions about different number systems and methods of performing operations.

Working with ESOL students, as with one group of Tigrean women, can bring about two-way learning, with the tutor expanding her own knowledge as much as the students. In such situations, where discussion may be limited by language differences, mathematical conventions can be explored (for example, the western use of the decimal point and the Arabic use of the comma, where 1.267 might be written in some notations as 1,267, thus causing confusion among westerners who read "one thousand two hundred and sixty seven")

Conclusions

From our experience as tutors in these groups, it is clear that there is scope for parent numeracy courses, both for those whose motivation is primarily to help their children, and for those who would like help with their own maths. We would envisage a course that combined learning through discussion and investigation with learning through direct teaching, and which would provide parents with the sort of understanding of maths skills that would enable them to transfer their learning to their children. Placing such courses in a community, rather than a school, setting allows participants to establish their own environment and make demands from their own needs, and thus to build up confidence in learning and teaching independently of the school.

Who asks the questions?

Alison Tomlin
RaPAL Bulletin No. 35, Spring 1998

Alison Tomlin is a part-time adult literacy and maths tutor working in South London, and a research student at King's College London.

Much of 'teaching' in adult basic education is structured as asking questions, though often questions to which the tutor knows the answer. We work within traditions in which **negotiating the curriculum, responding to students' needs and learning from the students** are valued, and we often work on topics chosen by students (for example, letter-writing, or metric measurement), but within that the questions are commonly posed by the tutor: "Ok, can we think of any other words spelled like *dear?*" "What did you think of the book?" "How long is this table, do you think?" This traditional structure has been the subject of much school-based research (e.g. Cazden 1988) summarises a range of research). **Some students and I have been trying out the use of students' questions instead, in both literacy and maths classes**.

My aim at the conference workshop and in this article is to ask you for advice and my own questions are listed at the end. Here I want to give you some examples of this work (using some of the work of the conference workshop participants) and point to some surprises, contradictions and difficulties. They are in three groups: literacy, maths and student interviews.

Literacy

Within literacy work I want to look only at some 'language experience' writing, using an example from a 'classic' of literacy work: writing personal histories. Taking dictation puts a great deal of control in the hands of the scribe – often the tutor. For discussions of this see Brown (1985), Moss (1995) and Duffin (1995). The control may be in choices of wording – whether to scribe 'ain't' or 'isn't', for example – in overall shape and, often, in content, because the scribe may ask questions to clarify and extend the writing. I was looking for ways to keep authorship in the hands of the students, to encourage critical reading and to help students read their own work with a sense of audience. I wanted to foster dialogue in writing as well as group talk – to make writing genuinely communicative. (I sometimes wonder if we tend to 'elicit' text (or extract it, like teeth?) from students, in order to get a text that can be mined for its 'key words', -ing endings, and the like.)

The example here comes from a class in which I took dictation from students, and checked it for accuracy. I didn't ask them questions for clarification or extension. Each piece was typed on a separate sheet with a box at the bottom:

> What question would you like to ask Mitchell?

Some students wrote their questions in the box. Some dictated them to me during a group discussion after reading each piece.

Here is Mitchell's piece:

> I come from Ghana.
> My mum and my daddy are dead.
> So I lived with my uncle,
> and I had to learn carpentry,
> because he was a carpenter.
> When I finished the carpentry,
> I started to put myself about sports,
> so I started boxing.
> Through boxing,

I came to Britain,
at 27 years.

At the conference workshop, participants suggested what they might have asked Mitchell if they had been the tutor taking dictation.

An amalgamation of the tutors' questions follows in the first box. Compare these questions with those asked by Mitchell's co-students, listed in the second box.

What are you doing now?
Are you still boxing?
What were your feelings when you left Ghana?
How old were you when you went to live with your Uncle?
How old were you when your mum and dad died?
What did you think of Britain when you came?
What was it like in Ghana?
Different from here?
Did you get hurt when you were boxing?
Have you been back to Ghana?

Were you good at boxing?
Did you make money at boxing?
How far did you reach?
Did you have a promoter?
How long did you box for?
Why did you stop?
Did you like carpentry?
Is it a good job for getting work?
How old were you when your mother and father died?
How old are you now?

The questions asked by co-students were used by the authors as a basis for extending their writing. Some questions were answered on the spot in the group discussion; the authors were able, of course, to ignore questions they thought inappropriate, but most of them dictated sentences in answer to each question, and then edited the first draft to include the answers. These second drafts were returned for group reading and we repeated the process of asking and answering questions. This means that the final pieces were quite different from what would have resulted had I asked the questions.

There are various 'benefits' from this in terms of group literacy work: the readers' interest is kept through reading several versions of the text; people as they think up questions seem to re-read the text more closely than usual; the group pays attention to (and, by implication, values) each person's contribution. I think the process also opened up wider issues of control and discourse. Tutors asked about feelings, students about money. Tutors focused on cultural/ geographical change, students on boxing and carpentry. It's likely that the tutors were aware of their own interest in students' feelings and the need to produce texts which can be used, as a basis for group discussion. Many of us are experienced in organising discussion about migration and loss, for instance, and Mitchell's response to the tutors' questions would have been a 'useful' text. But presumably the students' questions do reflect the group's interests and they are not those predicted by tutors.

Maths

I want now to look at very different questions: the 'problems' of formal maths education. I have always invited students to bring real 'problems' to do with maths into the classroom. These are often insoluble, for a variety of reasons (examples are 'budgeting' when in debt and living on benefit, planting a present of apple trees in a small garden and understanding the national budget). They have the obvious benefits of (potentially) helping students solve the kinds of problems that may have contributed to their decision to study maths, providing a context for a range of maths techniques and using the students' strengths and knowledge. My concern here however is with the more usual 'problems' of maths, those designed for practice or testing of particular skills or techniques.

Marilyn Frankenstein's book *Relearning Mathematics* (1989) is directed to individual adult readers and asks readers to write their own review quizzes. It is dense and lengthy, and I re-wrote the chapter on fractions for students who may have difficulty with reading. In doing so I changed many of the examples from a US to UK context, but left the structure and some of the wording the same and used many of the original diagrams. I will give some examples here from students' work on writing their own review quizzes.

Here is one of Marjorie's (one of the class) fractions problems:

4 children have a bar of chocolate between them.
How will you divide it?

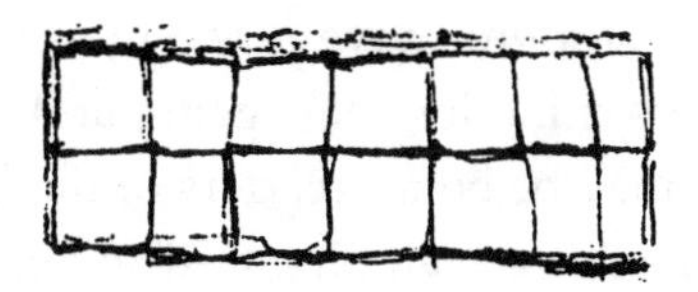

This question was beyond anything the group had come across in their work on the Frankenstein based fractions pack. In group debate they decided to ignore the fact that the chocolate pieces are different sizes, and assume they were all the same. Answers proposed included 14/4 each, 3 each and 2 left over, mother eats two pieces before dividing the bar, $3\frac{1}{2}$ each and $3\frac{1}{2}/14$ each. These solutions led to discussion about how they could all be right (you have to specify whether you are talking in terms of pieces or bars of chocolate). There was much laughter at how hard the problem was (Marjorie had miscounted; she intended there to be 12 pieces in all) and no-one downed tools and gave up.

Q5 Name each right fraction

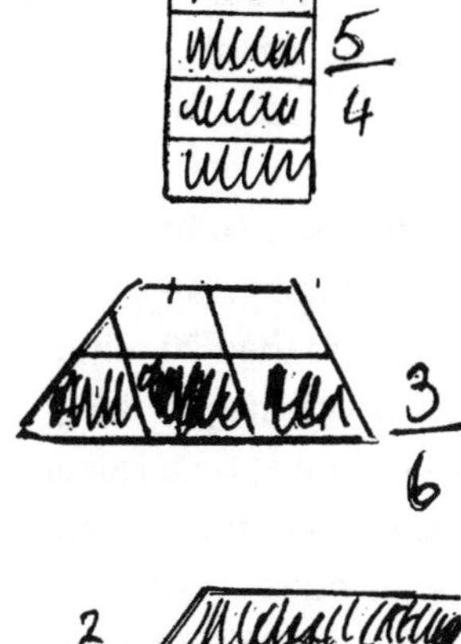

Toni, like Marjorie, modelled her questions closely on Frankenstein's examples. Here is one of her questions:

This arises from Frankenstein's use of *incorrectly solved* problems; she asks the reader to find the error. It is not clear in Toni's question whether the '3/6' diagram is a deliberate error or not. Three out of six pieces are shaded, but Frankenstein's examples included errors based on not knowing the pieces have to be the same size. One student suggested the 3/6 design showed a flat tile held at an angle from the viewer and drawn in perspective. Others challenged some who thought 3/6 was ok: "It doesn't look like half, though".

As far as I know the students had never before written maths questions; most maths students have probably *read* some during their school lives. They are writing in a new genre and I hope that by getting inside it they may be better readers of the questions set by others (that is, they may be developing critical language awareness in the context of maths study; see, for example, Clark & Ivanič (1997)).

Here is another of Toni's problems, written when the group topic was 'money':

> There are 12 people waiting to use the phone. Each person spends 28 p a minute and 3 minutes and 45 seconds on each call.
>
> a) how much does each call cost?
> b) how much do all the calls cost?
> c) how long does caller number 9 have to wait before it's his turn to use the phone?
> d) how long does caller number 11 have to wait?

This is as far from 'real life' as any textbook would dare to go, but Toni has caught the genre exactly and the group worked on the problem for more than 20 minutes. Again it was very difficult (for the person who wrote the problem, as well as others – no-one knew how to handle 45 seconds on a calculator).

Students seem to write, sometimes, problems that are 'too hard' for 'their level' but nevertheless find solutions to them. That suggests that much of the work we more usually do, with carefully written worksheets or choice of materials geared to exactly the individual's 'needs' at that time, may work to hold them back; they may be barriers rather than scaffolds.

Asked to set questions about fractions, Sandra found it difficult to work directly with the questions from the Frankenstein-based pack. So she wrote her own questions on a range of topics, including this one:

> Put these numbers in order of size with the smallest first:
> a) 305 b) 35 c) 350 d) 530 e) 503 f) 53

This is nothing to do with fractions, and I asked Sandra why she wrote it. She said this question was very hard for her (she is dyslexic and has problems with place value and zeros in big numbers) and she put it at the start of her set of questions in order to:

> Get them over and done with. I thought it would catch some people out, including myself, which it did.

It is clear here that Sandra is writing both for herself and for an audience of fellow students, and wants it to be hard. Writing questions gave her the opportunity to work out what she finds difficult, practise it, and get authorship of the problems rather than be defeated by them.

Interviews

A third kind of question is the sort tutors and/or researchers ask about students' experiences of their education. I intended to interview Pat and Cathy, for my research project, about their maths diaries. I couldn't find a way to fit this into their group's work, and asked them to interview each other instead. I gave them a few questions, with one catch-all on the end:

> Any advice for tutors or other students?

They both answered all the questions, but stopped to think (turning off the tape meanwhile) about this last question. I meant the 'advice' to be related to writing in maths classes (my research topic); Pat and Cathy interpreted it much more widely. Pat advised other students about the importance of regular attendance and the value of studying at college. Then she reverted to talking to Cathy (rather other students 'out there'), imitated my voice asking them to do the interview, and joked about students' not coming because they don't want to be interviewed. Suddenly, as I transcribed the tape I heard myself as others hear me – or at least as Pat hears me. Then Pat introduced Cathy, talking 'about how she feels about her studying', and Cathy said:

"Well, I've very much enjoyed the course, but sometimes I just think it's a *little* bit *wishy-washy*, that you sort of get told that you've done well because you're almost right, or you're on the right tracks. But in maths you're either right or wrong, *I* think."

There was more – an extremely coherent critique of both the teaching and the course content, followed by Pat intervening to defend the course. Again it seems that students are seizing a bit of time that is less directly controlled by the tutor to debate and work on issues that I understood quite differently. Being called 'wishy-washy' without warning was a horrible experience but the issues raised are much deeper than that; what I thought was a view of maths as socially constructed was understood, at least by Cathy, as tutor kindness (or patronage). Pat proposed the tape should be played to the group; it was not meant to be a private word in the teacher's ear. The recording quality was too poor for that to be successful, but some of the issues were raised. Had I been asking the questions in person, I would have had different answers.

Questions

Students' questions and answers in each of the areas – literacy, maths and research interviewing – merit much more consideration than I have given them here. Our questions might include:

- are students learning to ask textbook or teacher questions, or different and new kinds of questions, or all of them?
- when we offer work at the 'right level' are we limiting the students' opportunities for learning? who decides what level is 'right'?

- is using student-posed questions a signpost towards more democratic teaching or a top-dressing which does nothing about the basic power relations in a classroom or research project?
- how do we negotiate the teacher's emotional roller-coaster (up when the students take more of a lead, down when they use the lead it to criticise the teaching....)
- does writing your own questions help you deal with someone else's questions?
- how can we make more space for students to be critical of us?

Acknowledgements

Thanks to all the students who have given permission for me to use their work in research (some of the names have been changed) and to the participants at the RaPAL Conference workshop. Work discussed here took place in Southbelt Community Education classes and at the Bede Education Centre.

References

Brown, W. (1985). *The Student is the Expert*, in Frost, G. & Hoy, C. (eds) Opening Time. Manchester: The Gatehouse Project.

Cazden, C. B. (1988); Classroom Discourse: The Language of Teaching and Learning. Portsmouth, NH, USA: Heinemann.

Clark, R. & Ivanič; R. (1997). The Politics of Writing. London: Routledge.

Duffin, P. (1995). *Writers in Seatch of an Audience: Taking Writing from Personal to Public.* In Mace, J. (Ed.), Literacy, Language and Community Publishing: Essays in Adult Education (pp. 81–96). Cleveland, UK: Multilingual Matters Ltd.

Frankenstein, M. (1989). Relearning Mathematics: A Different Third – Radical Maths. London: Free Association Books.

Moss, W. (1995). Controlling *or Empowering? Writing through a Scribe in Adult Basic Education* In Mace, J. (Ed.), Literacy, Language and Community Publishing: Essays in Adult Education (pp. 145–170). Cleveland, UK: Multilingual Matters Ltd.

Time for "a great numeracy debate"

Dave Tout

RaPAL Journal No. 52, Autumn 2003

Dave is one of Australia's key adult numeracy workers and has been teaching mathematics to adults in Australia for over 20 years. He has worked as a teacher, trainer, university lecturer, curriculum and resource developer and researcher.

Introduction

The impetus for this article was the plenary address by Dr Ursula Howard, Director of the National Research and Development Centre in Adult Literacy, Numeracy and ESOL (NRDC) at the 2003 RaPAL Conference, *Celebrating Literacies? Language, Text and Power* at the University of Wolverhampton. In her talk Ursula explicitly stated that numeracy was an emerging priority and foresaw the need for "a great numeracy debate". She posed four main questions:

- is numeracy greater or smaller than maths?
- is it just skills and application?
- is it counting and sums, or algebra, reasoning and risk?
- how do we meet the need?

I shall address these by identifying different concepts of numeracy, suggesting where and how they fit into the *Skills for Life* agenda and by exploring some of the emerging issues.

What is Numeracy? And How Does it Relate to Maths?

There has been much discussion and debate about the word "numeracy" (O'Donoghue, 2003; Tout & Schmitt 2002 and Tout, 2001). The word itself was first coined in the UK Crowther Report (Crowther, 1959), where it was established to parallel literacy. Since then different countries have developed different meanings and usage. Interestingly, numeracy is still not recognised as a word by spell checkers in word processing packages such as *MS Word*.

The implications and consequences of using the term numeracy vis-a-vis mathematics (even if used with a defining term such as "everyday" mathematics) were frequently discussed. Numeracy was sometimes seen as the pretender – the junior, inferior partner to mathematics - because it was seen as just dealing with numbers and the four basic arithmetical operations. This view was commonly held in the school sector and also by the general public, the media and government. Practising ALBE (Adult Literacy and Basic Education, in Australia) teachers also questioned its use as an appropriate term to describe their work because of their perception that it was a lesser discipline than mathematics. Some teachers preferred the use of the term mathematics because it incorporated number skills and also included other strands of mathematics such as data and statistics, space and shape, measurement and even algebra.

On the other hand, the use of the term mathematics for many of the adult students brought with it the negative aspects related to traditional teaching of mathematics in schools. There it was often taught by rote, not related to any application or context, and not understood. Most adult numeracy students failed mathematics under this system and they return to mathematics education with much trepidation. Mathematics for them is associated with feelings of failure. It is a competitive and abstract subject filled with lots of work out of text books, stressful tests and little explanation of why and how many of the skills are used in society.

So there is a need for clarity here. What appears in most early descriptions of adult numeracy is that it is about the application and use of mathematics in everyday situations a functional view of mathematics education. More recent descriptions extend this and include a wider range of mathematics skills. They relate numeracy more explicitly to the needs and interests of the student within different contexts (home, community, workplace, study and training, etc.) where it is also seen to be dependent on the student's beliefs and attitudes. These descriptions also recognise that it is important to be able to communicate about the mathematics involved.

In recent years, the concept of 'critical' numeracy has also been introduced. Betty Johnston, in Australia, has argued that numeracy in fact incorporates, or should incorporate, this critical aspect of using mathematics:

> *'To be numerate is more than being able to manipulate numbers, or even being able to 'succeed' in school or university mathematics. Numeracy is a critical awareness which builds bridges between mathematics and the real world, with all its diversity.' (Johnston, 1994).*

She continues:

> *'In this sense . . . there is no particular 'level' of Mathematics associated with it: it is as important for an engineer to be numerate as it is for a primary school child, a parent, a car driver or a gardener. The different contexts will require different Mathematics to be activated and engaged in.' (Johnston, 1994).*

So the view of numeracy and mathematics which has developed in the adult literacy and basic education sector in Australia is one that sees numeracy as making meaning of mathematics and sees mathematics as a tool to be used efficiently and critically for some social purpose.

The Adult Literacy and Lifeskills Survey

One of the most comprehensive descriptions of numeracy that has been developed is in the numeracy component of the International Adult Literacy and Lifeskills (ALL) Survey. In their most recent paper, *Adult Literacy and Lifeskills Survey Numeracy Framework* (Gal et al, 2003), the authors fully describe the concept and meaning of numeracy. They, too, have taken the view that numeracy is the bridge between mathematics and the real world, and the authors have arrived at a definition for adult "numerate behavior" rather than of numeracy:

> *'numerate behavior is observed when people manage a situation or solve a problem in a real context; it involves responding to information about mathematical ideas that may be represented in a range of ways; it requires the activation of a range of enabling knowledge, behaviors, and processes.' (Gal et al, 2003).*

These views of numeracy are very different from earlier ideas of being just about numbers or about functional everyday maths. It is about using and understanding mathematics to make sense of the real world; using maths critically; and being critical of maths itself. It acknowledges that numeracy is a social activity. This is why it is okay to use the term "numeracy" to describe what we do – it is not downgrading, it is not inferior to mathematics – and as stated in the introduction to *Adult Numeracy Teaching,* "numeracy is not less than mathematics, but more" (Johnston and Tout, 1995).

What is the place of numeracy in policy and programmes?

Numeracy should have an equivalent focus to literacy and language within adult basic education. To achieve this, adult numeracy education must be supported by research, embraced in practice, and clearly communicated in policy at the national and local levels.

Unlike countries such as the USA, the UK has embraced numeracy in the language of policy making, where it has been named in programmes such as the *Skills for Life* and specified and supported in the core national curriculum. However, this needs to be supported in a number of ways. For example, numeracy needs to be viewed as a core essential skill by the public and the media. The public needs to see the importance of numeracy as a personal resource that can benefit the community at large, and it needs to acknowledge that it is not OK to fail or to be no good at mathematics.

Secondly, although numeracy may be named in policies and programs, practice in the classroom, in teacher training and professional development, in research and the development of teaching materials, can lag behind, especially in relation to literacy and ESOL provision. In these circumstances, numeracy needs to be given special attention and made a priority to make it a more equal partner in practice.

What is the role of teacher training?

The key to the development of comprehensive and successful numeracy teaching and practice is the support to teachers through teacher training and professional development. A major concern here has been that a large segment of ALBE teachers lack the pedagogical and content knowledge adequate to teach adults mathematics. Many are literacy or ESOL teachers with little or no training in teaching mathematics. On the other hand, school trained maths teachers may want to teach numeracy to adults in a more traditional way than that described for numeracy above. The approach, feelings and attitudes about mathematics education is entrenched in its own history and some have suggested (see Siemon 1989, Boomer 1986) that teachers need to be challenged and provided with different theories of mathematics education which will change them away from their traditional view of teaching mathematics. This will only happen through teacher training or professional development. So in both cases, any change in practice needs to begin by equipping potential teachers with both pedagogical and content knowledge of numeracy issues, classroom practice and assessment strategies.

The UK has recognised this need for adequate training and support of adult numeracy practitioners. Diana Coben and Noyona Chanda (Coben & Chanda, 1997) described the *ad hoc* nature of adult numeracy training in the UK, and listed a range of reasons why existing training was unpopular with practitioners: lack of experienced or qualified numeracy staff to act as leaders or mentors, lack of funding and lack of well-developed training materials. The authors believed that a programme for teacher training in numeracy should be developed that is based on articulated theory and research: and went on to recommend an Australian programme for teacher training developed by a university Adult Education Faculty and a teacher resource and information service in 1995. The programme, entitled *Adult Numeracy Teaching: Making Meaning in Mathematics* (Johnston & Tout 1995), was designed to establish a link between theory, research, and practice. Hopefully the Skills *for Life* agenda and the NRDC's support for its implementation will encourage such developments.

The role of curriculum, teaching materials and assessment

Another crucial aspect of improving practice is the writing of innovative curriculum, teaching materials and related assessment tools. It seems that much of this has already been established in the UK through the national core curriculum and the *Skills for Life* national strategy, including the development of national testing. However, there are important concerns with what has been described and specified in the core curriculum, especially in terms of a set of hierarchical maths skills which do not match adults' experiences, knowledge and development (see Ciancone & Tout, 2000 for further discussion of this issue). And there are major consequences in terms of assessing through a national testing scheme; teaching to the test can dominate the curriculum.

Yet, if the view of 'critical' numeracy described above is taken on board, then this can drive classroom practice and the curriculum and assessment can be made to fit what teachers see as best practice. These approaches need to be encouraged and supported through the teacher training and professional development described above.

The importance of research

Research in adult numeracy is sparse and so a research culture needs to be encouraged and developed. The UK has as strong a history as anywhere else in this respect: numerous reports and articles have described the development of adult numeracy research and practice in the UK (for example, Benn, 1997; Coben, 2000; Coben; O'Donoghue & FitzSimons, 2000). The key to useful and valuable research is to connect researchers with practitioners and to make sure the results are communicated to the field.

The UK is well situated to make this happen. First, it has been mainly behind the establishment and support of the group, *Adults Learning Mathematics – A Research Forum,* which is mainly a network of adult maths and numeracy researchers and practitioners from across the UK, with international connections. Second, RaPAL itself connects researchers and practitioners, including some in the area of numeracy. Third, with the establishment of the NRDC it appears that there will be a strong developmental strategy in relation to research in practice. So there is a good base for future work and development.

Conclusion

In terms of the four questions posed at the outset, if a 'critical' concept of numeracy is adopted as the philosophy and approach to the teaching of adult numeracy in the UK, then the answers to the first three questions are:

- numeracy is not less than maths, but more;
- numeracy certainly incorporates more than just skills and applications;
- it covers more than counting and sums, or algebra, reasoning and certainly involves taking risks, both by teachers and students.

Given the Skills *for Life policy* drive in the UK, including the establishment of the NRDC, and the existence of groups such as RaPAL and ALM, now is the perfect time to have the "great numeracy debate". This should ensure that all the issues and concerns noted above are addressed and that answers are provided to the fourth question of how to meet the current needs.

References

Benn, Roseanne (1997) *Adults count too: mathematics for empowerment,* Leicester, UK: National Institute of Adult Continuing Education (NIACE).

Boomer, Garth (1986). From Catechism to Communication: Language, Learning and Mathematics. In Bell, D. and Guthrie, S. (eds.) (1994), *An Integrated Approach to Teaching Literacy and Numeracy.* Sydney: NSW TAFE Commission.

Ciancone, T. & Tout, D. (2000). Learning Outcomes: Skills or Function? In *Proceedings of the Seventh International Conference of Adults Learning mathematics A Research Forum, July 2000,* Tufts University, Massachusetts, USA. Cambridge, MA: National Centre for the study of Adult Learning and Literacy, Harvard University.

Coben, Diana (2000) Numeracy, Mathematics and Adult Learning Practice. In: Gal, Iddo (Ed.) *Adult Numeracy Development: Theory, Research, Practice,* Cresskill, NJ: Hampton Press.

Coben, Diana & Chanda, Noyona (1997) Teaching 'Not Less than Maths, but More': an overview of recent developments in adult numeracy teacher development in England, with a sidelong glance at Australia. In *Teacher Development, Vol. 1,* No. 3, 1997, pp. 375–392.

Coben, Diana; O'Donoghue, John; & FitzSimons, Gail E. (Eds.) (2000) *Perspectives on Adults Learning Mathematics: Research and Practice.* Dordrecht: Kluwer.

Crowther Report, UK (1959) 15-18: *report of the Central Advisory Council of Education (England) vol.* 1, London: HMSO.

Gal, Iddo; van Groenestijn, Mieke; Manly, Myrna; Schmitt, Mary Jane; & Tout, Dave (2003). *Numeracy conceptual framework for the international Adult Literacy and Lifeskills (ALL) Survey,* Ottawa: National Center for Educational Statistics and Statistics Canada.

Johnston, Betty. (1994). Critical Numeracy, in *FinePrint, Vol.* 16, No. 4, Summer 1994, Melbourne: VALBEC.

Johnston, Betty and Tout, Dave (1995). *Adult Numeracy Teaching: Making meaning in mathematics,* Melbourne: National Staff Development Committee.

O'Donoghue, John (2003) Mathematics or Numeracy: Does it really matter? in *Adults Learning Mathematics Newsletter,* No. 18, March 2003, London: Adults Learning Mathematics (ALM).

Schmitt, M.J., (2000). Developing Adults' Numerate Thinking: Getting Out From Under the Workbooks in *Focus on Basics, Vol.* 4, Issue B, Sept 2000 Boston, MA: National Center for the Study of Adult Learning and Literacy, Harvard University.

Siemon, Dianne, (1989) Knowing and Believing is Seeing: A Constructivist's Perspective of Change, in Ellerton and Clements (Eds.), *School Mathematics: The Challenge to Change,* Geelong: Deakin University.

Tout, Dave (2001) 'What is numeracy? What is mathematics?' in Gail E. FitzSimons, John O'Donohue & Diana Coben, *Adult and Lifelong Education in Mathematics: Papers from the Working Group for Action (WGA) 6, 9th International Congress on Mathematics Education, ICME 9,* Melbourne: Adults Learning Mathematics (ALM) and ARIS, Language Australia.

Tout, D. & Schmitt, M.J. (2002). The inclusion of numeracy in adult basic education. In J. Comings, B. Garner, & C. Smith (Eds.), *Annual review of adult learning and literacy (Vol.* 3, pp. 152–202). San Francisco: Jossey-Bass.

Numeracy, literacy and mathematics

Richard Barwell, University of Bristol
RaPAL Journal No. 53, Spring 2004

What is numeracy? In his recent article, Dave Tout (2003), contributing to a debate about numeracy, offers an Australian perspective. For Tout, numeracy is seen as 'making meaning of mathematics' and mathematics as a 'social tool' (p. 10). Tout is, therefore, arguing for a distinction between numeracy and mathematics. I was not, however, completely clear of how Tout's distinction might work. 'Making meaning of mathematics' for example, is, ideally, what happens in mathematics classrooms. Does that mean that such classrooms are engaged in numeracy? Or mathematics? Or both?

If definitions are to be useful, they need to draw useful distinctions. Tout's conclusion that numeracy incorporates more than skills and applications, and that numeracy covers more than counting, sums, algebra and reasoning and involves taking risks could all apply to a definition of mathematics or mathematics education. Indeed Tout's brief outline of a critical approach to numeracy is very similar to Skovsmose's (1994; Alrø & Skovsmose, 2002) development of the idea of critical mathematics education (in which he proposes the term 'mathemacy'), which draws on a Freirean perspective to argue that mathematics is an important means of understanding an increasingly technologised world.

In this article, I want to argue an alternative position, that numeracy should be seen as a distinctive form of literacy. This approach distinguishes numeracy from mathematics and is derived from one of the origins of the term numeracy in literacy research. Let me start, however, by looking at some of examples of activities that involve numeracy.

What counts as 'numeracy'?

The Basic Skills Agency's Adult Numeracy: Core Curriculum (BSA, 2001, pp. 10-11) includes a number of images under the heading 'Numeracy', including a cheque, a payslip, a diet plan, a menu and a road sign. Do these images suggest activities involving the application of mathematics or mathematics itself, or literacy, or both?

The numeracy curriculum is based around three areas or 'capabilities':

- understanding and using mathematical information
- calculating and manipulating mathematical information
- interpreting results and communicating mathematical information

It is notable that these capabilities concern 'mathematical information' rather than mathematics, implying a connection between literacy and numeracy. Examples from the curriculum itself, however, suggest some confusion, with some items being more literacy-related (reading tables, recognising shapes) and other more mathematical (discussing mental calculation methods).

In their research on literacy, Barton & Hamilton observed many examples of activities involving numeracy (1998, p. 177; I have abbreviated the list):

- gardening, cooking, making clothes, knitting
- following current affairs with charts and diagrams in the newspaper
- health, medicine, contraception, dieting
- doing finances, dealing with bills and back accounts
- house repairs

- astrology
- arranging travel
- shopping
- map-reading
- using household technology, such as video machines.

Again, it is striking that these are all also examples of literacy activities. Following a recipe or a map involve making sense of texts of different kinds. These texts could be seen as having something in common, such as the use of numbers or charts or other 'mathematical information', but they are still text genres that must be interpreted.

Numeracy in literacy research

The term 'numeracy' makes several appearances in ethnographic research into literacy (see, for example, Barton, 1994, pp. 101–103; Barton & Hamilton, 1998, pp. 176-180; Street, 1995, p. 136). This work is based on a social view of literacy, which is seen as concerning the social practices involved in making meaning from or around text. Barton & Hamilton (1998), for example, see literacy as:

> *primarily something people do; it is an activity, located in the space between thought and text. Literacy does not just reside in people's heads as a set of skills to be learned, and it does not just reside on paper, captured as texts to be analysed. Like all human activity, literacy is essentially social, and it is located in the interaction between people. (p. 3)*

Tout's view of numeracy parallels a social idea of literacy by, in effect, replacing 'text' with 'mathematics', resulting in the formulation of numeracy as the social practices of 'making meaning from mathematics'. I propose a different approach: consider the above quote again, but with 'literacy' replaced with 'numeracy'.

> *Numeracy is primarily something people do; it is an activity, located in the space between thought and text. Numeracy does not just reside in people's heads as a set of skills to be learned, and it does not just reside on paper, captured as texts to be analysed. Like all human*

activity, Numeracy is essentially social, and it is located in the interaction between people.

This statement seems to me to usefully describe the process of interpreting the texts involved in the activities mentioned above. Indeed formulating this statement raises the interesting question: what kind of text? The texts involved in numeracy could be described as 'numerate' or perhaps 'mathematical'. Such texts must be interpreted, and this interpretation is a social activity. Numerate texts include the examples given above: recipes and cheques and so on. They also include texts such as rulers or protractors, coins and credit cards, calculators and computers, fingers and thumbs. It may be that, in order to interpret these texts, people draw on mathematical practices. Mathematical practices, for me, involve generalising and formulating abstractions, identifying and describing patterns and relationships. Thus, whilst following a recipe, the cook may double the quantities in order to produce a larger pudding (see Barton & Hamilton, 1998, p. 178). This halving involves drawing on patterns and relationships between abstract numbers. The cook, however, is not making sense of mathematics; she is making sense of a *text*, using mathematical *and* literacy practices to do so. For me, however, the literacy practices are primary in this situation. The cook must relate her interpretation to the recipe genre, to her experience of cooking, of discussing or learning recipes from friends and relatives or of watching cookery demonstrations on television. Within this frame, she draws on mathematical practices as part of the process of interpreting the text.

Discussion

I have argued that numeracy is best seen as a form of literacy, the literacy of 'numerate' texts. From this starting point, the idea of numeracy practices can be defined as the social practices involved in making meaning from such texts. In this light, numeracy overlaps with mathematics. Neither numeracy nor mathematics entirely encompasses the other. Mathematical practices may be involved in making meaning from numerate texts, but not always. If I slow down at a 30 mph speed limit sign, I have interpreted the sign (numeracy), but probably not used any form of mathematics. My response is part of the social practice of driving. On the other hand, if I argue with a shop assistant over the change she has given me, going over an arithmetic calculation as part of the discussion, mathematics has been used to develop a shared meaning for a set of numerate texts (price tags etc., coins etc.). Conversely, mathematics may be conducted without reference to any text at all, such as when discussing mental methods of calculation. The inter-relationship between numeracy and mathematics, as I have defined them, makes it difficult to separate them for the purposes of a curriculum. The overlap perhaps explains the nature of BSA adult numeracy curriculum, for example, which includes both literacy and mathematical elements. Finally, the close relationship between literacy, numeracy and mathematics, and the nature of the current debate, suggests a need for greater discussion between practitioners and researchers working in literacy research and teaching and mathematics education.

References

Alrø, H. & Skovsmose, O. (2002) *Dialogue and Learning in Mathematics Education: Intention, Reflection, Critique*. Dordrecht: Kluwer.

Barton, D. (1994) *Literacy: An Introduction to the Ecology of Written Language*. Oxford: Blackwell.

Barton, D. & Hamilton, M. (1998) *Local Literacies: Reading and Writing in One Community*. London: Routledge.

BSA (2001) *Adult Numeracy: Core Curriculum*. London: The Basic Skills Agency.

Skovsmose, O. (1994) *Towards a Philosophy of Critical Mathematics Education*. Dordrecht: Kluwer.

Street, B. V. (1995) *Social Literacies: Critical Approaches to Literacy in Development, Ethnography and Education*. Harlow: Longman.

Tout, D. (2003) Time for 'a great numeracy debate.' *Rapal Journal* 52, 9–13.

B2 ESOL

Further RaPAL references about ESOL

Ade-Ojo, G. 'The New National ESOL Curriculum: Is it Determined by Convention?', *RaPAL Journal* No. 53, 2004.

Loh, P.P.Y. 'Creative Respect For Learners, People of Diverse Cultures and the Earth', *RaPAL Journal* No. 54, 2004.

Gujarati literacies in Leicester

Arvind Bhatt
RaPAL Bulletin No. 25, Autumn 1994

Arvind Bhatt works on the Multilingual Literacies project and is based in Leicester

Introduction

My interest in Gujarati literacies began when I was appointed as the country's first full-time Gujarati teacher at an inner city community college in Leicester. The purpose of introducing Gujarati in the mainstream curriculum at the college was to facilitate communication in Gujarati between the generations in the home and the community which the college served. Later, CSE and GCSE courses were also instituted. Realising that such an initiative would require an extensive network of support, the Local Education Authority encouraged other schools and colleges to consider similar courses; this, in turn, required those who were involved in teaching Gujarati and other 'community languages' in the community to work together to evolve new methods of teaching these mother tongues and to share the meagre resources. Thus many adults from different backgrounds came together and formed formal and informal support groups, conducted their affairs in English and in their mother tongues and thereby renewed their interest in their own languages.

I am now working on a research project which is exploring the literacy practices of Gujarati speakers in Leicester. Before describing the project and discussing this aspect of adult literacy, a brief background of Leicester's multilingual make up may be useful.

Multilingual Literacies in Leicester

Successive settlers have brought their languages to Leicester over a period of the last 80 years or so. Thus after the first and the second world wars, Polish and Jewish settlers added their languages to the English of the residents. Also after the second world war there began a trickle of migrants from the Indian sub-Continent and from East Africa. This process accelerated as the former colonies and protectorates became independent states, culminating in the exodus of south Asians from Uganda in 1971–72. As there were many Gujaratis and other Asians in and around Leicester (for example, in Loughborough, some 12 miles to the north) and as the new settlers could occupy an economic niche there, the Gujarati speaking community particularly set about establishing themselves permanently, despite some opposition from the residents. Most of the Uganda Asians were literate in Gujarati and English. Marett (1989) has described the process of this settlement and the role of various welfare organisations in facilitating the process. Although the role of both Gujarati and English literacies are not fully explored by Marett, it is evident these literacies played a central role in mediation and communication both between and among communities in Leicester, and continues to do so today in the daily lives of Gujarati speakers.

Gujarati is now, in fact, the second most widely spoken language (after English) in Leicester with some 15% of its population identifying it as their first language; this estimate is based on the 1983 Survey of Leicester. The Gujarati speaking community includes groups with different religious affiliations: Hindus, Muslims and Christians. Unfortunately, there has been no linguistic survey since then (and the 1991 Census did not seek this information either), but recent estimates suggest that this figure is much higher now. Leicester today is a multilingual and a multiliterate city.

A Research Project in Progress

The project that I am involved in is entitled *Multilingual literacy practices: home, community and school*. It has been funded by the ESRC from May 1993 to January 1995 and is based at Lancaster University. Other members of the project team are David Barton, Marilyn Martin-Jones and Mukul Saxina.

The aim of our project is to study the uses of languages and literacies of a small number of Gujarati speaking households in Leicester. Our approach has been an ethnographic one. We have conducted in-depth interviews with members of these households about their literacy histories and their everyday literacy practices, including their attitudes and values. We have observed Gujarati speakers in the community, and in the home, and have followed some of the children to their primary, secondary and community schools. We have completed the data collection and are now engaged in the data analysis. Much of what follows is based on our observations during the data collection stage of the project.

Literacies Across the Generations

There are important differences across the generations, among the Gujarati communities, in terms of access to English and Gujarati literacies and in terms of the way in which they draw on their literacies in their everyday lives in Leicester. In our first project paper (Bhatt, Barton and Martin-Jones, 1994), we described the changing literacies in the lives of two men in one generation. In this older generation – those who arrived in Britain in the 1970s – some had the opportunity to develop their literacies in both Gujarati and English, whilst others were literate in English but only had oracy in Gujarati. Women generally were less fluent in English than in Gujarati which was used in both Gujarat and in East Africa in the household and the community domains. English literacies were used primarily in the official and political and public domains where women played a relatively minor role. The status of English literacy was thus of a higher status than that of Gujarati. This power imbalance between the public, official language (English) and the private, household and community language (Gujarati) still remains.

When the 'older' generation arrived in Leicester, they concentrated on settling down and developing new literacies to cope with the new situation. They thus learnt to cope with the literacy demands of Social Security, Welfare Benefits, school correspondence, forming associations and clubs, getting planning permission, using the media. They also had to deal with the hostility of some sections of the resident community. Some learnt the new technologies, some entered new work environments and took on new literacies. Some kept diaries in Gujarati, wrote letters to Gujarat, and to those left in East Africa, in English and in Gujarati, and printed wedding invitations in English and Gujarati.

In the religious and cultural spheres, Gujarati was predominant, at least among the Hindus. In the temple, three languages and literacies were (and still are) used: Hindi, Sanskrit and Gujarati. Later, English also began to be used, in notices and newsletters and in instructions about removing shoes and keeping the place tidy. Sanskrit was used for reciting *slokas* (religious verses) and readings from the Gita and other religious texts. In the mosques, both Urdu and Gujarati literacies were used, Arabic being confined to Quranic readings. Among the small number of Gujarati Christians, hymns, translated from English, were written or printed and sung in Gujarati. A Gujarati Bible was in regular use, some minutes of meetings were written in Gujarati and some correspondence was in Gujarati. As there were other Christians who spoke Hindi or Marathi, much writing and talk during meetings was in English.

In one aspect of community life, when cultural practices were being re-established in Leicester, women took a leading role. They brought with them recipe books in Gujarati and added new recipes using vegetables available then in Leicester market. They began collecting *garbas* and *raas* (songs to accompany dancing) for the annual *navratri* festival. Wedding songs, too, were collected and written down to be sung in competition with Hindi film songs at engagements and marriages. Some older women also filled ordinary exercise books with *bhajans* and other devotional songs to be sung in the temples.

For those Gujaratis who had oral fluency but were unable to develop literacy in Gujarati either in East Africa (where the language was replaced by Kiswahili in the curriculum) or in Britain (where Gujarati was seen more as a home language and as an obstacle to acquisition of English), the only recourse in learning songs, *doohas* (couplets recited during a *garba* or *raas* at the navratri festival), hymns or *dua* (Muslim prayers) was to transliterate the Gujarati in the Roman script, devising their own codes to transfer the Gujarati phonetics into English. There is an example of Yusufbhai, in our Working Paper, who used his Gujarati literacy to read the Muslim prayers and other texts written in Urdu, a language he cannot read. The ease with which these multilingual individuals moved between literacies in different languages is striking.

Thus Gujarati played a central role in fostering community cohesion across all the three religions in the older generation.

The Younger Generation

As the Gujarati community became established and accepted by the existing residents, older Gujaratis became aware that the younger generation used less and less Gujarati as they progressed through the English education system. Gujarati literacy was gradually being washed out of the community. The older generation felt both the generational and linguistic distance between themselves and their children and grandchildren. A recently coined saying in Gujarati expressed their fears well: when we moved from Gujarat to East Africa, we lost our *sagpan* (network of relations), when we were driven from Africa, we lost our *sampati* (wealth, acquisitions) and now in England we are losing our *santan* (children). In response to this situation, the Gujarati Hindu community organised Gujarati classes in the evenings and week-ends. The Muslim community instituted *Madressas* (religious schools) where Urdu and Arabic is taught. The Christians have not yet organised similar classes, probably because they are so few in number and scattered across Leicester. Moves were afoot to introduce the teaching of literacy in Gujarati in the state schools but this was resisted by most British educationists.

Voluntary teachers have done their best to support Gujarati literacies among the younger generation. Some mothers tried to teach their children to read and write Gujarati but were discouraged from doing so by some teachers, and in one case by a health visitor sent by the school, arguing that teaching of Gujarati will slow down the acquisition of English by the child. Even so, some state schools such as the college mentioned earlier in this paper, did manage to introduce Gujarati in the timetable. Teaching of Gujarati in the community and the state schools led the adult teachers to form associations to advance their cause. They thereby acquired new literacies in terms of adopting modern foreign language teaching methods for Gujarati and designing syllabi to conform either to the state norms or to the norms in Gujarat. New type of resources were also devised and Gujarati owned book shops imported text books from Gujarat and Bombay. More interestingly, information technology began to be used for word-processing in Gujarati and several versions of Gujarati software began to be designed as the market for these products were recognised. The adults helped to design such software, trained themselves, or did this through mutual help, and insisted on print equality at least with English. It is a measure of some advance that one of the adults and his daughter in our sample used this new technology to design and print a Gujarati-English bilingual wedding invitation (a *kankotri*) for the occasion. The Gujarati print media already use this modern technology to improve the presentation of the newspapers.

Some Everyday Literacy Practices of Selected Individuals

A good way to illustrate the literacy practices which I have outlined above would be to briefly describe the everyday literacies of a few individual Gujaratis. The names have been changed to preserve confidentiality.

Sarlaben

Sarlaben is in her 40s, married with two children, and is a full time teacher in a primary school. She has come directly from Gujarat to settle in Leicester with her husband who is from East Africa. She prepares her lessons in English and Gujarati, sometimes using a Gujarati typewriter (which is awkward to use). Most of her Gujarati work is hand-written. She is compiling a Gujarati text book for the younger age pupils and hopes to publish her work in the near future. She is also in charge of a community school where she teaches Gujarati on Sunday mornings. In her spare time, she sings, professionally, for entertainment and for *sanji na geet* (pre-wedding songs). These songs she either knows by heart or she selects from her collection both printed and hand written. She also helps her sons with their homework in English and in Gujarati. Like many other families, they write letters jointly to their friends and relatives in Gujarat or East Africa in both Gujarati and English. She is also a regular reader of Gujarati books.

Rajivbhai

Rajivbhai is in his 50s and came to Leicester from Uganda where he was a teacher in a secondary school. For a time he pursued the family trade of goldsmithing but also did some part-time evening teaching in Gujarati. Later he was appointed as a teacher of Gujarati at a secondary school where he eventually took charge of the Gujarati GCSE examinations. In his work, he uses both Gujarati (for teaching) and English (for administration). He is a regular reader of Gujarati newspapers and magazines from which he culls out teaching materials for use in the class room, in GCSE examinations and for training other teachers of Gujarati. His filing cabinet at home and in his office at school bulges with such materials. He is writing a text-book for the Gujarati GCSE course. He is an avid reader of Gujarati novels, which he borrows from the local libraries, and maintains contact with educationists in Gujarat by writing letters in Gujarati. He is also adept at using the Gujarati software on his computer. He is a member of the Gujarati Teachers' Association, for which he sometimes takes notes of the meetings, writes up the minutes and conducts some correspondence using both English and Gujarati.

Ramaben

Ramaben describes herself as a housewife. She is in her 40s and has three children. Over the last two years she has been very busy with arranging first the engagement and then the wedding of her eldest daughter. She began to make a list of all her relatives in Gujarat and in East Africa, as well as here in Britain, with the help of her husband and other close relatives. She often used informal contacts and the regular newsletter, which is circulated within her *gnati* (caste group), to track down distant relatives and friends of the family. There were some 1800 names last time I inquired. This list is now fast becoming a resource for other families in the *gnati*. She likes to read Gujarati magazines for women and sometimes novels if she has time. Although she knows English, she prefers to speak, read and write in Gujarati. She maintains her links with Gujarat and East Africa.

Yunusbhai

Yunusbhai, a Muslim, came to Leicester from Gujarat about four years ago. He married Firozaben who was brought up in Malawi. Yunusbhai is trilingual, at least. As well as English, he is fluent in both Urdu and Gujarati which he describes as his mother tongue (*matrubhasha*). He is an accountant by profession and uses English to write official letters. On the telephone, he starts with English but quickly changes to Urdu/Hindi or Gujarati if the client speaks one of these languages. Sometimes he takes notes in Gujarati. He writes for a newsletter for his *kom* (sect) in Urdu but uses Gujarati to proof-read and make amendments. He is involved in work for his mosque and uses all three literacies to make lists, notes, hand write notices and correspondence. He writes letters to his friends in Urdu and Gujarati. He can read Arabic and has studied Farsi.

Rev. Jansari

Rev. Jansari is a Christian minister in an inner city church. He was a teacher in Uganda but later took orders and settled in Leicester together with other Asians from Uganda. As a minister of the church, he deals with mainly English speaking parishioners, but he is also a *dada* (grandfather) for the Gujarati speaking Christians in and around Leicester. He reads the Bible, sings hymns and says prayers in Gujarati when he takes a service for Gujaratis who come from other cities in the Midlands. Realising that many of the younger generation may not be able to read Gujarati, he uses the Roman script to phonetically transcribe the Gujarati prayers and hymns. He writes letters in both English and Gujarati but makes notes in his diary in English. In his house he is surrounded by books, newspapers, newsletters and magazines in English and in Gujarati.

Support for Gujarati Literacies in Leicester

Gujarati is now taught in some state schools in Leicester as a National Curriculum subject in the Modern Foreign Languages departments. Until recently (1992), these languages had central support from the Local Education Authority but now the teachers rely on self-help (in contrast to European languages). Community schools also run Gujarati classes for both adults and children. The numbers attending these classes fluctuate and hard information is difficult to collect. Some Colleges of Further Education also run Gujarati classes for beginners, for GCSE examinations, for literacy only or just for oracy. Bi-Lingual Skills courses are also organised. For this year (I994/95), the Youth and Community Education Service listed 21 classes relating to Gujarati as compared with 183 for French, 100 for German, 45 for Italian and 49 for British Sign Language, for example. There are also English as a Second Language courses for adults, and training courses for the unemployed in industrial skills and English combined.

There is thus some support from both the state and the community itself to support multilingual literacies. However, support for literacies in English and European languages take up most of the resources. Literacies in Gujarati for both adults and young people is *ad hoc* and unsystematic. This is very much my own impression, but this paper suggests that these literacies are greatly valued by the community and provides an element of social and family cohesion together with religion and cultural traditions. It is an area where further research is urgently needed.

References

Bhatt, A., Barton, D. & Martin-Jones, M. (1994) Gujarati Literacies in East Africa and Leicester: Changes in Social Identities and Multilingual Practices, Working Paper Series No. 56, Centre for Language in Social Life, Lancaster University.

Leicester City and County Councils (1983) Survey of Leicester: Initial Report of Survey, City and County Councils, Leicester.

Marett, V. (1989) Immigrants Settling in the City, Leicester University Press.

The literacy practices of Welsh speakers

Kathryn Jones
RaPAL Bulletin No. 25, Autumn 1994

Kathryn Jones is a post-graduate student at Lancaster University

This article describes my MPhil/PhD research on the bilingual literacy practices and values of Welsh speakers living in an urban and a rural community in North Wales. I'm going to begin by describing the context of this research and how the concerns of various language groups have shaped the scope and focus of this project. I'll then give a brief account of the way that I'm conducting this research and describe some of the literacy practices that I have observed so far.

The context: A change in status and a change in texts

The status of the Welsh language in the public domain and Welsh speakers' opportunities for speaking, reading and writing Welsh in the activities of their everyday lives have undergone many changes in the latter half of this century. Changes in the status of Welsh in different domains are described in some detail in The Language Forum's document "*Strategaeth Iaith*" (Language Strategy) published in 1991. These changes have come about as a result of many campaigns by various pressure groups (again see "*Strategaeth Iaith*" for details). The efforts of these groups, and the Welsh Language Acts of 1967 and 1993, have extended the role of Welsh in public domains, particularly with regard to literacies. The 1967 Welsh Language Act gave documents in Welsh "equal validity" with English ones and the

1993 Welsh Language Act now requires that every public body in Wales treats the English and Welsh languages "on a basis of equality" in "the conduct of public business and the administration of justice in Wales" (HMSO: 1993). As part of providing a bilingual service, public bodies have been producing bilingual information leaflets, posters and signs. This means that there is an increasing range of Welsh and bilingual texts used in the public and, to some extent, private sector that were previously only produced in English.

Defining the project

I am Welsh and my commitment to research on language use in Wales is that it informs the efforts of groups and organisations who are concerned in their various ways with 'promoting' the use of Welsh in Wales. I have discussed this research project with the members of various organisations. Each group has its own specific agenda and particular concerns but they all share an interest in discovering how people use Welsh and the factors which affect people's language use. Many of them have a specific interest in the way that the policies and practices of various public bodies are changing as a result of the new Welsh Language Act and how this is affecting the way people use language in their interaction /contact with these institutions. It is in the light of these interests that the aims of this project are:

- to investigate the language use and literacy practices of a sample of Welsh speakers, both in general and with a focus on institutional or official literacies in particular
- to investigate how their language use and literacy practices are culturally, historically and institutionally shaped
- to investigate language use in particular events in terms of literacy practices and the talk around texts.

My research is located in a small market town in north-east Wales; called Rhuthun, and the outlying rural areas of Bontuchel and Cyffylliog. Cyffylliog and Bontuchel are agricultural communities and most of the Welsh speakers living here have some connection with farming or the local forestry commission. In Rhuthun, apart from locally born Welsh speakers there are also a lot of 'professional' Welsh speakers (teachers, librarians, civil servants) who have moved to Rhuthun from other parts of Wales. I have been living in Bontuchel with a farming family since the beginning of March 1994. I have become involved as much as possible in the social life of both communities. Most of my observations of people's literacy practices have been carried out as a participant in a range of events. I am collecting data by taking photographs, collecting examples of the texts people use and making oral recordings of events, by talking to people about the things they do and keeping extensive field notes.

Literacy Practices – some preliminary observations

So far in this research, I have been participating in the social events of these communities and observing literacy practices in terms of the event as a whole and the activities of particular individuals. Many of people's literacy practices in Welsh are connected to their participation in Welsh social events and the role they play in those events. I will now identify some of these events and some of the literacy practices connected with them.

Chapel and Church

The Welsh chapels in Cyffylliog, Bontuchel and Rhuthun are a central focus of many people's Welsh literacies. Chapel services are structured around written Welsh texts. Apart from the minister's prayers, the occasional congregational recitation of "*Ein Tad . . .*" (the "Lord's Prayer") which people know by heart, some announcements and people's whispered greetings to other members of the congregation, all of the service involves some kind of reading. For most members of the congregation this just means reading and singing the hymns – both the words of the hymn and the sol-fa or old notation (most of the congregation use sol-fa) and listening to someone else read a prayer or something from the bible. In a *cyfarfod gweddi* (prayer meeting) when a minister is not present, two or three members of the congregation take the service. They usually choose two hymns which they read out before the organist plays the music and the rest of the congregation stands up to sing. They also read something from the bible or another religious text and they say a prayer. The prayers they say are also read aloud. A few write their own prayers but most read from a book of prayers that they have at home or a book they have borrowed from the library.

In Sunday School, both adults and children read out parts of the bible, and answer questions in the lesson book about the piece they have just read. The Sunday School organiser writes in his or her book which member of the congregation begins and ends the Sunday School each week. She or he also writes down the order of next week's services which she or he announces at the end of the service together with other announcements written on pieces of paper and given to one of the chapel deacons. A register is taken in each Sunday School class. The money everybody brings with them is written down in the register. The Sunday School secretary collects the registers and the collection and, at the end of the service, announces how many have attended, which class has learnt the most lessons and how much money was collected. The secretary also announces who will begin the service next week.

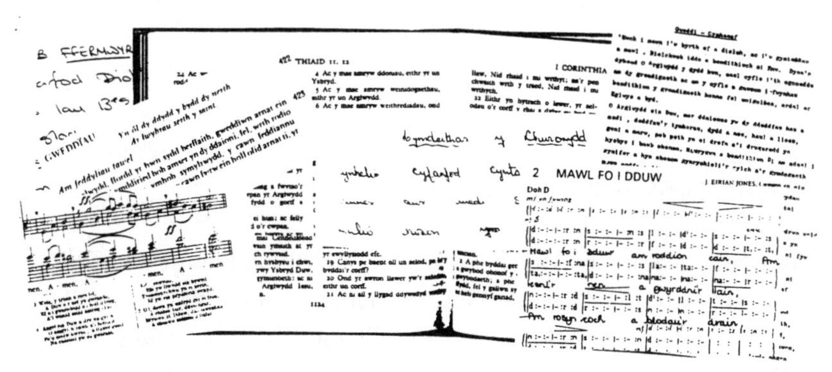

Figure 1.

Some Welsh speakers go to church. Although most churches have some Welsh or bilingual services, the majority of the services are in English. While the administrative texts of the chapel are in Welsh, in church they are in English. Chapel and church porches have notices, hand written and typed: notices of forthcoming events and lists of who is playing the organ each month, who is responsible for the flowers and so on. In chapels these notices are in Welsh while in church they are in English. Both church and chapel produce some bilingual notices. These are often for events to which people outside the church or chapel are also invited. The church in Cyffylliog, for example, had a bilingual notice posted on the church gate and village notice board notifying all the villagers of the church's harvest festival services in both English and Welsh. The chapel Sunday School secretary produced two posters advertising the chapel's annual Sunday School trip. The Welsh version was put up on the chapel notice board and the English version was displayed in the Post Office.

Figure 2.

Chapel clubs and societies

For many people, the Welsh chapels in Cyffylliog, Bontuchel and Rhuthun are central to their Welsh social activities. These social activities involve people in particular literacy practices using Welsh rather than English. Chapel members organise various clubs and societies for all age groups. There are clubs for both primary and secondary school children. In these clubs the children have quizzes and games, they learn songs and they write and perform sketches and so forth. Most of these activities are prepared by adults who write the songs and sketches or choose ones written by other people. The adults also prepare their own quizzes, writing the questions themselves.

For the adults there are literary societies which meet once a month and have guest speakers or performers. The literacy practices of the *'gymdeithas'* (society) meetings depends on the activities of each particular meeting. Each society, like all the other clubs and societies have committee meetings to decide upon the programme of the year and organise events. The president or chairperson, secretary and treasurer each have their role in this organisation which involves them in particular literacy practices which include writing notes, letter writing, preparing posters and tickets, writing programmes, writing lists of things to do, keeping accounts, writing cheques and so on.

Other social events

Other regular social events in these communities which take place in Welsh and which involve people in Welsh literacy practices are, for children and young people, the *URDD* (Welsh Youth League) and *ffermwyr ifanc* (Young Farmers). *URDD* and *ffermwyr ifanc* meetings are conducted in Welsh and their various activities, treasure hunts, games, quizzes, competitions, discos and concerts involve a range of literacy practices in Welsh. Members of the *URDD* and *ffermwyr ifanc* take part in the *URDD eisteddfodau* (festivals) competitions for which they learn songs, recitations, write poems, plays, short stories, write and perform sketches and so on. Applications for the various competitions have to be made by filling in forms and most competition entries get a written adjudication of their entry.

Merched Y Wawr (literally "Daughters of the Dawn") is a nation-wide women's organisation in Wales. There are branches in Bontuchel and Rhuthun which meet once a month. Each month they have a guest speaker and each month the meetings begin with singing the *Merched y Wawr* song and conducting the branch business. Branch business is organised by the president and secretary who also refer occasionally to the treasurer. The secretary begins by reading out the minutes of the last meeting, the president then reads out letters the secretary has received during the week and the contents are discussed and acted upon by the group. They sell tickets for forthcoming events and make lists of group members who are prepared to take part in various activities.

There is also a men's group, a 'dinner club', which meets every month in Rhuthun. Men living in Rhuthun and in the local rural area are members of this group. The club meets for a meal and they also have a guest speaker. The men's dinner club meetings do not involve much formal business. However, like *Merched y Wawr*, they have a president, secretary and treasurer responsible for drawing up a programme for the year and organising speakers and events. Both groups have a clear aim to support the Welsh language in their constitutions and conduct all aspects of their events in Welsh only.

Figure 3.

These are just some of the social events that take place in Welsh in these communities. There are others that I have not mentioned. There are many social events, like WI (Women's Institute) meetings, which take place in English and in which some Welsh-speakers take part. Most of these take place in Rhuthun. There are social events which could be termed 'bilingual', such as the Bontuchel and Cyffylliog agricultural show, where Welsh and English are both used in the written texts and in the oral part of the event.

Some Welsh speakers seem to be involved in a lot of social events such as those that I've described here. Some play active roles as committee members and organisers of events, which involves them in particular literacy practices. Others do not seem to take part in these kinds of organised events. Some Welsh-speakers seem to conduct most if not all of their social lives in Welsh. For others, their social lives are a mixture of both English and Welsh activities. These seem to be just some of the factors which affect the bilingual literacy practices of Welsh-speakers.

This article provides only a brief description of my research and some of my observations to date.

Teaching English literacy using bilingual approaches

Monica Lucero and Jan Thompson

RaPAL Bulletin No. 25, Autumn 1994

Monica Lucero is a lecturer in the ESOL Department and Jan Thompson is Curriculum Manager of the Return to Study Department of Westminster Adult Education Service.

Introduction

This paper discusses the work of an ALBSU funded local development project, Teaching English Literacy Using Bilingual Approaches, for Arabic, Bengali and Spanish speakers, from 1989 to 1992.

The first aim of the project was to research and develop an appropriate pedagogy for the teaching of English literacy to bilingual adults (in the case of the Arabic speakers and the Bengali/Sylheti speakers, the groups were for women only, but for the Spanish speakers (from Latin America and Spain) the group consisted of both women and men). The second aim was to disseminate the findings of the project into the practices of ESOL and adult literacy across the field generally.

The paper attempts to clarify what is meant by bilingual literacy in the context of second language learning. It looks at the methodologies developed in the classroom to help students maximise their skills, knowledge and learning from previous experiences through a process of reflection and analysis of their current circumstances. The paper also considers briefly the way processes, begun within the curriculum through the work of the project, pre-empted the current thrust in adult education and training for student outcomes in terms of accreditation and access to the labour market.

What is of particular interest here is that immigrant/settlers and refugees/asylum seekers have always sought educational and training opportunities which lead to agreed outcomes. In the past, however, they have all too often been contained within an adult education provision premised on the grounds that English was an end in itself. It is useful to look back and evaluate previous practice – and we did this on the project – to achieve a continuity and coherence in educational thinking and practice.

From the point of view of the students, tutors and managers on the project, its real strength lay in the fact that it developed from grass roots work done by two voluntary organisations, the Migrants Resource Centre (MRC) and Voluntary Action Westminster (VAW). It was on the basis of research funded by the Central and West London Open College and undertaken by MRC and VAW into the training needs of voluntary organisations (which identified a demand among members of a range of communities for bilingual literacy classes), that a proposal for a joint initiative between MRC and the adult education institute was made to ALBSU for funding in 1987.

Students and Tutors: Learning from Each Other

The three tutors, all women, appointed to the project shared the cultural and linguistic backgrounds of their students. The Spanish speaking co-ordinator came from Chile, the Arabic speaker from the Lebanon and the Bengali tutor from the Sylheti district of Bangladesh. They all had community work experience and were trained and experienced English teachers.

Their briefs included doing outreach work, publicising the project and generally making contact with potential students for project classes. Project classes were also publicised through the adult education institute's recruitment and advising sessions. This meant that in the first instance many students who would otherwise have been referred to traditional ESOL beginner classes came to the project instead. Most of the students coming through this route were Arabic speakers, who wanted a women's only class, and Spanish speakers who expressed a preference for being taught by a Spanish speaking teacher. Most of the students in these groups were literate in their first language, had a secondary education, and some had studied at university. The Bengali/Sylheti group took longer to bring together and needed more outreach work through community networks. When they did come together, however, they also demanded a women's only group.

The three tutors had different concepts about teaching and their approaches were permeated with their cultural backgrounds and specific educational experiences. This was seen as a strength as each of them started using teaching styles familiar to the students, thus establishing a context for discussing and exchanging views as to what the appropriate methodology for the project might be. They also explored the possibilities of using traditional ESOL/literacy teaching approaches.

The value of the tutor sharing mother tongue with students quickly became evident. It provided the team of tutors with in depth knowledge about the students and their educational needs. For example, when asked why they had not learnt English before or why they had abandoned previous attempts at learning English, the answers were

> *"because I felt as a child", "I felt ridiculous", "I could not give opinions", "I did not like the other students", "another student was making life difficult for me", "the teacher had wrong ideas about my country", "why should 1 go back to school? It's like being punished", "What's the point? I forgot everything I learned, I am useless" and finally "the teacher did not have an English accent".*

The use of mother tongue also helped to break down barriers of communication, and made it possible to begin building a sense of solidarity within the group of students who, although they shared language and culture, had different social, educational and life experiences. These were inevitably reflected in casual conversations among tutors and students and became the best source of what Paulo Freire would call "generative themes". Comments on the advantages and disadvantages of living in London reflected common preconceived ideas and stereotypes about British people and other ethnic groups, and myths about the English language.

Some of the questions raised were answered or discussed on the spot. Cultural identification was a positive force at play and tutors managed to control debates when they became heated, as they frequently did. The more controversial issues raised became part of the syllabus and were confronted and analysed in a didactic manner (this will be explained later in this article). Mother tongue also played an important role in facilitating the exchange of information between more experienced students who had lived in Britain a long time and those who had arrived more recently.

Uncovering Themes: Learners and Protagonists

One of the findings of the project concerning teaching approaches was related to the adult education practice of negotiating learning content with students. The findings showed how easy it can be to discourage and de-motivate students when they do not have enough language to express what they want to talk about in class. A widely used resource by teachers of students who speak very little English is to teach new language by incorporating elements from students' own countries and cultures. For example, if the weather is the topic, comparisons will be made with the weather in Bangladesh or Columbia. Project tutors recorded discussions with students with the aim of negotiating the syllabus.

Below is a transcript from a discussion in Bengali about shopping, one of the subjects the Bengali women wanted to learn about. Such comments raised a number of pedagogical issues and became the key to developing the Project's methodology. Discussions about the learning needs and content took place in mother tongue and preceded the designing of course content and materials.

Transcript

What is the shopping system like in Bangladesh?
It takes longer there because you have to bargain for prices and you have to pay with money in cash, not with cheques.

Sometimes the shopkeepers cheat when they have the chance. It is more peaceful here, we don't have to bargain. But forget about Bangladesh, talk about this country.

We want to prepare a lesson about your ideas about shopping . . .
Yes, in this country I go to any shop. I'll buy shoes, jewellery, clothes . . . In Bangladesh I couldn't go shopping and I didn't bother, there was always somebody to do the shopping for me.

In Bangladesh those who went shopping would put things in the rickshaw, here we have to carry it ourselves. We are talking again about Bangladesh . . .

Is there a reason for not wanting to talk about Bangladesh?
Our country could be the same as this country. Here men and women work but in our country only men can work. Women don't.

Yes, in this country everybody works, that's why they make progress . . . this country has many good things. When we buy things, if you don't like what you bought you can exchange or return things. In Bangladesh you sometimes spend a lot of money and you don't get a good thing.. In this country it doesn't happen. My heart is not in Bangladesh any more.

From these discussions tutors were able to abstract two main aspects of syllabus design. One aspect was straightforward and had to do with linguistic needs. The bilingual syllabus designed informed a methodology which encouraged students to use English skills actively without feeling constrained or unconfident because they were afraid of making mistakes. Mistakes were seen as "first attempts" or evidence of "partial knowledge which needed improving". Negative transfer of features of mother tongue was discussed and explained and areas of similarity of usage and feature were identified and taken advantage of, quickly establishing a firm base from which students could build their knowledge. This was particularly advantageous for the Spanish speaking group (due to the influence of Latin).

Proof reading was very effective for contrastive and comparative analysis. Pieces of writing were transferred onto an OHP and were corrected and improved collectively with contributions from everyone in the group. Critical analysis of language and how it works was based on mother tongue and then tried out in English. Students lost their fear of taking risks, and informed guesses were continually assessed and adjusted as they became more in control of English.

The other, major aspect of syllabus design related to students' prior learning and socio-cultural perceptions of themselves and the other students in the group, and their perception of the position of their cultural groups in British society. Although trying to deal with these extremely complex issues seemed to be outside the tutors' professional capabilities and the scope of English teaching, it was felt that these issues had a direct impact on student retention and motivation, and in developing group skills which allowed students to take full advantage of the learning situation.

The tutors felt that they could not shy away from these very real problems and that part of applying bilingual methodology was dealing with these areas organised in a syllabus. This showed that negotiating the syllabus was not only about responding to students' explicit requests about what they wanted to learn; it was also about helping them to acknowledge areas of learning which they had not identified but which had emerged as teaching points from the 'generative themes' of the background discussions. If controversial issues had been avoided they would have become obstacles to learning. Students were presented with suggestions for topics from which they could select. The topics proposed in this way were normally accepted as they had clear connections with improving the students' understanding of their present circumstances. Teaching materials were based on content provided by the students, as discussions were taped and transformed into items for language learning, and language was quickly absorbed as the linguistic material had direct relevance to students and helped to shape their identities in English.

Some examples of the topics included:

Reasons for Migration: why are we here?
Tutors and students had left their homelands for different reasons but migration has roots in an economic system which provoked it. Contextualising the reasons that brought tutors and students to Great Britain counteracted distorted images about countries and communities in Britain and asserted the value of students' language and cultures in British society. This discussion led to looking at different ethnic communities living in London and to discussing varieties of English and accents.

Learning and Mental Health Issues
A group of women who had suffered repression discussed how several physical symptoms like headaches and pains, and also lack of memory, affect their learning. They explained how discussing the causes of their problems helped them to lose fear and to look for solutions.

Banks and Interest
Students wanted to learn the language of banking, including how to apply for loans, how to keep savings and earning interest. A discussion was included on students' views on earning interest, as some students were against this for religious reasons.

Developing critical awareness

The idea of developing English literacy skills with a range of students (some who had little or no literacy in mother tongue, others who had some secondary schooling and a few with higher education experience) seemed daunting to the tutors at the start of the project. Gradually, as real learning needs started to be uncovered and common ground was found, the tutors felt confident in exploring a wider definition of 'literacy', an approach benefitting all students in the group, preparing them for life in Britain by building bridges between past learning experiences, including language, and emerging needs. English learning aspects, such as linguistic structures, grammar patterns, register, intonation and pronunciation became more relevant to students as they realised they could achieve meaning and communicate in a context they could control. Although English language learning was not always in strict relation to the content of discussions, the fact that students were already motivated and felt part of the group facilitated learning and language interaction. They could actively intervene as they felt free to use their own language if necessary. The framework that put together students' and teachers' understanding of each other, and of the learning context, was the critical awareness raised about life in Britain and students' expectations of it as presented and analysed in class. Paulo Freire quotes a woman's description of her literacy class:

> *"I like these discussions because they show how I live. But while I am busy living, I cannot see myself. Now I can observe how to live." (Own translation, Pedagogia del Oprimido, Paulo Freire, Siglo Veintiuno editores 1979).*

Conclusion

The value of using bilingual approaches in developing literacy skills in English emerged from the work of the project in a number of ways. The most immediate benefits could be seen in the way that students were able, through discussion in the mother tongue, to set their own agenda and contribute to the construction of a curriculum which reflected the complexities of their experiences, both in their countries of origin and Britain, in a material way. The use of mother tongue provided a framework for all that went on in the classroom and was a means of establishing a sense of solidarity within each group. This was important because the groups, with the possible exception of the Bengali group (all of whom came from Sylhet district in Bangladesh and were facing similar experiences as they joined their husbands in Britain) were by no means socially – or even linguistically – homogenous.

For example, the women in the Arabic group came from a range of Arabic speaking countries and from diverse social and cultural backgrounds. The group consisted of both Muslims and Christians and religion was a major difference and a potential source of conflict. The tutor used discussion in the mother tongue, however, to confront and resolve conflict and to help the group to bond and to establish shared objectives. In this group, which became the most cohesive and stable of the three, the focus was very much on learning as a process and they developed a conscious commitment to a particular, formal learning style.

Although the overall benefits were similar for all three groups there were significant pedagogical differences in the way each group worked. These reflected differences in previous social, cultural and educational experiences of the students and their tutors in their countries of origin and the specific ways they were experiencing being migrant/settlers in Britain in the late eighties.

In effect, each tutor used a combination of techniques from a range of English language teaching disciplines and the field of adult literacy, albeit a somewhat different combination in each case.

For example, the Arabic speaking group started, drawing upon traditional ESOL methodology, by developing students' oral skills in English in a systematic way before moving on to work on basic literacy skills. These were then developed through the introduction of grammatical patterns which were presented in a highly formal and structured way. The methodology used here was reminiscent of both EFL teaching practices and those used in teaching formal grammar to native speakers. This was followed by systematic practice (in the classroom and at home) of graded exercises, and combined with opportunities to read abridged but complex texts in English of up to a thousand words, and to write assignments.

The Spanish speaking tutor, who acknowledged the influence of the radical theories of Paulo Freire, borrowed language experience approaches from adult literacy practice and used discussion to uncover issues relevant to the students, particularly those around health and work, and problematised them in the context of the classroom. This problem posing/solving approach reflected the real life experiences of the students in the group (who were migrant workers, or refugees) for whom access to services such as health and education had long been problematic, (see Out of the Shadows, Migrants' Resource Centre, 1985). The students developed their literacy skills through reading and writing about these issues but also worked on grammatical patterns in a formal way. Once again a combination of formal and informal techniques were used.

In the Bengali group the approaches used initially were highly formal. The students, none of whom were literate in Bengali, were introduced to the alphabet and small units of meaning (short sentences) through phonics and techniques such as copying. Discussion was used to communicate with students in a variety of situations, e.g. to develop confidence, to negotiate the syllabus, to support learning through checking for understanding, and to give information about access to services such as health and education.

The tutors met regularly to discuss the various methodologies they were using, to share ideas, discuss problems, explore common ground and develop materials that could be adapted for use within each group. As time went by a more unitary pedagogy began to emerge from the various methodological syntheses being deployed in each group. Unfortunately, the task of disseminating the findings of the project was disrupted by the abolition of ILEA, in 1990, and the subsequent re-structuring of adult education in London.

Nevertheless, the advantages of using bilingual approaches to develop English literacy skills were clearly demonstrated by the speed with which students moved from basic to high level literacy skills (this was most notable in the Arabic group, where even students illiterate in the mother tongue made tangible progress) which enabled them to make choices about employment or further study.

The combination of the radical pedagogies of critical literacy theory, involving discussion and problematisation, with more formal teacher centred approaches seems to have provided a rigorous framework which allowed students to develop their literacy skills from a soundly grounded basis, through a process of critical analysis, and to take control of their learning. The use of the mother tongue in the classroom unlocked the learning process by giving students access to knowledge about language and literacy; they were able to discuss grammatical categories, to label parts of speech and to discuss tenses in a way which is not possible in an ESOL class with an English tutor.

Perhaps the most outstanding lesson of the project was the need for flexible approaches which use a combination of methodologies in a pragmatic and creative way and which acknowledge the essential heterogeneity of groups, ensuring that the articulated outcomes of individual students are achieved.

ESOL through sport

Barbara Hately-Broad

RaPAL Journal No. 52, Autumn 2003

Barbara is a Senior Lecturer in the School of Education & Professional Development at the University of Huddersfield. At the time of this Summer School project, she was employed as Basic Skills Curriculum Co-ordinator with Wakefield Adult & Community Education Service. Her research interests include Post-16 Traveller Education and ABE in the British Armed Services.

Introduction

One of the perennial issues for both English for Speakers of Other Languages (ESOL) and adult literacy providers is how to attract young male learners. This article describes the way in which the Wakefield Adult and Community Education Service devised an approach to this problem through their annual summer school provision for asylum seekers.

For a number of years the Service, in cooperation with the Wakefield Metropolitan District Council Asylum Seekers Team, has provided a week-long Summer School programme including a combination of both English and vocational classes. The curriculum catered primarily for families but an analysis of actual arrivals in 2000-01 showed that many of the asylum seekers arriving in the area were, in fact, young single men. Although this group was catered for in the day-to-day, term-time provision, no provision had been made specifically for them in Summer School.

To address this problem a new Summer School programme was devised predicated on a common theme of 'sport' – a topic which crosses language barriers – and which combined traditional English classes with practical sports activities. By utilising activities within both the cognitive and psychomotor domains, it was possible to accommodate the greatest variety of learning styles and provide a wide spectrum of learning opportunities.

The implementation of the programme, initially devised by the Basic Skills Curriculum Co-ordinator, was made possible by the availability of a basic skills tutor who was also a qualified football coach. However, we were also fortunate in being able to call on the support of a Community Development Worker based in Castleford High School who was able to negotiate the use of their facilities for these classes. The school provided us with a teaching classroom for the morning classes, use of their outdoor sports facilities and, perhaps most importantly, use of their indoor sports hall. Without this co-operation, the course would have proved much more problematic as the weather during the week of the Summer School proved changeable to say the least!

The Curriculum

The usual pattern of the day was for the learners to be provided with classroom sessions during the morning and then sports sessions in the afternoon. The creative challenge was to embed ESOL learning within the sports sessions in a productive and enjoyable way.

The curriculum helped learners to become competent, quickly, in a range of English skills necessary for their everyday life. Learners were

also given the opportunity to undertake accreditation via a West Yorkshire Open College Network (WYOCN) Regional Languages Framework module at Entry Level which included familiarity with a number of everyday situations: the recognition of a number of common signs (e.g. open/closed); ordinal numbers as used in everyday life (e.g. first floor); asking for and following directions; recognition of a variety of shops and simple vocabulary associated with shopping; and ordering of food and drinks. Although the learners targeted had varying levels of competence in spoken English, none had previously undertaken any accreditation and many were hesitant in employing their existing skills. The accreditation was, therefore, seen primarily as a confidence building exercise. Additionally, this particular accreditation was achieved through the production of a portfolio of evidence which ensured that learners could demonstrate competence by providing evidence in a wide variety of ways and allowed learners to work at their own speed. If necessary, the compilation of evidence could be carried forward into the term time group provision.

In each of the morning classes, the fifteen learners enrolled were divided roughly into three groups – beginners who had no existing knowledge of English, improvers who a small knowledge of English and intermediates who had a working knowledge of English. Working under the overall direction of the Basic Skills Coordinator, and facilitated by a tutor or volunteer, each group worked on the same topic each morning at their own speed. For example, in the session dealing with asking for and following directions, although all learners were provided with a map of the Castleford locality in which they were all living, the different groups were provided with maps of differing complexity. Similarly, towards the end of the week when learners were undertaking tasks to demonstrate their competence, they were all asked to complete the same type of task but at different levels of competence. For example, when undertaking a simple 'shopping' role play activity, learners with basic levels of English were only asked to greet the shopkeeper, ask for specific items and then close the conversation, whilst the tutor acted as the shopkeeper. Those learners with a greater knowledge of English worked in pairs and were required to be able to take the part of both the customer and shopkeeper, thus demonstrating their ability to employ a wider range of vocabulary.

The afternoon session fell into two parts. First, the learners undertook general warm-up activities to reinforce the learning from the morning session. For example, in one exercise laminated cards were made showing 'ground floor', 'first floor', 'basement' etc. and learners were asked to run to specified 'floors' whilst in another, cards were produced showing a bus with a large, clearly visible number and learners were asked to 'catch' certain buses. Initially these activities were carried out as a group activity to support the less able members of the group and build their confidence in their own knowledge. As the week progressed, learners were divided into teams and the activities completed as relays. These activities not only helped to reinforce learning through experiential activity but also proved to be extremely popular with the students who sometimes became highly competitive.

During these sessions two supervisory staff were always present – in this case the tutor who also acted as the coach and the Basic Skills Co-ordinator. This was necessary, not only for health and safety reasons but also to check learning. During the warm-up session the Co-ordinator completed check sheets for learners providing evidence that they were able to understand and use the necessary vocabulary and, during the 'game' part of the session, made a note of the relevant vocabulary introduced which was reviewed at the end of the session.

After the warm-up, the learners then actually played a sport – five-a-side football was the most popular, though badminton and basketball were also tried. During this part of the session little attempt was made to provide formal learning. Instead learners were provided with the necessary vocabulary as and when the need arose. This also proved successful. The learners were already familiar with the rules and vocabulary of the game in their own language and initially tended to choose to be in teams which had a common first language. However, as the week progressed, we began to insist that instructions such as 'pass' and 'offside' were only to be given in English and, as learners became more familiar with the English vocabulary, more mixed nationality teams began to form – based often on perceptions of each other's skills rather than nationality.

General comments, often about the skill or parentage of the referee, were also only accepted in English.

For the last afternoon, a visit had been arranged to Castleford Tigers – the local Rugby League team. The team's ground is only a short walk from Castleford High School, through Castleford centre itself and this provided an ideal opportunity to check the week's learning in a practical activity. Learners were provided with a checksheet covering all the sight vocabulary they had learnt during the week, such as entrance, exit, Ladies, Gents, open, closed, which they had to find during their walk or at the ground. This experiential consolidation activity demonstrated clearly to the learners the relevance of their week's learning and its application in everyday life.

Once at the ground, the learners were given a complete tour including a general introduction to the game, a demonstration of the training equipment (which some tried out) and an explanation of a training session. At the end of the tour, learners were delighted to be given tickets to the next match. This was a very generous donation on the part of the club – although some of the learners had previously expressed interest in the game, it would have been almost impossible for them to have been able to afford the £10 entrance fee.

Evaluation

In overall terms, this pilot Summer School proved a very successful venture in a number of ways. First, we succeeded in attracting a number of young male learners who often prove elusive. Of the fifteen men who enrolled on the programme, twelve were between the ages of twenty and thirty-two. This had the additional long-term benefit of introducing these men, who were often placed in single bed-sit accommodation, to others from their own communities and so reinforced the work of both the Asylum Seekers Team and their own religious communities in helping them to forge social contacts. Second, we were able to further these 'social' outcomes by providing them not only with an insight into local culture in the form of Rugby League, but also with an appropriate vocabulary which would enable them to participate in informal sporting situations. Third, through this intensive provision we were able to ensure that all the learners quickly became familiar with the English vocabulary and structures necessary to enable them to cope with basic day to day situations. Eight of the learners completed the accreditation during the week and, for a number, the confidence this gave them encouraged them to undertake further accreditation including the Pitman's English tests. Finally, all of the students continued to improve their knowledge of English by regularly attending ESOL classes in the following autumn term, effectively demonstrating that this initial, specifically targeted provision had successfully provided them with an initial positive learning experience, which encouraged them to undertake further learning.

In conclusion, we feel that this programme serves as a potential model of good practice, not only for young male ESOL learners, but also as a means of encouraging young male learners to join more general literacy and numeracy classes.

B3 Dyslexia

Further RaPAL references about dyslexia

Lobley, G. and Millar, R. 'More on Dyslexia', *RaPAL Bulletin* No. 3, 1987.

Herrington, M. 'The Leicestershire Dyslexia Study Group. A Student-Centred Research Group', *RaPAL Bulletin* No. 8, 1989.

Herrington, M. 'Dyslexia: Old Dilemmas and New Policies', *RaPAL Bulletin* No. 27, 1995.

Lankshear, P. 'Dyslexia and ABE – where now?', *RaPAL Bulletin* No. 27, 1995.

Walker, M. 'Specific Learning Difficulties: A Three-Pronged Approach', *RaPAL Bulletin* No. 27, 1995.

Whitehouse, G. 'An FE Student's Experience of Assessment or "It's a Big Shock Finding Out You're Disabled!"', *RaPAL Bulletin* No. 27, 1995.

Hodges, N. 'Arrows, Cycles and Spirals: Time in an Adult Dyslexia Class', *RaPAL Bulletin* No. 37, 1998.

Chappell, D. 'Dyslexia Study Workshops', *RaPAL Journal* No. 50, 2003.

Summerfield, S. 'Through the Maze of Strife: A Personal Case History of Dyslexia', *RaPAL Journal* No. 53, 2004.

Dyslexia: The continuing exploration. Insights for literacy educators

Margaret Herrington
RaPAL Bulletin No. 46, Summer 2001

Margaret Herrington is Director of the Study Support Centre at the University of Nottingham

Introduction

Dyslexia continues to be 'problematic' for many adult literacy educators in post-16 education (Herrington 1995; Kerr 2001). In part this stems from continuing uncertainties about definition, terminology, origins, boundaries, perceptions, methods and resources. But it can also stem from a lack of knowledge about the developing debates about dyslexia and thence any acknowledgement of the areas of agreement among dyslexia researchers and practitioners. This 'not knowing' also seems implicit in the new adult literacy and basic skills initiatives in the UK. The recent Moser Report on the future of basic skills provision (1999) signalled that for policy purposes, general literacy problems are best separated from the literacy difficulties experienced by dyslexic people. Dyslexia is seen as a specific learning difficulty and guidelines for provision are included within policies for disability. Yet such a separation does not appear to have a strong theoretical basis and the day-to-day realities are that literacy educators in all sectors encounter and respond to dyslexic students. Clearly, both policy and practice require some 'joined up thinking'.

In this paper I shall attempt to challenge one of the arguments which has obstructed the development of further understanding in this field among adult basic education staff (and I write from the position of having worked in adult basic education for sixteen years). I shall then suggest that the most useful way of considering dyslexia (Herrington & Hunter Carsch 2001) is one which integrates the unfolding picture about 'in person' dyslexic differences with the broader concepts of literacy incorporated within New Literacy Studies (Barton, Hamilton, & Ivanič 2000; Gee 1988, 2000) and the broader visual communication context described by Kress (2000). The traditional and crucial role of the literacy educator as a co-explorer of literacies with their students will be re-emphasised, with reference to some recent research with dyslexic HE students in the University of Nottingham Study Support Centre. This research reveals that when dyslexic students and their tutors generate certain types of 'conversation' (Herrington, forthcoming 2001), important insights are revealed about how literacies are experienced by those deemed to be dyslexic and about precisely how institutional and cultural conventions continue to exclude and marginalise.

Dyslexia, the Middle Class Disease. A Lingering Myth?

It has often proved difficult to have dispassionate discussions about dyslexia with fellow educators in the UK. A particularly destructive myth has been that of dyslexia as a 'middle class disease'. The negative power of this phrase stems from the suggestion that dyslexia has somehow been appropriated by one social class at the expense of another and that this has been done by conceptualising it as a special disease. Teachers in the UK have often expressed concerns that middle class parents of children with literacy difficulties have sought the 'dyslexia' label in order to avoid less acceptable labels of 'not very bright' et al.; and that they have also attempted to access more educational resources for their children, sometimes at the expense of children with literacy difficulties from poorer homes.

In part, parents have been able to sustain this position because of the method of identifying dyslexia used by educational psychologists. This is based on a 'discrepancy' principle. The verbal and performance IQ of the student is determined using standardised IQ tests [WAIS] and a further battery of tests is used to assess literacy performance. If a student is deemed to be of average or above average IQ and yet there is underachievement in particular areas of literacy performance (a discrepancy), then the student is identified as having a specific learning difficulty (e.g. dyslexia). If IQ is below average then low literacy performance is deemed to be connected with a general learning difficulty and is not termed 'dyslexia'.

This method has been heavily criticised, not least by educational psychologists themselves (British Psychological Society 1983 , Stanovich 1991 etc):

> *" It is more educationally and clinically relevant to define reading disability without reference to IQ discrepancy . . . problems with the IQ concept are endemic . . . researchers cannot agree on the type of IQ score that should be used in the measurement of discrepancy . . . changes in the characteristics of the IQ test will result in somewhat different subgroups of children being identified as discrepant . . ." (Stanovich, ibid. p. 130)*

IQ measures have been challenged both in principle (cf Gardner's Multiple Intelligences 1993) and in practice. Varying scores on different days for the same student are not unknown. Literacy tests too have been challenged. They do not provide an accurate snapshot of the broader reading performance of the adult in adult settings.

In practice, psychologists are often more interested in the variations in IQ subskills associated with dyslexia and, along with colleagues in other disciplines , in pinning down areas of relative cognitive strength and weakness. A vast body of research has developed which is devoted to exploring the defining weaknesses of dyslexia (Snowling 2001 ; Fawcett 1995 ; Stein 2001). However, this association of dyslexia with average or above average intelligence has fuelled the myth described above. Concerned middle class parents who recognised that their children were reasonably intelligent despite difficulties with aspects of literacy refused to accept the rather sloppy connection teachers tended to make between literacy and intelligence (slow reader= slow thinker). By obtaining IQ scores they could challenge this thinking and show that their children were intelligent/educable.

Because middle-class parents appeared to be suggesting a distinction between those who were 'bright' with literacy difficulties and those who were not, their actions encouraged a continuation of the situation in which testing for dyslexia became a process of separating out those who had average or above average IQs from the rest; and even then only for those with parents who either had their own resources to pay for testing or who had sufficient power in relation to a school to insist on formal testing. This contributed to the accusation that they were really only interested in securing additional resources for the most 'intelligent'. Yet for those educators who are aware of dyslexic patterns, all the practical evidence is that these difficulties/differences occur across the ability range and though it may be far easier to spot them in someone who has a high and obvious intelligence, it cannot be acceptable to limit the identification to such people. Nor can it be acceptable to secure additional resources solely for this group.

Further Exploration is Necessary

If literacy educators were right to entertain some suspicions about unsubstantiated claims regarding dyslexia, they were wrong if they responded to such confusion by refusing to explore dyslexic patterns. There was clearly something to explain here: if people of all abilities could have difficulties with literacy, then the assumed relationships between literacy and intelligence/educability were being challenged. It was clearly important to explore the implications of this for educational conventions dominated by literacy-based modes of assessment.

They were also wrong if they did not ask further questions about the reasons for sustained difficulties with accessing written text. The educators' professional responsibility requires a continuing investigation into why some literacies are inaccessible. Literacy researchers and practitioners have produced a growing body of evidence about literacy practices and conventions in an array of contexts Freire 1972; Street 1984, 1990; Barton 1994; Barton, Hamilton and Ivanič 2000; Crowther, Hamilton & Tett 2001; Wolfe 1998, Mace 1992, Houghton 1995/6, Lillis & Ramsay 1997) and through close analysis of how practices are used by the powerful to maintain their status, they have exposed the key sources of inaccessibility. However, when 'in person' characteristics are also implicated, for example, when a student cannot remember words they want to be able to read and have tried a range of methods over a considerable period, there is a case for asking questions about underlying cognitive processing as well as about confidence and motivation etc. To argue for such investigation does not undermine the case for maintaining a 'critical literacy' stance about how literacies are implicated in exclusionary social practice. Rather, it enriches our understanding of excluding processes. It is only problematic if the findings from such work are used to reinforce models of literacy which focus solely on individual cognitive performance and which view 'literacy context' as essentially unproblematic.

There is evidence that the refusal by some tutors to investigate has resulted in literacy students feeling that their difficulties have not always been properly identified (Walker forthcoming 2001) and literacy tutors have not always had adequate training about the array of specific learning difficulties which can undermine literacy achievements. This delay in a fuller analysis and articulation of types of difficulty with literacies may also have delayed recognition of the enhanced cognitive skills which may accompany such difficulties. Whilst literacy educators have achieved remarkable outcomes with literacy students over many years (Ivanič 1988, 1996, 1997; Moss 2000; Mace 1979, 2000), relatively few have investigated dyslexia explicitly (Klein 1989, 1994, 1995; Morgan 1995; Morgan & Klein 2000; Herrington forthcoming 2001).

When this was done, students revealed powerful descriptions of literacy processes. The adults in a dyslexia research group at Leicester University in the 1980s described what they saw on the page. They identified visual discomfort [blurring and moving of letters etc] and demonstrated how coloured overlays/paper had helped with some aspects of the discomfort. This did not seem to be related to any problems with sound/symbol correspondence or phonemic segmentation and yet was experienced as a barrier to reading. When they described reading and writing, they used dynamic, visual, three dimensional analogies. These descriptions were more valuable for developing effective teaching strategies than test scores and three examples are provided below:

> *"I have to read each night-its like keeping down the undergrowth on country paths. If I don't keep going through, they get overgrown"*
>
> *"there definitely seems as though there is a veil over part of my brain . . . my thought pattern is not at all the same"*
>
> *"my problem is I cannot form letters,so I cannot write. Although this is very extreme I can copy if I see the words but if the words are taken away from me I cannot know where to begin to start to form a letter. To me it seems like having a television set inside my brain with the letter on it and that set is spinning very fast. But to make things even worse, there seems to be a curtain in front of the set and I get very fast peeks at the letter. The letter can appear at any angle, at 45 degrees or even backwards. I don't know which is the right way when I try to put it onto paper". (Herrington, forthcoming 2001)*

They also described visual/kinaesthetic learning strengths when describing their particular vocations/skills: photography, hair styling, coaching swimmers, engineering etc.

It has taken some inspiring dyslexic teachers such as Susan Parkinson of the Arts Dyslexia Trust and most of all dyslexic writers themselves (West 1991) to point us even more forcefully towards the possibility of enhanced cognitive strengths which accompany some literacy difficulties. They have encouraged the start of a second tradition within the literature which does not view dyslexia as **a neurological disorder with associated cognitive processing deficits and literacy behaviours** (tradition 1 in Figure 1) but as **a different thinking/learning style** (tradition 2). In particular they have highlighted the enhanced visual spatial and global thinking skills which appear to accompany literacy difficulties. See Figure 1 overleaf.

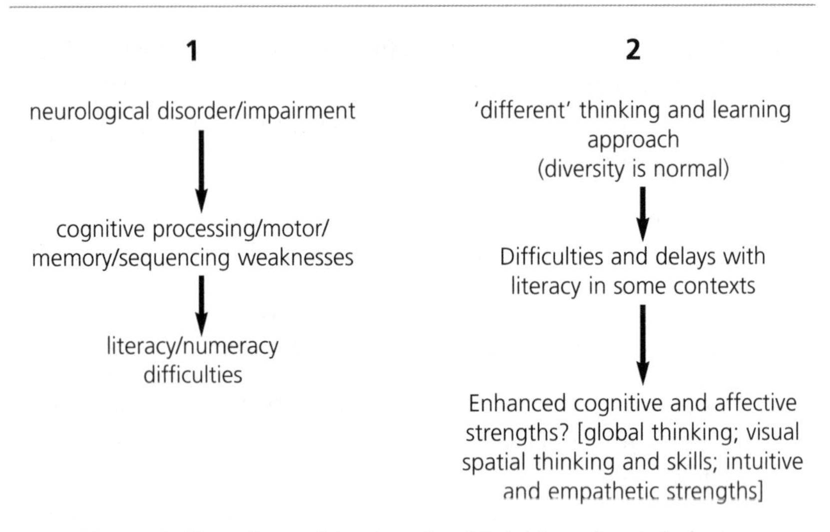

Figure 1: Two discernible strands of thinking about dyslexia.

Dyslexia: a different cognitive style and a socially constructed disability?

The second tradition is now beginning to gather pace and particularly in higher education. It suggests a social rather than a medical model of disability. Dyslexia is still designated as a disability in higher education and a medical model conceptualisation still prevails. This view is of dyslexia as an 'in-person' impairment which can be designated as a disability. In contrast, the social model of disability acknowledges impairment but focuses on how social and economic values, conventions and practices construct 'disability'. Disability researchers have pointed out that so called non-disabled people construct the disability by creating inaccessible structures, whether physical, social, or attitudinal (Barnes & Mercer 1996). Whilst this is relatively easy to see in the case of physical and sensory impairments, the ways in which the disability is constructed in the case of learning difficulties/differences are not usually rendered explicit. Yet in the case of dyslexia it is possible to argue that certain literacy practices and conventions and values create the 'disability'. In contexts which do not require the rapid processing of large amounts of written information and in which the learner is not expected to write intensively and accurately to time, it is possible to argue that the dyslexia may not be experienced as a disability at all.

'Critical literacy' theory can help to explain the disability creation process. Though it is clear from the above that dyslexia is not only concerned with literacy, it is often seen as primarily about this and the model of literacy usually invoked is an 'autonomous' one. This focuses on a technicist view of literacy as involving letters, sounds, words etc which are the building blocks of literacy 'acquisition', and usually sees the learner who has difficulty with these simple elements as having a specific or general learning difficulty. In contrast, the 'ideological' model (Street 1984) views literacy as social practice and a 'critical literacies' perspective reveals how certain ideas about literacy and standards lead to educational exclusion and how certain non-school based literacies are ignored (Barton and Ivanič 1991). The autonomous model allows the disability to be constructed whereas a 'critical literacies' model enables dyslexic learners to understand how attitudes to literacy affect educational and social policy, to reframe past schooling experience and to reject prior judgements made about them.

In an academic context, as well as in ABE, it is vital for students to be aware of the different, co-existing literacies and of their respective status- the social class and use of language and literacy issues are alive and well for some students in British universities This enables students to move from a personal deficit paradigm to one which contextualises each literacy task , which clarifies their power within their situation and which empowers them to challenge if they see fit. The recent classification offered by Lea and Street (1998) which distinguishes between study skills deficit/academic socialisation/and academic literacies is helpful in this area: the academic literacies approach allows us to identify the contest between literacies for status and to help students to locate themselves within this (Carys Jones 1999).

This theory also allows the interesting question of whether the literacy of dyslexic learners is a distinctive literacy. Many writers have pointed to the particular areas of literacy with which dyslexic people have difficulty (Gilroy and Miles 1996; Steffert 1996) and traditionally this has been seen within a deficit paradigm despite the authorial strength of many dyslexic people. Although the arguments above reveal that the social perception and valuation of difference is key, they also encourage us to look carefully at the cognitive and affective profiles associated with dyslexia. Several dyslexic writers have attempted to describe the nature of the difference and what may be involved in the enhanced skills (Hetherington 1996, Stacey 1997). The Arts Dyslexia Trust (ADT), too, has focused on identifying some of these and has recently noted what it believes to be a common feature of dyslexia:

> *"the inability to concentrate on the linear, sequential aspect of words, figures and time; together with a better than average ability to absorb and understand the three dimensional information which all our senses, most importantly our visual sense, brings us every day" (Preamble to the ADT symposium at Green College, Oxford, November 2000)*

In HE where the dominant paradigm remains a deficit one, there is growing evidence of the recognition that there may be some valuable elements in this learning style and some recognition that particular kinds of vocational skill rest heavily on such elements: dyslexia is common among artists, engineers, designers etc (Steffert 1999). The twin themes of visual processing difficulties and visual strengths run throughout the literature on dyslexia The difficulties were evident in the earliest recognition of 'word blindness' (Pringle Morgan and Hinshelwood) and of 'twisted symbols . . .'strephosymbolia' (Orton) and have continued through to the currrent work by Stein on the biological bases of certain kinds of visual perception problems.

The theme of potential visual 'strengths' has also long been evident. Winner et al (1999) have identified the recognition of spatial talents from Orton in 1925 to Rawson in 1968, to Geshwind and Galaburda (1982 and 1987) and more recently in the 1990s from West (1991). Whilst acknowledging the evidence which has been gathered around this theme and in particular that which shows that

> *" 'spatial' populations have a higher than average incidence of reading or language difficulty" (p4)*

and that

> *"Six of the twenty world class mathematicians studied by Bloom (1985) reported some difficulties learning to read and none of the twenty learned to read before school, despite their undoubtedly high IQs. All thirty four of the inventors studied by Colangelo et al (1993) reported difficulties in writing and in verbal areas, along with strengths in mathematics" (p4),*

Winner et al (ibid.) have also identified some of the limitations in the research evidence: case study research with individual dyslexic people does not consider the incidence of those characteristics in the population at large; the focus on the relative skills of dyslexic people who do better on spatial than on sequential or verbal tasks may give misleading results; and there are inconsistent findings when absolute spatial strengths are investigated with dyslexic and non dyslexic groups. Their own research (study 1) revealed that their dyslexic 'subjects', "performed either equivalently to or worse than the control participants on a range of spatial tasks". However, Parkinson has shown the limitations in the quality of the tools used: three dimensional tests designed by two dimensional thinkers produce highly questionable results.

Kiziewicz,M. (2001) also acknowledges that *" No good tests of visual spatial ability yet exist to fully confirm this theory-we are still waiting for them to be developed by visual spatial thinkers" (p39)* and this remains a major barrier to investigations in this field.

Even so, Steffert (1999) has used a batch of currently available visual spatial tests in her investigation of the relative incidence of dyslexia among art and non art students. She did not find more dyslexia than in the population at large among art students, if the phonological deficit model of dyslexia was used. She did find 40–50% of art students with problems with reading and writing; and difficulties with syntax were the most marked (a *syntactic dyslexia).* She also found evidence of a visual learning style among the art students which contrasted with the style of the non art students. Her additional, smaller, sample of students who had already been designated as dyslexic before they arrived on their art course revealed more clearly this mix of enhanced skill with shape rotation and innovative creative style with higher scores on the dyslexia questionnaire and poorer reading, spelling and writing. For her they were a "more extreme form of art students in general". (p.156)

Brigden has investigated differences between dyslexic students on art and non art courses and found that there were few differences between the pattern of difficulties experienced by dyslexic students on both types of course and she also offers support to Steffert's view of a dyslexic syntactic rather than phonological difficulty. Further, she found from her sample of students at the Surrey Institute that

> *"a minority of art and design students with dyslexia demonstrate the most severe literacy difficulties and yet show excellent originality and creative talents" but that for the whole group one could not "assume that an art and design student with dyslexia necessarily has visual spatial strengths" (p.35) because different learning preferences were recorded.*

There is still much to unravel in these proposed relationships but if further research does reveal hard evidence of enhanced visual processing strengths, it will be important to locate the discussion about dyslexia and literacies in a broader visual communications framework. This may require a 'critical communications' rather than a 'critical literacies' perspective; one which embraces ideas about relationships between visual and literary modes of communication in an increasingly visual culture (Kress 2000); the shifting location of power in the new arrangements; and a focus on whose modes of communication are included/excluded. West frequently argues that the new emphasis on rapid visual communications will benefit dyslexic people and that the balance of power will shift in favour of visual thinkers (2000). Though this may be a welcome realignment and one more in tune with the longer term view of human communications, this view may well underestimate the power of certain literacies to retain the high ground and certainly the powerful interconnectedness of visual and literary forms.

At the moment however, we do not know enough about the intellectual landscape involved in the enhanced skills, nor about whether these apply to all dyslexic learners in some degree, nor even whether they are exclusive to dyslexic learners. However, it is important to pursue the investigations both to find effective methods for the individual learners and to expand our understanding of the multiplicity of learning styles. Though it may appear to be part of a continuing separation of dyslexic literacy and other literacy problems, I would argue that it is part of opening up the bigger debate about literacy difficulties in general. As a literacy educator I have been grateful for the insights from dyslexic students which I can now use with all. In terms of the general arguments, I suspect that all intractable difficulties with aspects of literacy involve some elements and degree of dyslexia. This may appear to bring us back to the ABE starting point where tutors may feel that there is little point in distinguishing between dyslexic and non dyslexic but there is now a deeper understanding of how literacy is experienced and a new set of tools for the literacy educator to explore with their students.

In the remainder of this paper I shall pursue these questions further with reference to dyslexic learners in HE. The following report is based on my practice in 'supporting' dyslexic students in the Study Support Centre at Nottingham University. This Centre provides individual and group tuition to any student who wishes to explore some aspect of their learning and literacy; and dyslexic students account for over half of those using SSC at any one time. At the heart of this work are the students' own descriptions, within interactive 'conversations' (Harste 1994) with tutors and other students. These release illuminating accounts of both the 'in person' experience and of the reasons why certain institutional constraints are particularly difficult to counter.

Student Descriptions

This small scale, reflective, generative piece of 'research within practice' is based on two particular sources of data:

- notes from the dyslexic students' monthly discussion groups (8 meetings during one academic year). These regular monthly meetings started in 1997 and the size of the group varies between three and eighteen students. The form of the meetings involves some core activities such as an 'information exchange' about ICT, about ways of handling the tasks set by tutors, and

about ways of dealing with inappropriate teaching and learning styles. The group also works on selected themes: making changes in systems at the university of Nottingham; and exploring ways of describing dyslexia to people who do not understand/recognise it. Notes are taken and circulated before the following meeting in a 'dyslexia friendly' form (12–14 font, sans serif, coloured paper, succinct style, left justified only)
- case notes from selected individual student files (6 in all). These files record the main content and the decisions made within the individual study support sessions, which are provided for approximately four hundred dyslexic students who use the Centre each year. The selection in this case was done on the basis that there had been more time for informal explorations in sessions with these students. Students studying a range of disciplines were included but all personal details are anonymised.

The limitations in the research methodology are acknowledged at the outset. No generalisations about all dyslexic learners can be derived from this evidence and particular descriptions are quoted if they reflect recurring themes.

The two sets of insights discussed below are:

- 'visual' references in students' descriptions of reading and writing
- students' descriptions of time

In our practice, we always explore with students how they see the page and how they attempt to describe visual 'disturbance' and 'discomfort' with reading text. We also often hear powerful 'visual' descriptions about literacy and about ways of thinking e.g. enhanced 3-dimensional visual spatial thinking: *"I can design an engine in my head and turn it round, looking at it from all sides/angles"*.

On a daily basis we hear comments about literacy, learning and time. These are important because they can show us a non-deficit view of 'difference' and yet may offer additional reasons why some literacy practices are more problematic than others.They may of course reveal a spectrum of experience with text of which dyslexics may be at one end but which is not unfamiliar for other readers. Ultimately they may help us to open up discussion about how everyone experiences text.

Emerging Insights and Questions

Reading

1 Switching modalities, between words and visualisation, to make meaning

The first description I would like to consider is from a postgraduate HE student: she was concerned about the volume of reading she had to undertake for her thesis and so we discussed the possible reasons for her slow reading speed. I asked her to tell me what happened when she read text. She had

> *"to look at each word to make it connect to the next one . . . I draw pictures in my mind of what people are saying [in the text] in order to understand it . . . I have to read and re-read . . . I have to go slowly and have to re-read to fill in . . . create a visual image and then back to the words to fill in and to check it" (1)*

The reason for the delay was now evident. The student needed to find pictures in order to understand words and it is possible almost to sense the effort and energy required to do this. Though an intellectual and imaginative task, it had a physicality about it. She appeared to be switching between modalities to make sense of text. In the next example, this switching appeared to be an essential prerequisite to understanding.

> *"I need to convert what I read either into a visual image or a metaphor.If I can't convert the text, I find that I cannot convert the meaning" (6)*

This appears to differ from the enhanced appreciation of text which any reader with a visual imagination can obtain.

2 *Extra consciousness/feelings of visuals alongside text*

For a further student it was not a switching of modalities but a running of the two together.

> *"When I read text, pictures leap off the page"* (7)

This is possibly nearer to what non dyslexic people experience but it has an unusual dynamic visual spatial quality.

Another student reported that:

> *"it is not a picture as such but a feeling of an image which is a three dimensional structure. This shows all the relationships and interconnections within the content."* (2)

This description indicates that reading is a process which leads to a 'full' visual spatial experience, as all the stimulus from the text is translated into visual form. There is not a switching of modalities here but a rapid construction of a complex dynamic visual image. The same student also said that,

> *"reading takes all my brain to do it. But I can listen and speak at the same time"* (2)

This suggests that the mass of interconnected visual images uses up neurological space, that there may be a perceived physiological limit to it. This, too, may limit a student's volume of reading and account for reading fatigue. For this student, speaking and writing did not have this effect but I have known other students for whom this is the case.

3 *Reading as creating visual frameworks which are filled in with successive re-reading*

Frequent reference has been made in the discussions about reading as involving several stages. All the students in the discussion group agreed that the process of reading any new and unfamiliar text involved at least two stages: processing the text and then processing meaning on the second reading. It appears as if the stages which others may do automatically have to be separated out and done one at a time [perhaps part of what is usually described as an automaticity problem in dyslexia].

Further conversation revealed that reading was a process of creating and filling in a frame:

> *"the first read creates a structure or framework. The second starts to fill in the content and the third completes the process . . . (2)*

Others noted that far more than three reads were needed with unfamiliar text. There was often a complete blankness about new text, with no idea about the kind of document it was and what to do with it.

In the example below, the process of filling in is described graphically:

> *"Have you ever turned an abacus on one side and seen the balls fall down . . . some very quickly and others very slowly . . . thats what it is like" (3)*

The abacus provides the frame and the movement at different rates suggests that reading meaning comes through in this way. There is a delay before all balls have dropped down.

The question of whether the type of text affected this experience suggested that students felt that some sorts of fiction were experienced in a different way. The re-reading was less problematic in the case of poetry,

> *"because there is a small amount of text and you have to keep re-reading it anyway. Visual imagery is important for full interpretation." (4)*

Adverts were described as providing a multi-sensory experience including 'visual poems'. This suggested that text within media visuals was absorbed within a much bigger visual experience.

One student suggested that the layout of the text was important because

> *"I need corners to find a place on the sides. I need all the corners to find out whereabouts it is" (5)*

This may explain why some dyslexic students always seek boxed text for ease of reading.

4 Experience of colour in relation to text

Students regularly discuss the effects of colour in relation to text. Two main categories are explored: first the impact of colour on visual discomfort. If a student experiences the white spaces between words as' torchlights shining out '[as a student outside this sample recently reported], then coloured overlays can serve to dim this. Certain colours also have the effect of bringing the text into 3D or of quietening down the movements they experience when looking at text. However, dyslexic students within this sample have also spoken categorically about the intense emotional messages carried by colour.

It is clear from this material that textual forms which are dominated by black text on white paper will 'disable' readers with this kind of reading experience. Particular conventions of text production, governed largely by economic considerations, contribute to the 'disability'.

These descriptions of reading both reveal a high level of activity [switching,expanding ideas,filling in meaning] and reasons why delay in accessing text may occur. I shall now turn to some descriptions of writing.

Writing

Three dimensional references emerged during discussions about writing, too. Ideas can be compressed into jumbled words

> *"I can have three ideas at the same time and the three ideas are like three dimensions which become compressed into one jumbled word. I have no way of unencoding them. When I produce malapropisms it is because I am having three ideas at the same time and they are being transformed into one 'three dimensional' word."* (2)

Ideas are also described as being experienced in dynamic multi-dimensional matrices or as contained within some visual shape. Some get in the way of some sorts of writing. Students who have had extreme difficulties in organising a piece of writing describe their thinking in visuals which do not help to produce linear academic stories.

Students at Leicester University (1993–94) first alerted me to this with these descriptions of thinking about an essay question:

'everything is connected . . . it does not matter where you start'

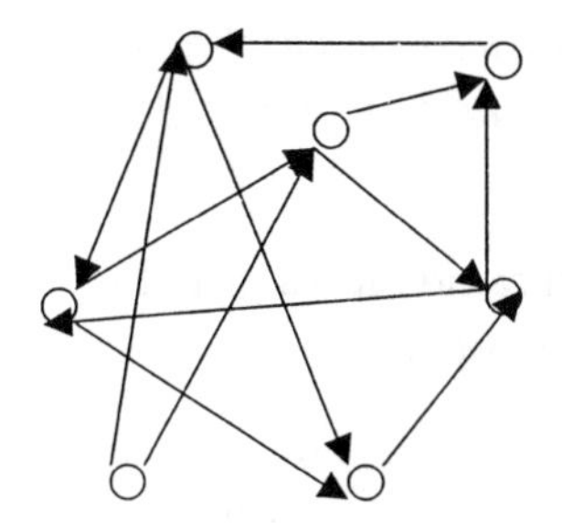

'all my ideas are contained in here. I cannot get them out'

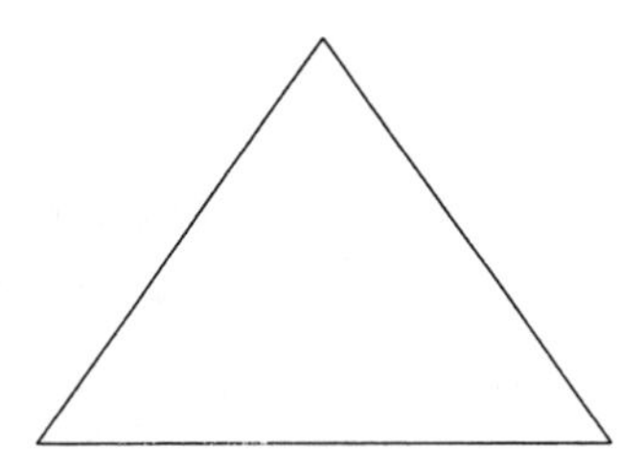

Within this sample, the first description was repeated and the implications explained:

> *"my supervisor says my dissertation has lots of repetition. The problem is that I am writing about each bit and the connections from each. How can I avoid repeating the links?" (8)*

Further discussion revealed that a student:

> *Dislike[s] writing because it feels two dimensional (2)* [compared with the thinking and if no interpersonal interaction]

Another student described the process of collecting his ideas and material for an essay as creating a ball of string:

> *"I cannot really start writing until I've got the ball of string and I can find the starting thread." (9)*

Once again, visual spatial/kinaesthetic terminology is employed.

Learning and memory

Traditionally, dyslexia does not just involve 'weaknesses' in reading and writing. Certain aspects of memory are also implicated (McLoughlin 1994). The descriptions here emerged during discussions on retrieving information.

> *"my mind thinks in right angles" (10)*

"like a giant buffet table with lots of dishes and bits falling off the plates. If I chunk them I can gain something" (2)

". . . it is like a detailed room . . . you see the light but then it is switched off. You've to remember what is in there. If the light is on long enough, we can remember." (4)

For some it appeared as if the information came onto a visual blackboard and they had to get it down or lose it:

"It is like a carousel of visual slides. You have to write them down as you are going along...but you can't keep up with it" (8)

"The words are coming through like a tape going past-with a small window for me to see – I can write down the words I see in the window. The ideas are going faster than this." (2)

Students also described the use of colour for exam revision. One student created her summaries of courses in visual form; not using mind maps but creating detailed drawings depicting parts of the course, organised around one key word. She found this easy to memorise and would always draw it out once she got into an exam so that she had a course reference tool to hand. This student had a weak short term auditory memory (10).

Speech

This creation or experience of visuals did not just occur during reading and writing. Two students reported regular visual interference during speech

"I am getting an 'electric circuitry' picture as I am talking (11)

"As you are speaking, I am getting a picture in my head

"What of . . ."

"I am putting my head inside a professional hat and am looking round inside the hat at the different elements. I can turn my head around inside the hat" (9)

The experiences described here may contribute to the explanation for some of the reading and writing delays which dyslexic readers report. They also suggest a kind of rich visual experience with reading, writing, speaking and listening which can be communicated by visual analogies and metaphor.

Time

Students may also be having to switch between dimensions in terms of time. Time is deeply implicated in the identification of dyslexia and dyslexic students are renowned for difficulties with 'time management' or for being so compensated that they are hypervigilant in this respect. Institutional time acts as a disabling contextual constraint for many dyslexic students. Students commonly complain of too much writing being required in the time available and of practices which require all their modular essays to be in on the same day. Negotiating over time becomes a way of life for some, against a background in which the cultural valuation of speed in relation to reading, writing, information processing, and memory sets the broader 'disabling' framework.

But I began to be interested in how time was being experienced rather than simply why it wasn't being managed. The student descriptions in this sample reveal two possibly related insights in this respect: that time can be experienced as a total blank and that students can experience time as if it is a separate dimension.

Despite practical reasons to be aware of time, this student experienced the following blank:

> *"I can arrange to see someone at 5.30 about a flat . . . and I do need the flat . . . but something else will have attracted my attention and I have no sense of time passing. When I come to I am way into the evening"* (12)

A mature student in the sample, recently described himself going into his garage to spray paint a motor bike.

> *"I had gone in to do the job one evening and thought it would take about three hours. I came out at 7 a.m..I felt I had been there a couple*

> *of hours but in fact I had been there all night. I hadn't been to the toilet or had a coffee in all that time. I was quite taken aback . . ."* (2)

He mentioned that when he needs to concentrate that this can happen [almost the crowding out theory of dyslexia] but even more interesting was the description of the experience within the time blank:

> *". . . when I was doing the job...it's like I was in this wonderful world of creativity . . . you get stimulus and satisfaction . . . you get all these lovely feelings flooding through you . . . and then you come out and its like black and white. The dog needs taking for a walk and its like everything catches up with you . . . tiredness and thirst etc."*

Most people experience 'losing track of time' when they are fully engaged with an activity but in these cases the blank seems to be so total as to prevent warning signals which would alert many people . . . the need for the new flat . . . the fact that the night has been experienced; and the nature of the experience can seem qualitatively different: creative and in colour with powerful emotions versus a flat, routine, black and white experience of real time.

This experience of different visual spatial dimensions seemed very different to this author's experience and so I contacted the Arts Dyslexia Trust to discuss the issue further. In private communications with Susan Parkinson, an experienced educator and Secretary of ADT, she suggested that for her, "time is one dimensional because it goes one way only. If you are a 3D thinker, you have to move out and into the one dimensional area of time . . . it is like moving from one key to another". She also felt that "time is unreal . . . the essence of it is not being experienced". Clearly, as a researcher/practitioner, I would need to explore the available research on how time is experienced among the general population before being able to draw any conclusions about dyslexia. Even so, at the individual level, these 'dyslexic' descriptions cast a new light on time management work in learning support settings.

Conclusion

The qualitative evidence provided here shows dyslexic students using visual references and analogies when describing their experience with reading, writing, speaking and listening. In some cases the visual dimension impinges on the reading process and can slow it down, as the reader can only really make sense through pictures. In others the visual 3D framework enables us to make sense of what the reader is experiencing. The evidence also invites us to look further into how time is experienced by individual learners and thence why certain values about time, learning and assessment play a particularly strong role in creating disabling contexts for dyslexic students.

This direct evidence from learners has provided useful insights and should inspire literacy educators in all sectors to investigate further with their students. Though the limited sample size prevents any deductions about how all dyslexic students experience literacy or even about whether these are exclusive to those deemed to be dyslexic, these kinds of descriptions occur sufficiently often to warrant an additional strand in the discussions tutors have with students about literacy. The practical challenge of discerning patterns of difficulties among literacy students should be part and parcel of literacy practice. If students struggle, the task is to find out how they are operating with text and to identify the internal and external barriers etc . . . Tutors engaging with such an investigation will find many different patterns of working with text and these will inform the necessary further thinking about delivering the national curriculum in ABE.

This 'opening up' of thinking and practice does not in itself resolve the continuing problems with 'diagnosing' dyslexia in relation to IQ scores and with allocating extra resources to this group if similar resources are not also available for other students whose thinking learning styles are not currently catered for. The labelling in relation to IQ remains unacceptable in principle. The rationale for separating out those with dyslexic type difficulties [at all levels of ability] for additional resources would have to stem from findings that distinctive methods were required with additional resource implications. A case can be made for this in some adult literacy education settings. An in depth student centred approach will at its best cater for most patterns of approach to text, provided that the tutor has an appropriate 'gaze' (Herrington forthcoming 2001). But in the day to day rush of adult literacy practice it cannot always be at its best and the discussions about language processing which are at the heart of effective practice, and are vital for dyslexic students, may not occur. The allocation of some resources specifically for them may be essential for any learning to occur. A case can also be made in higher education for additional resources to ensure that the dyslexic students have staff with whom they can work to deal with the vast increases in information processing and the added time constraints. However, there are other learners who are disadvantaged in HE and they have an equally valid case for additional resources. Ideally there would be more resources for all.

The insights revealed here are important but it should be acknowledged that this is still relatively new territory and we are not yet at the point where we can create a research design which could produce strong quantitative evidence in these respects for all dyslexic learners.

However, this is an important area of professional development with which ABE tutors should engage. The exploration should not be left to HE students, art college or otherwise. Intractable literacy difficulties should lead to suspicion about some degree of dyslexia and the nature of the discussions with suitably aware tutors may lead to important breakthroughs for some students. What is required is that ABE tutors work with an open mind about thinking and learning styles and that they are aware of the evidence which is now emerging from dyslexic adults.

References

Barnes, C. & Mercer, G. (1996) *Exploring the Divide. Illness and Disability.* Leeds: The Disability Press.

Barton, D. (1994) *Literacy: An Introduction to the Ecology of Written Language.* Oxford: Blackwell.

Barton, D. Hamilton, M. and Ivanič, R. (2000) *Situated Literacies. Reading and*

Writing in Context, London:Routledge

Barton, D. & Ivanič, R (1991) *Writing in the Community*, London: Sage

Brigden, A. (2001) Dyslexia in Higher Education Art and Design.Summary of the HEFCE SpLD [Dyslexia] Special Initiative at the Surrey Institute of Art and Design, University College, *SKILL Journal*, No 69,March 2001,pp 32–37

Brigden, A & McFall, C (2000) *Dyslexia in Higher Education Art and Design: A Creative Opportunity,* The Surrey Institute of Art and Design University College, Farnham, Surrey, GU9 7DS (Available from the Academic Registrar).

British Psychological Society (1983) Division of Educational and Child Psychology, *Occasional Papers,*volume 7, No 3, 1983

Crowther, J.,Hamilton, M.& Tett, L. (2001) Powerful Literacies, Leicester: NIACE

DfEE (1999) Improving Literacy and Numeracy. A Fresh Start.

Fawcett, A.J. (1995) Case Studies and Some Research in *Dyslexia and Stress*, ed, Miles,T.R. & Varma,V.1995 London: Whurr

Freire, P. (1972) *Pedagogy of the Oppressed.* London: Penguin.

Gardener, S. (1985) The Development of Written Language within Adult Fresh Start and Return to Learning Programmes. ILEA Language and Literacy Unit, *Occasional Paper* No **2**. Published in 1992 as The Long Word Club. Lancaster University: RaPAL.

Gardener, S. (2000) Student Writing in the 1970's and 80's, RaPAL Bulletin No 40 Winter,pp 8-11

Gardner, H. (1993) Multiple Intelligences. The Theory in Practice. New York:Harper Collins,Basic Books

Gee, J.P. (1988) The Legacies of Literacy: From Plato to Freire through Harvey Graff. A Review Article of Graff. H.G. (1987) The Legacies of Literacy. Continuities and Contradictions in Western Culture and Society. Bloomington: Indiana University Press. In *Harvard Educational Review* **58**, No 2, May 1988.

Gee, J.P (2000) The New Literacy Studies: From Socially Situated to the Work of the Social, in Barton, Hamilton and Ivanič, Situated Literacies, London: Routledge pp180–196

Gilroy, D.E. & Miles, T.R. (second edition 1996) *Dyslexia at College.* London: Routledge.

Harste, J.C. (1994) Literacy as Curricular Conversations about Knowledge, Inquiry and Morality. In R.B. Ruddell, Ruddell, M. R. & Singer, H. ( ),

Theoretical Models and Processes of Reading. Newark, Delaware, USA: International Reading Association

Herrington, M. (1995) New Policies: Old Dilemmas. *RaPAL Bulletin* **27**. University of Lancaster: RaPAL.

Herrington, M. (forthcoming 2001) Adult Dyslexia: Partners in Learning, in Hunter Carsch, M, & Herrington, M. (eds) *Dyslexia and Effective Learning.* London: Whurr

Herrington, M. (forthcoming 2001) An Approach to Specialist Learning Support in Higher Education. ibid

Herrington, M. and Hunter-Carsch, M. (2001) Dyslexia: A Social Interactive Model, in M. Hunter-Carsch ed. *Dyslexia: A Psycho-social Perspective.* London: Whurr.

Hetherington, J. (1996) Approaches to the Development of Self Esteem in Dyslexic Students. In *Conference Proceedings; Dyslexic Students in Higher Education, Practical Responses to Student and Institutional Needs.* SKILL/University of Huddersfield

Houghton, G. (1995/96) Who says so . . . Who?. *RaPAL Bulletin* **28/29**. University of Lancaster: RaPAL

Ivanič, R. (1996) Linguistics and the Logic of Non-standard Punctuation. In N. Hall & A. Robinson, *Learning about Punctuation.* p148–169. Bristol: Multilingual Matters.

Ivanič, R. (1997) *Writing and Identity. The Discoursal Construction of Identity in Academic Writing.* Amsterdam/Philadelphia: John Benjamin

Ivanič, R. & Simpson, J. (1988) Clearing Away the Debris: Learning and Researching Academic Writing. *RaPAL Bulletin* **6**, Lancaster University: RaPAL.

Jones, C. Turner, J. & Street, B ed (1999) Students Writing in the University. Cultural and Epistemological Issues, Amsterdam/Philadelphia: John Benjamins Publishing Company

Kerr, H. (2001) Dyslexia and Adult Literacy: Does Dyslexia Disempower? in Crowther, J. Hamilton, M. & Tett, L op. cit. pp 69–79

Kiziewicz, M. Do Art and Design Students have strong visual spatial skills? in Brigden and McFall, op.cit. pp.109–110

Klein, C. (1989) Specific Learning Difficulties. *ALBSU Newsletter* No.32. (A summary of the ALBSU Project).

Klein, C. (1994) *Diagnosing Dyslexia.* London: Adult Literacy and Basic Skills Unit.

Klein, C. (1995) *Demystifying Dyslexia.* London: Language and Literacy Unit. (Now based at the South Bank University, London.)

Kress, G (2000) The Futures Of Literacy , RaPAL Bulletin , Summer 2000 pp 1–19

Lea, L & Street, B.V. (1998) Student Writing in Higher Education: An Academic Literacies Approach, *Studies in Higher Education* 23(2)

Lillis, T., & Ramsey, M. (1997) Student Status and the Question of Choice in Academic Writing. *RaPAL Bulletin* 32 . Lancaster University: RaPAL

Mace, J. (1979) *Working with Words. Literacy Beyond School.* London: Writers and Readers Publishing Cooperative in association with Chameleon.

Mace, J. (1992) *Talking about Literacy. Principles and Practice of Adult Literacy Education.* London and New York: Routledge

Mace, J. (2000) The Significance of Student Writing, RaPAL Bulletin No 40,pp3–6

McLoughlin D., Fitzgibbon, G. & Young, V. (1994) *Adult Dyslexia: Assessment, Counselling and Training.* London: Whurr.

Morgan, E & Klein, C. (2000) Adult Dyslexia in a Non- Dyslexic World. London: Whurr

Moss, W. and Richardson, S (eds) (2000) Student Writing: Past, Present and Future. *RaPAL Bulletin* 40. Lancaster University: RaPAL

Snowling, M (2001) From Language to Reading and Dyslexia, *Dyslexia* vol.7, no 1, Jan-March 2001, pp37–46

Stacey, G. (1997) A Dyslexic Mind A-Thinking. Dyslexia, *Journal of the British Dyslexia Association* 3, no. 2. London: Wiley.

Stanovich, K. (1991) The Theoretical and Practical Consequences of Discrepancy Definitions of Dyslexia in Snowling, M. & Thomson, M. *Dyslexia. Integrating Theory and Practice*, London: Whurr,pp 125–143

Stein, J. (2001) The Magnocellular Theory of Developmental Dyslexia, *Dyslexia* vol. 7,No 1, January-March 2001, pp.12–36

Steffert, B. (1996) Sign Minds and Design Minds: The Trade-off between Visual Spatial Skills and Linguistic Skills. In *Dyslexia in Higher Education, Learning Along the Continuum.* 2nd International Conference. Conference Proceedings, University of Plymouth, p.53–69.

Steffert, B. (1999) Visual Spatial Ability and Dyslexia, Part 1 pp8–49 & Part 2 pp127–167, in Padgett, I. (ed) *Visual Spatial Ability and Dyslexia*, Central St Martins College of Art and Design, The London Institute, Southampton Row, London WC1

Stephens, C. (1996) Developing Awareness of Learning with Individual Students. In C. Stephens, (Ed), *Dyslexic Students in Higher Education. Practical Responses to Student and Institutional Needs,* p 46–50. Conference Proceedings, SKILL/University of Huddersfield.

Street, B. (1984) *Literacy in Theory and Practice.* Cambridge: Cambridge University Press

Street, B. V. (1990) Putting Literacies on the Political Agenda. *RaPAL Bulletin* **13**, University of Lancaster: RaPAL.

Walker, M. & Chappell, D (forthcoming 2001) Effective Support for Adult Learners in Hunter Carsch & Herrington (eds.) op. cit.

West, T. (1991) *In the Mind's Eye. Visual Thinkers, Gifted People with Learning Difficulties, Computer Images and the Ironies of Creativity.* New York: Prometheus

West, T. (2000) Dyslexia in Art and Design: A Creative Opportunity, pp9–18, op. cit. Brigden, A & McFall, C. (2000)

Winner, E. et al (1999) Dyslexia and Spatial Talents: Is There a Relationship? Discussion Paper, Boston College and Harvard Project Zero

Wolfe, M. (1998) Shake Rattle and Write, RaPAL Bulletin No35, Spring, pp17–20, Lancaster University: RaPAL

Letter to the editor

Dear Margaret

Here are some comments from one of my Study Skills students (GCSE English) who had been assessed as dyslexic. I asked him how he saw words in print when he was reading, and wrote down what he said.

'I see white (glistening) around the black print and a line (sometimes) under individual words'

'like a three dimensional comic book'

'Like a stage without an audience . . . I see perspective . . . it's like a movie screen. It's not like you control them but when you read the story, you sort of bring it to life'

and

'Starting writing is like being in a Japanese restaurant where the counter thing constantly rotates. The difficulty is that it's gone past too quickly for you to capture it'

He also drew to illustrate his points as he went along.

I had not said anything about the 3D effect mentioned in the article in the Dyslexia edition of the RaPAL Bulletin (Summer 2001) and so was fascinated by the way that this was implied by what he said.

Best wishes
Kate Tomlinson, Stroud College

Dyslexia: Moving forward in further education

Doreen Chappell

RaPAL Bulletin No. 27, Summer 1995

Doreen Chappell is an experienced ABE Area Organiser, who has worked extensively for twelve years with dyslexic students in Leicestershire, providing specialist support for a wide range of students and delivering in-service training.

Wigston College is fortunate in having a strong Language Support Department and especially in having acquired two established ABE schemes from a former county service. Therefore, in these sectors of the college, there is extensive staff knowledge and experience of dyslexia, as well as keen interest and a desire to know more from less experienced basic skills tutors. Support for students with dyslexia is still in its infancy in college, but it is becoming established and there are now three main routes: individual support, workshops and short courses.

Individual Support

Students refer themselves, or are referred to the special needs tutor, with requests for discussion/extra support; these are all people enrolled on college courses who experience difficulties or who have previously undergone a dyslexia assessment. These students are interviewed by me and a joint decision made to arrange individual support appointments, or to continue as before, with the knowledge that help is available if needed. Students may also be referred for formal assessment, funded by the college. Currently, students from the college Access course make up the bulk of the clientele.

Support for individual students is essentially student-centred. It can include tuition to recognise spelling errors and methods to improve this, help with organising the content of essays, teasing out ideas, discussing information about dyslexia, and lending an understanding ear when dyslexic difficulties, and feelings about them, seem insurmountable.

Dyslexia Workshops

A dyslexia workshop has run weekly for just over a year. This draws students from college and from the community. It is also intended to provide a bridge between considering and beginning a college course, such as GCSE.

Students involved in the group include: those whose dyslexia is undiagnosed but who have been wrestling alone with the difficulties involved; those who have had a previous assessment but very little contact with other dyslexics; and those who have come to terms with the label but wish to improve in specific areas.

The group enables students to discuss and unload their feelings about dyslexia, to work on difficulties that arise, and to prepare the way to developing the confidence to enrol on a college course, assured of continued support if needed.

Short Courses

We have also developed a short course which enables students to explore dyslexia for themselves. This is open to all students whether 'diagnosed' or not and is running currently. The course empowers students to explore literature about the subject, to take the initiative in explaining the sort of problems/frustrations that occur, and to consider open college accreditation. The Leicestershire Open College programme 'Investigating and Coping with Dyslexia' allows students to gain accreditation for their own research, for case studies, and for ways of dealing with their particular area of difficulty. A similar course is planned for the summer to take place in a local village. We believe that this is an important step forward in ensuring that students do not passively accept diagnoses by others.

Those students who have taken advantage of this menu of support have described what dyslexia feels like for them and how it affects their study practices. The comments on the following page are from students who identified their need for support; there will be many more who still feel they're unable to cope because they lack intelligence, or through a conviction that their F.E. course is not right for them. Unfortunately, by the time these students are referred, they are often so demoralised by their failures that they do not keep their appointments and subsequently give up the course.

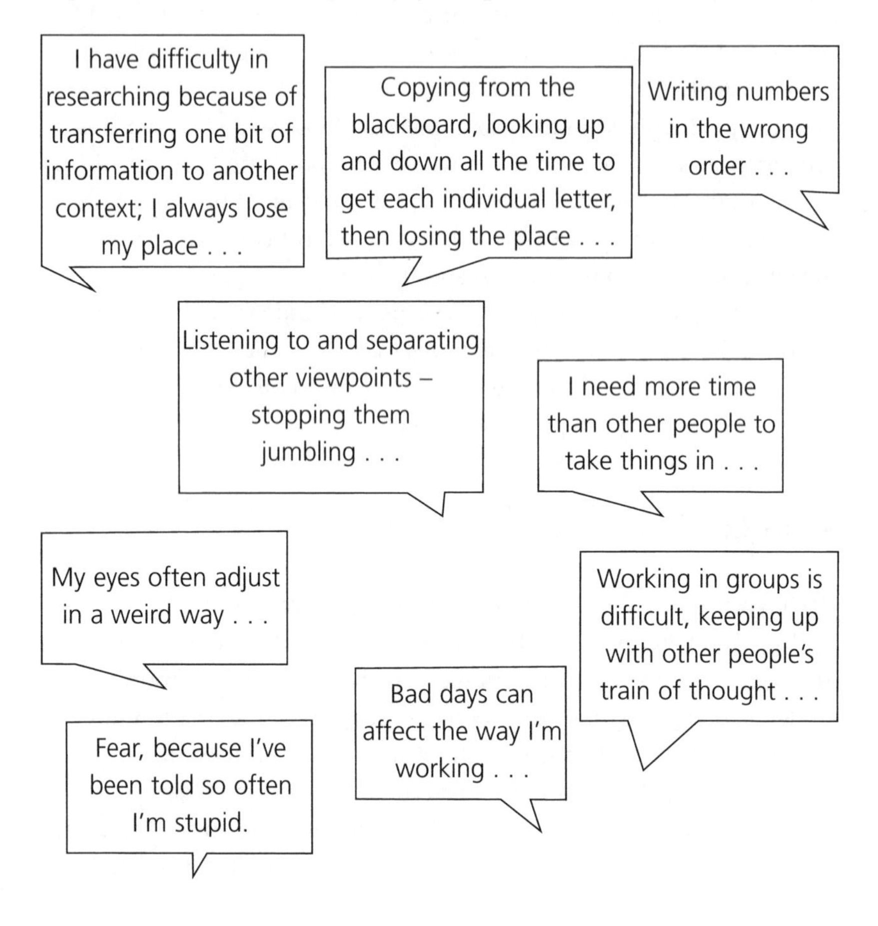

Staff Development

This support work has revealed the need for all college staff to be more aware of the implications of dyslexia in the F.E. context. They will need to recognise the struggle of the students and also the particular effects of some of their teaching methods. However, there is still some resistance and lack of knowledge in many departments and evidence of misconceptions:

The last comment was made to a very bright student heading for

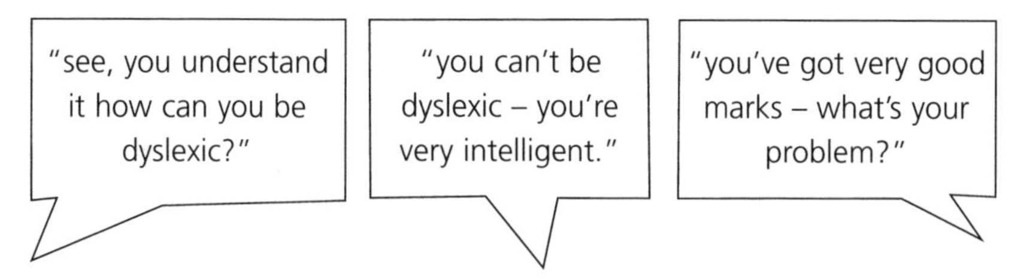

university, who had taken hours and hours and hours over an assignment, with support and encouragement from Link into Learning tutors and her husband; it was good, but it had been a very difficult process to get there.

The staff development implications are therefore considerable and a training session, in the regular staff development slot, has recently taken place. This aimed to explain what dyslexia is, the particular problems encountered by dyslexic students on college courses, and ways in which students can be better supported – for example, by use of coloured paper, dictaphones, consideration of the size and type of print in reading materials, and a reduction in the need for laborious copying.

My colleague and I welcomed the opportunity to deliver this session, but were disappointed, though unsurprised, by the small response. We hope, however, that by initiating interest, at least a glimmer of awareness will develop and eventually lead to further requests for training.

As far as Wigston College is concerned, there are real steps forward in acknowledging the problems associated with dyslexia, and towards providing appropriate support. I feel that the college could benefit from the development of a specialist service which could take on board issues for students and tutors and supply both individual help and necessary staff training and advice. However, the difficulty of accessing appropriate funding, and some misunderstandings, are still hampering this move. Open minded attitudes are essential and greater recognition and acceptance of ways to smooth the path of the dyslexic F.E. student are needed. After all, these students simply aim to achieve equality with their peers, without a superhuman struggle, or without feeling so embarrassed about the difficulties that they 'become afraid of admitting it.'

Dyslexia and higher education

Chris Rishworth

RaPAL Bulletin No. 27, Summer 1995

Chris Rishworth, a second year Archaeology undergraduate at Leicester University, identifies his preferred way of learning and shows how the academic environment has both stimulated and, at times, undermined his writing confidence.

I was brought up on a farm. This allowed me to have access to tools, equipment and open space. In this world I found skills, and expressed myself, by stripping engines down, working with tools on metal and wood, and watching things live and grow on the land. Literacy skills were a bit harder to focus on and I went along with them on one level but I only found it easy to read things in the landscape around me. The ideas I had about our world I had in my head; I didn't think about writing them down.

In the world of work I have been ambitious to try new things. I have been fortunate in that most of my employers thought me bright . . . as I had many skills . . . but I always wanted literacy skills to complete the circle, so that I could work in more areas with more competence and confidence. I was offered the chance to manage part of a museum before I came to university, but at the time I felt I could not even write a letter without including a thousand grammatical errors.

Learning to study at university has had its positive and negative sides. I began to see the beauty and function of language in expressing ideas and thoughts, but at the same time it has been very painful. Sometimes my head has ached; it seemed as if new connections were being made in my brain. At the start of my course I experienced a great feeling of de-skilling and no longer felt in control, because I no longer had time for the practical, productive, and creative activities which everyone needs; and writing had not yet become creative for me. In the academic environment I also discovered an isolation from others as I concentrated on writing skills, and could not contribute as you normally would in a work environment by interacting with other people, and addressing practical challenges. I have, therefore, during some periods, felt that my contribution is low and this affects my confidence with writing. I have, however, been fortunate to have worked in the field for my course work; a recent placement with The Forest Service in America has provided renewed confidence. This organisation regarded my drawing skills highly and I was able to assist in directing a project where I could teach my creative skills to others. My humbling experience at university allowed me to understand how difficult it was, for some of the team I directed, to grasp the drawing and recording skills that I had found so easy.

My preferred way of learning is still through what I touch and see – through mechanical, tangible, connections. For example, I find it difficult to learn about a topic in archaeology just through reading; seeing the site and handling the material are crucial to my understanding. However, my reading skills have improved and I am beginning to feel what it is to understand the ideas of others through interpreting their written language; and during my course I've been stimulated greatly by ideas and concepts in written text. In terms of my own writing, I still find some aspects very difficult. However, I have become very competent in using computers, despite being technophobic at the outset (due to the old forms of computer language). Modern icon and menu driven software is so much more interactive and friendly for a dyslexic student, and spell checkers are invaluable. My handwriting varies with my confidence and so I use the computer to give a print out which has constancy and clarity; then I can feel more confident about presenting my written ideas to others.

At each stage within the higher education system, I have come across a great deal of misunderstanding about the difficulties I have faced, from those who have no way of understanding what the problem is. I, of course, also have great frustrations and some ignorance of how to overcome the hurdles and so cannot be too angry with them. My answer has been to press on with grim determination, only possible because of the confidence key people have had in me; people who understand some of the problems students like myself face within the education system. This has egged me on to success.

I'm learning to read again

John Karlik

RaPAL Bulletin No. 26, Spring 1995

John and Jill Karlik are directors of a Christian English Language School in Whitby, but make periodic visits to Guinea-Bissau and The Gambia to assist in adult literacy projects in a local language. John wrote this article in 1992, four years after he suffered a stroke.

We all possess abilities which become so much a part of our daily lives that we take them for granted. I want to recount an experience which taught me to value my literacy skills more highly.

In my sitting room, everything seemed different. The armchairs, table and pictures were all there as before, yet something seemed to have changed. Then my eye fell on the television. The sound was turned down and there were lines of letters on the screen, but they were only lines of shapes, with no meaning and no relationship to one another; they might as well have been Chinese characters. *"I can't read,"* I told my wife, Jill, with as much composure as I was able to muster.

I felt confused all that day. I wanted to communicate, but my sentences were disjointed, unfinished, jumping around from one subject to another like knights in a chess game. I wandered about the house, and spoke about people and happenings nobody could recall.

Jill called the doctor, who gave me a few simple memory tests. *"I'm going to tell you an address,"* he said. *"Try to remember it. I'll ask you in a little while."* Then he went on, *"How old are you?"* – *"Forty-one,"* I said. But no, that was wrong. *"Eighty-one,"* I ventured. – *"Can you remember your telephone number?"* I couldn't. – "Can *you tell me that address? The one I told you before?"* I couldn't. The doctor diagnosed a stroke and told me to go and see my regular doctor the next day.

A stroke is, in most cases, a symptom of old age. The veins thicken, lose their elasticity and become brittle or blocked. One of the veins carrying the vital blood supply to the brain bursts and blood enters the brain tissue, putting the cells out of action. If possible, the internal mechanism of the nervous system tries to re-route the electrical signals to other cells in the neighborhood. These cells have to accept an additional load and, in fact, learn to do a new job. They can in some cases completely replace the cells which have been damaged and the lost functions are gradually restored.

I found I had to re-learn the use of some everyday objects. I came to our front door and could not remember how to open it. *"There must be a lock somewhere," I* thought. But at least such everyday skills were easy to regain. I had Jill's constant loving attention, and that made a big difference, because in many ways I was like a child, having lost some of the most elementary memory store.

I had also forgotten certain recently acquired vocabulary. English was not my mother-tongue and I had learned it in Australia, so I was not really familiar with words like "heather" and "moors" before we moved to the North of England. Now I had to re-learn these words, but they came back quickly as Jill drove me round to "re-introduce" me to the area.

My sudden dyslexia – the loss of reading – presented me with a problem of more serious proportions. It was not that I had forgotten how to spell the words. I could still write and type, but I was unable to read back something I had typed just minutes before. Because of long-standing Parkinsons Disease, my handwriting was very poor, so I always typed everything. Now I could not read it back to check my many typing errors. *"It will be a long, slow job,"* the specialist told me. *"You may never read again. But the only way to find out is to try. "*

Inwardly, I shrank at this verdict. To learn to read again? Now, at sixty-one? The prospect appeared as daunting as an invitation to climb the Himalayas. How fast would I be able to learn? As it happened, for a few years Jill and I had been missionary volunteers, working together on linguistics and literacy projects. The analysis of a previously unwritten West African language had formed the basis of my doctorate thesis and we had produced an alphabet and reading materials to enable the people to learn to read in their own language. The psychologists at the National Hospital in London were very interested, even excited, at having a linguist as a new dyslexic, able to describe his own literacy problems. They called me in for one of their multidisciplinary seminars and examined my symptoms in some depth, but they could not offer me any strategies for re-learning. So we set out to apply all our own so-called "expertise" in literacy to my own needs.

Acquired dyslexia, according to medical writing, is a condition inhibiting the recognition and recall of letters with resulting loss, in varying degrees, of the ability to convert the written symbol into a mental concept. If caused by a stroke, it is very often accompanied by a "blind spot". Since the damage is to the brain, not to the eyes themselves, it makes no difference which eye one looks through: the blind spot is still there. The area can be small or large, up to a total loss of sight. In my case, it was so small that my optician, who happened to examine me a few days after the stroke, did not discover the defect at all. But even though small, there was the blind spot, sitting just to the right of the centre of my vision, getting up to all sorts of mischief: looking at my wife, it was unnerving at first to find that she had a blank spot instead of her left eye. Of course, this happened every time I looked at someone, and I saw myself one-eyed in the mirror.

The existence of my "blind spot" proved to be less amusing as I applied myself to reading: the blank space moved along the line ahead of my eyes, blotting out part of the next word. Sometimes I assumed it was the end of the line and moved to the next, missing several words. I also found that it did not help when I tried to use a book printed in bigger letters because the bigger the characters, the wider the spaces, and therefore, the greater the likelihood of missing the beginning of the next word. Walking down the street, I stopped, surprised, at the shop sign, "Imperial Cancer", and I had to read it twice to discover the full title: "Imperial Cancer Research Fund".

If several short lines left a blank space, there appeared to be some indistinct words filling the gap. When I tried to read them they would disappear over to the right. *"The mind does not like to have unexplained blanks"*, the doctor said, when I told him about this strange phenomenon. *"It tries to explain the sudden appearance of an unexpected blank by providing something to fill it, something which could logically be expected to be found there"*. In this case it was an indistinct line of letters. After a few weeks I lost this interesting illusion because my brain accepted the legitimacy of my "blind spot" and stopped trying to provide excuses for empty spare when it happened to be superimposed on a genuinely vacant spot.

My form of dyslexia is sometimes called 'word-blindness' or 'letter-by-letter reading'. The only way I could make sense of a word was to work it out letter by letter. Normally, we do not read letters in isolation. The fluent reader takes in a whole word at a glance and interprets the meanings of the letters in relation to each other. But when I attempt to read a word like ***'cautious'*** for instance, I don't know whether the ***'c'*** will sound like ***'k'*** or like 's The next letter, ***'a'***, can have many values. The word could be *'car'* or ***'cat'*** or ***'cake.*** It is only when I get to the ***'u'*** that I know the syllable sounds like ***'caw'***. As for the ***'tious'***; it has to be read as a unit, and somehow I could not recognize these units. Within a few days of the stroke, I was able to tackle syllables and short words, but I found some letters quite unfamiliar. I was intrigued by the wiggliness of ***'g'*** and wondered what it sounded like. When told, I was able to re-learn this particular letter quite quickly, but others were not so easy. More difficult were the letters which looked alike, such as ***'b'*** and ***'d'***; or ***'s'*** and ***'c'*** , and I often mix up *'ch'* and ***'sh'.***

Then there were sounds symbolised by different letters in other languages. In Greek, the sound ***'n'*** is written by a letter which looks very much like ***'v'.*** Although I know very little Greek, for the first few weeks after the stroke I kept reading ***'v'*** as ***'n'***: I experienced a whole array of symptoms. Fortunately, none of them lasted long: letters jumping; lines broken in the middle, overlapping and not lining up; attempting to read words back to front. Double figures appeared to be triple figures, so no matter how much I concentrated I could not decide whether the number was ***'88'*** or ***'888'***. Several times I had a distinct feeling that I should be reading an invisible line above the one which I was on, like a musical score line accompanying a line of lyrics.

Jill made up flashcards for me with words and parts of words. I entered wholeheartedly into the spirit of what she was doing but I was still not able to make the connection when she combined cards with ***'cook', 'er'*** and ***'ing'*** to form ***'cooker'*** or ***'cooking'*** It was very discouraging.

We turned for advice to the local Teachers' Centre, but all they could offer was easy-reading children's material. Try as I might, I simply could not recognise the word ***'pig'***, however many times I came to it in the repetitive story about the old woman and her pig. Then I picked up a copy of "Cry the Beloved Country" and found I could cope better with 'real' reading material. Jill helped the pace along by reading alternative paragraphs with me. I still had to pick out most words letter-by-letter but I was thrilled to find I could recognise a few distinctive words such as ***'Johannesburg'***. I began to make some progress. I found support and encouragement with my efforts from the speech therapist at our local hospital, whose help had been called upon because of an increased slur in my speech.

Wanting to re-visit old friends while I still had some health, we took time off for a round-the-world trip a few months later and we continued to read together daily. In Esperance, in the far south of Western Australia, the headmaster of a small Christian school showed us how he had introduced audio equipment to help his slow-readers. They would listen to their favourite book on cassette and read along with it. We realised that this was effectively what we were doing as Jill read aloud with me, but she had to read below normal speed if I was to follow it.

As I began to get faster at picking out words, I started trying to read books against time, with my stop-watch. The first few chapters were difficult: I kept losing the place and it took half an hour to read a page. But gradually the meaning of the letters and the arrangements of the syllables and words started to make sense and I reduced the reading time to 20 minutes a page. Unfortunately, if I stop practising, my time gets slower and slower again.

I found I was able to follow the words of hymns in church, especially the slow ones, but I was unable to follow the speed of a normal Bible reading. I skipped over words in my efforts to keep up and then found myself trying to read them backwards. It was a year before I could read fast enough to follow text being read at an ordinary speed.

At the beginning I was totally unable to make head or tail of handwriting. When a handwritten message was in English – or French, German or Portuguese – Jill quickly set the matter right by simply reading it out to me. But when I had a handwritten card from my friends in Czechoslovakia, I was helpless, because the letter combinations were too difficult for Jill to decipher them. After about three months, the ability to distinguish handwriting came back quite suddenly, though I can still only read it letter by letter, and only if it is neatly written. My family can leave me messages carefully written in a childish script, but most people do not write neatly so I still have to rely on others to read much of my correspondence.

When I tried to get back into normal work again, I found there are many different reading skills I had taken for granted. Have you tried to pick out one book from the shelf; or to find one name in a directory? The skill you need is called 'scanning' and it is not until you lose it that you come to appreciate its worth. I stumbled by accident on an excellent method to improve my scanning skill. I borrowed books and cassettes in the "Talking Book" series, in an effort to force myself to read faster and take in larger 'chunks' at a time. I discovered the additional challenge that large sections of some paragraphs had been left out to make the spoken text shorter. Trying to find a key word in the text to pick up the storyline again provided me with just the practice I needed in scanning.

Some young people from the local secondary school volunteered to come *to* my home and read aloud to me, to help me with research for some historical articles I was writing. I recorded them as they read and then was able to listen to the recording again later and try to read along with it. I found I had good motivation. It was easier to persevere with this material than with commercially recorded novels.

I struck more of a problem, however, when the material I needed was in Czech, because there was no-one in our small town who could understand Czech except myself. I managed to persuade the department of employment to allow me to have a Czech reader come over to help me for a while. Unlike English, Czech is written in a regular 'phonetic' form and Jill found she was able to learn to read paragraphs of Czech aloud to me, although she did not understand anything of what she was reading. This was a great help after my

Czech reader left, when I needed to skim through the text to check up on something.

It has been a long haul since the stroke just over four years ago. I know I am much better now by the fact that my wife no longer says: *"That* was *very clever of you, Darling".* Looking back, there were small 'triumphs' to lighten my way, like the first time I was able to read a Bible verse aloud as we read around a group, instead of having to nudge my wife to break the silence when my turn came. Occasionally I recognise a whole word in a hazy, indistinct way, as if I saw it through a veil, and then, when I go back to verify it and read it letter-by-letter, I am thrilled to find that my initial recognition was correct.

A lot of my reading practice has been with the Bible, because the content is familiar and word recognition is easier. It was a great blessing to me to rediscover certain passages of the Bible and now, as I keep reading it, it holds a new message for me. I know I am in the hand of the Almighty God and can say, as the Psalmist said, "My *times are in Thy hand."*

B4 Disability

Further RaPAL references about disability

'LASER – Language of Signs as an Educational Resource', *RaPAL Bulletin* No. 17, 1992.

Laugeson, C. 'Literacy and Deafness: Research and Practice Issues', *RaPAL Bulletin* No. 17, 1992.

Two Can Resource Unit. 'What is Two Can?', *RaPAL Bulletin* No. 17, 1992.

Herrington, M. (2000) Conference report from workshop 'Moser – Basic Skills Standards and People with Learning Difficulties and Disabilities'. SKILL Annual Conference, Policy and Practice 2000 and Beyond, 25–26 February. *RaPAL Bulletin*, No. 41, 2000.

A silent minority: Deaf people and literacy

Crissie Laugeson
RaPAL Bulletin No. 16, Autumn 1991

Crissie Laugesen is a lecturer in Adult Education in Preston, Lancashire. In 1990 she was asked to provide evening classes for Profoundly Deaf Adults in need of literacy/English help. She suffered hearing-impairment herself as a child and often still has reduced hearing and tinnitus.

What is Deafness?

Interpretations of the term "deaf" vary within the hearing community. In the Deaf Community, the word has a clear meaning: Deaf people do not hear at all in the accepted sense of the word "hear", and what little hearing they may have is not sufficient to understand conversation, listen to music, distinguish the normal sounds of everyday life. "Profoundly Deaf" is a term often used, for the benefit of hearing people, to identify those who are totally without hearing or have very little residual hearing. Of these, "Born Deaf" refers, obviously, to congenital deafness either through damage (e.g. Rubella) or heredity. "Deafened" people are those whose loss of hearing occurred through accident or illness either in childhood or after maturity: the writer and poet David Wright and MP Jack Ashley are examples of this.

Other less severe degrees of hearing impairment are apply to those known as "Hard-of Hearing" and "Partially Hearing". These people have enough hearing to be helped significantly by a hearing-aid, though their hearing remains imperfect. Hearing-aids, whilst a wonderful invention, lack the sophistication and selectivity of the fully operational human ear, and amplify all sounds without discrimination: conversation, and what to a hearing person is "background noise", can become confusingly blended into what Wright calls a "blurry bumble of noise". (1)

For the purposes of this article, I shall use the terms as described above, so "deaf" means having profound or total loss of hearing. Out of respect for colleagues and friends in the Deaf Community, I shall use the capital letter D as do they when referring Deaf people.

Reading, Writing and Deafness

Despite attending schools for the Deaf for the requisite number of years, many Deaf adults in Britain have more than their fair share of difficulties with the written word.

A survey conducted in 1979 (2) disclosed that many Deaf children were leaving school with shockingly poor levels of literacy; so poor, in fact, that tabloid newspapers, instructional manuals, safety regulations and government forms are beyond the comprehension of the Deaf school leaver. Even worse, the results concerning the most profoundly Deaf revealed similar reading ability to that of hearing eight-year olds: unable, therefore, to cope comfortably with Teletext and sub-titled TV News. The surveyed children are now approaching thirty years of age, and we can conclude that there are at present in Britain a great many Deaf adults whose level of literacy prevents them from coping easily with the most common everyday reading tasks, whether at work or in their leisure time.

Aids for Deaf people may be useless for a Deaf person who also has difficulty with reading and writing. Tele-writers – special telephones fitted with letter-printing screens, upon which Deaf users can send and receive telephone calls – are irritating or embarrassing for the slow reader. Even the favourite leisure activity of the average Briton, watching television, presents problems despite the latest advances in Teletext. The very latest development is Simultaneous Sub-titles on the TV News: these, confusingly, have to be read from the bottom line, and disappear rapidly upwards into the lap of the Newsreader, so the struggling reader has to try to un-learn the painfully acquired reading habit of starting from the top and reading downwards. Subtitles for the Deaf are a positive step towards equality of access; unfortunately, they themselves are virtually inaccessible to many of the very people they were created for.

A History of Discrimination: Education the Deaf

Throughout history, Deaf people have been a minority subjected to discrimination, low status and marginalisation, regarded at best as curiosities, at worst, as freaks. Such is the status of speech in the hearing world, that those unable to produce it according to normal standards have always been mocked; those unable to respond to it, ignored or institutionalised. In Mediaeval Europe, at a time when illiteracy was the norm rather than a matter for social concern, the spoken/heard word was all-important, and in order to be able to put his/her mark or thumb-print to a legal document, the average person had to be able, to *hear* it read or recited. While the lives of the lower classes were difficult, and no one's lot was really happy, the status of the Deaf person in the community was particularly low: deprived of legal rights, regarded as feeble-minded and incompetent, the Deaf were, literally, without a voice.

Upper-class Deaf persons fared rather better than their peasant counterparts, especially in sixteenth-century Spain, where aristocratic in-breeding had resulted in many cases of hereditary deafness. Spanish law prevented the inheritance of land or titles by Deaf persons without speech, a situation which brought about an interest in the education of the Deaf and Spain became the first nation to organise education for Deaf people, albeit for aristocratic material interests. The first recognised teacher of the Deaf was Pedro Ponce de Leon; a monk from the silent Benedictine order, he was familiar with a form of sign language and no doubt used this in his work. He was expected to teach his pupils to speak, read and write well enough to comply with the law, and appears to have been successful for several years. His greatest achievement was in proving to a concerned and wealthy interested party – the Spanish nobility – that Deaf people were not gesticulating idiots but educable human beings with equal rights – *if they were wealthy enough to pay for tuition.*

In England, Deaf people had aroused interest by the sixteenth and seventeenth centuries. As in Spain, efforts at educating the Deaf were confined to the wealthy, not, in this case, to ensure inheritance, but rather as scientific experimentation. In an era of scientific advance, the educability of the Deaf became the subject of discussion by writers and physicians, but remained the preserve of the wealthy and was kept in the hands of private tutors.

Larger-scale Deaf education began in Britain in the 1760s, when Thomas Braidwood's School for the Deaf in Edinburgh became a flourishing and prosperous business, as wealthy parents sent their children to be educated by Braidwood after Dr Johnson wrote its praises in 1775. Success inspired a move to London and the opening of another private Deaf school (with similar high fees), followed by the first "charity" school for Deaf children of the poor, to be run by Braidwood's nephew, Joseph Watson. The Braidwood/Watsons were succeeded by the Asylum System, founded by private benefactors and maintained by public subscription, and by a variety of religious and charitable schools which used mixed-method approaches and both Deaf and hearing teachers. By the mid-nineteenth century a proliferation of schools and methods suggests some confusion and rather low standards:

> *"The wretched teacher of the Deaf had neither training nor professional status, nor a union, nor, indeed, any official recognition for his existence at all" (3);*

and some shocking discrimination: [schools] ". . . taught speech to paying pupils and allowed signing for the massed poor" (4).

The "massed poor" were "allowed" to sign while the wealthy Deaf children were the focus of teachers' attention and taught speech, creating a divide in the Deaf community and a speaking elite.

Methods of Teaching/Learning: Oralism versus Signing

Conrad in 1979 laid the blame for the poor standards of literacy and speech of many Deaf people firmly at the door of the particular teaching method which had dominated British Education of the Deaf for the previous century, to the absolute exclusion of all other methods: namely, Oralism (2).

In Britain, Braidwood and his followers had been secretive about their methods, but eventually admitted to using a mixture of speaking, signing and literacy – including finger spelling – to achieve their notable success. Sign was used alongside oral methods, albeit in the discriminatory way as mentioned above. In France, Sign Language was used freely in schools for the Deaf to very good effect, and French methods were exported to the United States of America by Thomas Gallaudet in 1816; here it flourished, giving Deaf teachers the opportunity to work alongside their hearing counterparts, and in many cases to run their own schools and colleges. American Sign Language dominated US Deaf Education in the nineteenth century, producing literate and educated Deaf persons well able to take their place in society.

Nevertheless, to rid Deaf people of the "dumb" or "mute" label by giving them the "gift" of speech would seem to be a commendable and worthwhile aim. In Milan in 1880, the second International Congress of Educators of the Deaf ruled that all schools were to adopt Oralist methods; Deaf children were to be talked to, made to answer with their own voices (though they themselves could not hear those voices), forced to talk and to respond to the spoken word, and to talk to and lip-read each other. Little matter that the Deaf themselves showed every inclination to avoid speech unless all else failed, for they had long learned from the reactions of hearing people that their voices often sounded strange. The fact that Deaf people chose not to speak to each other (and why should they, when they could not hear each other's voices?) but instead invariably chose to use another form of communication mattered not to the Milan Congress. The important thing was to *make the Deaf speak,* an objective which was to take on the proportions of missionary zeal.

In the years leading up to the Milan Congress, the idea that deafness was one of many social evils still held firm, and the Deaf were still looked down upon; the Deaf poor even more so, and in Britain, their language, Sign, identified them as belonging to a doubly despised group. That they gathered together, married, had families and formed communities, alarmed those with strong views on social hygiene, who feared that the development of a "variety" of Deaf people would pollute the human race. One of the most extreme of these was the Scots-born Alexander Graham Bell, who advocated preventing Deaf people from marrying each other (he was married to a Deaf woman) in order to eradicate deafness. His eugenic theories encompassed Sign Language, which he saw as a manifestation of degeneration. Teaching the Deaf to speak was seen as a step towards their normalisation, standardisation, *humanisation;* upon such views as this was the power of the Oralist argument based. It is ironic that Bell is celebrated in the hearing world for his invention of that modern communication device the telephone; the Deaf have other reasons to remember him.

In 1880, the Deaf education scene in Britain was a confused mix dominated by the ideals of the social elite, who could afford the intensive one-to-one instruction Oral methods demanded; in the more egalitarian and forward-looking USA, Deaf Education (Sign-led) was showing impressive results for all pupils. Despite this, the Oralists won. Bell, who feared that Sign Language united and strengthened Deaf communities and thereby prevented the eradication of deafness, put all his influence and financial resources behind the Oralists. Their views prevailed in the Milan Congress of 1880, which had only eight US delegates out of a total of 164: of these eight, any who might have supported Sign were outvoted. The ostensibly humanitarian aim of Oralism, that of "restoring deaf mutes to society through speech and lip-reading" (4), overlooked the fact that it was literacy and education, rather than the "power" of speech, which restored the rights of Deaf people. The right to remain equal but different, even if that meant using Sign Language and having an "abnormal" and therefore little-used speaking voice, was taken away from them by the decision of the Milan Congress.

The effect of the Milan decision was devastating. It gave strength to the argument that Oralism was the only valid way to teach the Deaf, and was used in Britain as vindication of a deliberate movement to suppress the form of communication Deaf people were using amongst themselves, in education and daily life: Sign Language. Oralism meant that Sign Language must not be used in schools for the Deaf, and one immediate effect of this was the enforced redundancy of teachers who were themselves Deaf, unless their own voice production met the exacting standards required of Oralist teaching methods. The Deaf were robbed of the teachers who knew their needs and problems from personal experience, and their education was put in the hands of hearing educators who imposed their own, supposedly superior, form of communication.

Oralism dominated British Deaf Education for ninety-nine years, until Conrad revealed the low literacy levels and the conspicuous lack of oral success. Since Conrad's survey in 1979 (2), appalling revelations have been made about the conditions and harsh treatment meted out to Deaf pupils: children being made to sit on their hands to prevent signing, being beaten if signs of any sort were used, tongues pulled to force speech . . . Oralist methods were hard on the children and brought out the worst in "educational" sadism and prejudice, as William Bury, a pupil from 1934–1945, describes:

> *"I came to this school when I was five and I could sign fluently . . . I kept winning things at maths 'cos it was visual . . . But I failed at speech . . . the teacher called in the Head and said, 'Billy is poor at speech and I have difficulty teaching him'. And the Head said,' . . . we'll just have to put him in with all the monkeys – the little gesticulators."' (5)*

One of my own students told me about his teacher asking him (he had to lip-read the question) how long a pencil was; when, like any person, he showed the length with his two hands, he had them sharply slapped with a ruler, and was forced to repeat the words "about eight inches", clearly enunciating the t's and the -ch- sound several times until he was thoroughly humiliated. Accounts of being slapped, beaten, punished in so many ways was attest to the determination of those Deaf people who struggled in secret to keep their language alive and now, when they come to me for help with written English, are so patient with my own dismal efforts at Sign Language.

Current Prospects

The 1980s saw British Sign Language regain its respectability in the education of the Deaf, and it is now being used more and more widely. Oralism still has its devotees, and is still used in some schools, but the "outlawing" of Sign is no longer practised.

Since the Warnock Report and the 1981 Education Act, many Deaf Schools have closed down as a result of the present policy of integration, which places Deaf and other hearing-impared children in normal schools. Some are in groups in Partially Hearing Units, others are alone in "individualised integration". The Department of Education and Science now classes Deaf children as "Children with Special Needs" and does not categorise them as a group – Deaf learners – in need of their own schools. Whether the true motives of this move are to aid the children's integration or to save the large amounts of money formerly devoted to Deaf Schools is a question for debate.

Many Deaf people reject the label "special needs", which carries with it the unfortunate stigma of "handicap" or disability. They class themselves, and I tend to agree with them, as a linguistic minority, whose second language is English; their first language is BSL, a rich and profound language which I am just beginning to understand. That it should have been suppressed, that Deaf people should have had their hands tied for so long, is a shameful tragedy. This suppression has resulted in so many Deaf adults having limited access to the literacy they need to function fully in our society that it is, indeed, time for change.

References

1. Wright, D. (1990) *Deafness, A Personal Account*. Faber & Faber.
2. Conrad, R. (1979) *The Deaf School Child: Language and Cognitive Function.* Harper & Row.
3. McLoughlin, M.G. (1987) A *History of the Education of the Deaf in England.* Ashford Colour Press.
4. Miles, D. (1988) *British Sign Language: A Beginner's Guide.* BBC/Open University.

5. Signs of Our Times (1990) *School Days.* Listening Eye/Channel 4 Publications magazine.

Other works worth reading

Kyle, J. & Woll, B. (1988) *Sign Language: the study of Deaf People and their Language*, Cambridge University Press.
Sacks, O. *(1990) Seeing Voices.* Pan Books

Lost for words: Aphasia and literacy

Susie Parr
RaPAL Bulletin No. 26, Spring 1995

Susie Parr is a speech and language therapist who has worked with aphasic people for twelve years. She is currently employed by the City University, London, and is working on a research project investigating the long term effects of aphasia.

It is difficult to imagine life without words. Communication, through the medium of written, spoken and gestured language, is of course the foundation of many aspects of human functioning: work, education, domestic and social life, leisure and religion. Yet 150,000 people in the United Kingdom suffer from aphasia and find themselves literally lost for words.

Aphasia can occur as a result of brain damage caused, for example, by a stroke, head injury or encephalitis. It is an impairment of language processing: aphasic people can experience difficulty understanding what is said and written, and in formulating language.

Words or names may be forgotten, even simple, familiar ones.
Or they may come out wrong:

Hallo er . . . This is my er . . .

The difficulties can occur on several levels. They can involve the use of word meanings, grammatical structures, the distinctions between speech sounds and the relationship between these and written forms.

The levels of severity can vary. Minimally aphasic people may find themselves able to pick up the threads of their lives with relative ease, occasionally troubled with word-finding difficulty or problems processing rapid and complex speech. Some severely aphasic people, however, are effectively cut off, unable to process any speech or writing and completely unable to express themselves.

This disabling condition can occur in the absence of any physical impairment. It is essentially 'invisible'. It is also chronic, in the sense that many people never recover fully from it. In the United Kingdom, people with aphasia are generally offered language therapy as in-patients in hospital, and as outpatients on their return home. However, for many, the provision of therapy is limited. This is due to a number of factors influencing 'service delivery', not least the recent changes in the health services and the rationing of health care.

Weak muscles may make
a person difficult to understand
They may also sound a little drunk . . .

The speech and language therapists who are able to offer language therapy to aphasic people have a number of assessment and therapy tools at their disposal. These reflect the different ways in which aphasia has been understood over decades of academic, medical and linguistic study.

Many different schools of therapy co-exist and directly influence the way in which the therapist will assess, interpret and deal with the impairment. However, these approaches are not always compatible, theoretically and practically. For example, two influential models of aphasia make unhappy bedfellows. The *functional* school addresses the language impairments of aphasic people in terms of their ability to carry out communicative activities of daily living, such as ordering a meal, or using the telephone. In complete contrast, the *cognitive neuropsychological* approach seeks to develop a model of language processing through the study of aphasia. Therapists adopting this approach use extensive psycholinguistically controlled testing to pinpoint the type of language breakdown. This leads to logical and specific therapy, which has predominantly focused on single word processing.

What has all this got to do with literacy? Aphasic people experience literacy difficulties, which vary with the type and severity of the aphasia. Thus, for some, writing their name and address may be overwhelmingly difficult, while others find themselves struggling with the intricacies of formal correspondence. A further dimension to these difficulties is, of course, added by the fact that aphasia is an acquired condition.

The aphasic person, formerly proficient in spoken and (perhaps) written language use, is suddenly fared with a chronic and disabling impairment. 'Coming to terms' with such a condition is not a rapid and easy process.

It can take many years for the aphasic person to take stock of linguistic losses and to evolve ways of dealing with them. The fact that people need to use language in their attempts to deal with chronic illness, enquiring about and explaining the condition, negotiating treatment and legitimising what has happened, can add to the sense of disempowerment.

As a speech and language therapist, working closely with aphasic people for many years, I started to feel that the ways in which I was attempting to address everyday reading and writing, using available asessments, was wide of the mark. The cognitive neuropsychological assessments seemed too mechanistic and reductionist, and the functional assessments too prescriptive and based on assumptions about what constitutes 'normal' everyday language activities.

Focusing on everyday literacy, I decided to try and develop an assessment which would be based on the findings from a survey of the everyday reading and writing of non-aphasic people. Through the process of semi-structured interviews, I managed to compile a list of everyday literacy practices, and to rank these in terms of the importance invested in them by the respondents. However, I was alarmed by the sheer number of the activities cited, the variability of practice, and by the fact that the 'top twenty' activities were not representative of the practice of all the respondents. In addition, a coincidental finding of this study was the way in which the respondents delegated and shared literacy practice with partners, family members, friends and even shop-assistants. Thus, people who were not aphasic, were not necessarily independent in literacy. In fact, literacy was coming across to me as a service, a social event, intrinsically connected with processes of negotiation and bartering. Reflection about my own everyday practice seemed to back up this, to me, startling revelation.

With illusions of constructing a neat, relevant assessment package abandoned, I turned to literacy research for help – and got it. Brian Street's (1984) distinction between autonomous and ideological models of literacy offered a framework from which to consider the tensions within the study of aphasia. I could at once see analogies with the autonomous model in many assessment approaches, particularly in the cognitive neuropsychological approach. Street also drew my attention to the social construction of literacy, the hidden skills which contribute to literacy, and to the rich potential of ethnographic investigation. Levine (1982) offered a radical critique of prescriptive literacy assessments and programmes, pointing out the political infrastructure of literacy teaching. And Fingeret's (1983) study of the use of social networks in literacy illuminated and consolidated my own incidental findings.

With a much firmer idea of literacy as a social, as well as a technical, cognitive and linguistic process, I set about trying to approach literacy in aphasia in a different way. Having read Barton and Padmore's (1991) account of the relations between social roles and literacy practice, I decided to try and study aphasia and its effects on literacy from this perspective. I interviewed twenty people about the extent to which their previous work, domestic, social and leisure roles had changed with the onset of aphasia, and how this had impacted on their literacy practices. I also asked them about social networks, and the kinds of back-up strategies they used, including personal help, procedures such as drafting and chunking written material, and technical aids.

Many interesting revelations emerged from this study. For example, some respondents did not see aphasia as the main cause for role loss or change. Other factors were often cited: changes in family circumstances; the death of a friend; being made redundant and so on. In addition, the change of role was not always perceived as a loss. Some respondents described the ways in which the quality of their lives and relationships had been enhanced following the stroke. Another revelation concerned the intricate back-up systems operated by aphasic people carrying out literacy practices in the implementation of various roles.

One of the most sophisticated examples involved a man who had maintained the mechanics of writing and the ability to recognise spelling errors, but who had lost the ability to produce syntactically correct sentences and to generate correct spellings. His wife was not a confident speller, but together they worked on developing a system of writing letters which worked very well. They would negotiate aloud what they wanted written. She would draft it into a written form. He would identify the mis-spelt words. Together, they would consult the dictionary and correct them. Finally, he would copy out the draft in mechanically perfect script.

Another key revelation emerging from this study was the fact that the respondents were clearly reacting to their aphasia in a number of different ways. Some were desperately keen to re-establish lost skills, and to resume previous roles. Others seemed resigned and fatalistic, others appeared to wish to forget the effects of the stroke and were doing this by withdrawing from a number of roles, others were depressed, and so on. It became clear that people were wresting significance from their common experience, but the significance for each person was completely idiosyncratic. Each way of coping, however, would clearly impact on the person's response to functional therapy, which generally operated on the basic premise that every person's aim was the recovery of lost skills. In addition to this, the dynamism of the coping process was evident in that the significance of aphasia was changing, over time, for each person.

This second study brought into question, too, the assumption that independence should be the unquestioned goal of therapy. Clearly, aphasic and non-aphasic people do not undertake literacy activities in a social vaccuum. Often literacy is part of an intricate and highly developed social behaviour, as much strategic as linguistic. The assumption that each aphasic person entering therapy should be tested on his or her ability to perform a set number of prescribed reading and writing tasks 'adequately' had to be questioned.

In the final study, I looked more closely at the contributions of different models of assessment to the reading and writing of three aphasic people. To do this, I compared an ethnographic assessment with cognitive neuropsychological testing. I tried to look at the type and quality of information yielded by each, the relationship between the two types of assessment, and the aphasic participants' interpretations of each. The ethnographic investigation involved interview, diary-keeping and seven hours of participant observation. The cognitive neuropsychological assessment yielded useful information about each person's language processing, where it was breaking down and what specific therapy might address particular problems. The ethnographic assessment yielded information about the range of activities and practices, the strategies developed, the significance of aphasia for each person, and its consequences. Again, I was surpised by the range of literacy activities undertaken by the most severely aphasic participant. He said he could not read, and yet was observed to undertake a wide range of reading activities. Similarly, the least impaired person, who technically should have been able to tackle complex and abstract writing, limited himself to certain types of writing in which spelling mistakes and grammatical difficulties would not be out of place and therefore would not embarrass him. These were puzzle activities and note-writing. He had long since abandoned all attempts at producing written correspondence, delegating these to his wife. This was simply not on his agenda. Whilst there were some correspondences between the results of the psycholinguistic tests and the findings from the ethnographic study, (such as difficulty reading non-words aloud being evident in struggles with saying holiday place names) generally it was not possible to use one model to predict difficulties which were brought to light by the other.

Each of the respondents reacted in different ways to the various procedures they underwent. Most interestingly, they construed the psycholinguistic assessments variously, as a game, a puzzle, a school-like activity, a source of information and explanation and as having curative powers. This seemed to confirm the idea that no literacy event, even the most ostensibly empirical, can be free of significance and reference for the participant. The allusion to the school-like nature of the activities corresponded with Street and Street's (1991) ideas about the influence of pedagogical conventions on literacy teaching and offered a new way of thinking about therapeutic discourse.

The project came to an end in a very different form from that anticipated at the outset. I became absorbed with qualitative accounts of literacy, their strengths and potential contribution to assessment. It became clear that *prescription* should be replaced by *description* and *consultation,* if functional needs and abilities are to be assessed in a way which is relevant to the aphasic person. Placed in the context of issues raised by the disability movement, the idea of *autonomy,* rather than *independence,* as the goal of therapy, assumed central importance. Associated with this was a growing concern about the way in which aphasic people are effectively excluded from research into their own condition.

Thus, the project raised more questions than answers. These are still being debated in academic circles concerned with the study of aphasia. The potential of the qualitative approach has yet to become acceptable within the hard science culture of aphasiology.

Nevertheless, undertaking the study was a hugely rewarding experience. It opened my eyes to a number of things: to some of the consequences of aphasia; to alternatives to a medicalisation of language therapy; to the limitations of research conventions; to the extent and nature of the literacy difficulties experienced by aphasic people, but also to the extent of their resourcefulness in dealing with them. The study has many implications for clinical work. It is being reconstituted in the form of academic papers, a book and a clinical manual, which will hopefully prove interesting and useful to therapists working with aphasic people.

References

Barton, D. and Ivanič, R. (eds.) **Writing in the Community** Sage Publications, London, 1991.

Barton, D. and Padmore, S. *Roles, networks and values in everyday writing*, in Barton, D. and Ivanič, R. (eds) 1991.

Fingeret, A. *Social networks: a new perspective on independence and illiterate adults*. Adult Education Quarterly, 33,1 33–146,1983.

Levine, K. *Functional literacy: fond illusions and false economies*. Harvard Educational Review, 52, 249–266, 1982.

Street, B. *Literacy in theory and practice*, Cambridge University Press, 1984.

Street, J.C. and Street, B. *The schooling of literacy*, in: Barton, D. and Ivanič, R. (eds) 1991.

Note: The illustrations accompanying the original article, which explain the effects of aphasia, came from the organisation **Action for Dysphasic Adults** *(see* Digest in **RaPAL Bulletin**, no. 26, *page 33)*. The terms *dysphasia* and *aphasia* are used synonymously.

Developing literacies within ABE to promote self advocacy work alongside adults with learning disabilities

Catherine Jamieson and the 'Sticking Up For Your Rights' Group
RaPAL Bulletin No. 38, Spring 1999

Catherine Jamieson has first degrees in both Education and Social Work and, since 1995, an M.Ed. at Stirling University. She has been working full time in ABE since 1990.

Background

From 1990 until 1997 a full time Senior Community Worker was seconded from ABE to work with adults in a Social Work Resource Centre for people with learning disabilities in North East Edinburgh. The worker involved was seconded to Social Work because of her experience in the field of learning disabilities. Her role included developing opportunities for service users to be active in the community, increase their confidence and develop effective literacies, numeracy and communication skills. Developing advocacy, and in particular self advocacy, for service users in a community context was an important part of the work ABE began to set up.

How the group began

ABE was represented on various joint working committees including a multi-agency Planning Team for people with learning disabilities. The ABE worker found it both embarrassing and disturbing that two people with learning disabilities were attached to the committee who were given no help before, during or after the meeting to make sense of the sometimes complex issues and arguments they were hearing.

Powerlessness

These individuals were **token** members of the committee with no power, and few opinions, who were unused to speaking even in peer groups and who could easily be intimidated in groups of professionals. They were being placed in a situation that was potentially both humiliating and patronising. Attending committees such as this seemed to point up some of the crippling effects of isolation and institutionalisation which had contributed to the difficulties faced by this group of people.

Along with the ABE worker, two of the other participants from the planning group, Social Work Department staff, also felt it was unacceptable having tokenistic involvement from people who badly needed to have their valuable opinions heard, and came up with the idea of offering to support a group of people with learning disabilities who might be interested in planning local services. They advertised the group around the community and to all day centre users in the area. They suggested that people might want an opportunity to be part of the planning process, to form the service users arm of the North East Edinburgh (multi-agency) joint planning group (Learning Disabilities subgroup). They arranged to run the group in the local Community Centre.

Response to the offer

There was an extraordinary response to the offer of this group to service users. Around twenty people with learning disabilities turned up and continued to attend weekly over the next few months. It wasn't only people attending Social Work Day centres, service users came from the voluntary sector, Further Education and from the community. A wide range of ability has always been represented on the group. Many of the people in the group showed signs of having very poor self-esteem and low confidence. Some used signing (Makaton) systems, some were already part of the member councils in their day centre, some were supported to attend by staff, some had lived in long stay hospitals for as long as 40 years, while others were not long out of school. All were ready to have a chance to speak about the kind of things that mattered to them.

Who is involved in supporting the group?

A range of staff, volunteers and service users are involved in facilitating the sessions. In the early days two ABE workers helped the group members to develop the structure, ground rules and goals which have, underpinned the work of the group as well as helping group members with all literacy and numeracy issues.

How the sessions were set up/the role of the staff

The sessions were developed initially around an existing Social Services report on day service provision as well as from the issues the service users took to the group. It was quickly established that the staff were there to provide structure, minute the meetings, set up graphic recording* for the non-readers. The participants were encouraged to develop ground rules and then went on to setting down the aims of the group. The group then formulated letters by getting together in small groups and referring to the larger group before sending them out. The facilitators helped the group members to put what was being said onto paper and helped group members to read the replies. All of the meetings were clearly minuted so that everything said in the meetings could be taken back to clients/members councils and discussed by a large number of people with learning disabilities. The group soon enjoyed having the power to establish what was talked about. They decided to call themselves the Sticking Up For Your Rights Group (SUFYRG).

Speaking as equals to professionals

One of the important areas that the group needed to address quickly was how to communicate effectively with professional groups in face-to-face situations. The group worked hard over several sessions, discussing various ways that this could happen. They needed to find a way to feed back to the joint planning group that would allow people to express their opinions without getting confused or tongue tied. Through discussion the group came up with the idea of deciding what to say, then making a tape together with a transcript to play to the professionals. A small group attends the planning group on a monthly basis and will then answer questions about the issues on the tape. The membership of this small group changes every month so that every one can have a turn presenting ideas to a professional committee.

SUFYRG campaigning against the cuts/speaking up

Edinburgh is one of the unfortunate councils that found themselves rate capped under the Conservative government, in addition to suffering from the lack of revenue that came about as a result of the poll tax. This has meant that the local services have been very hard hit by the lack of available money. Inevitably, the local government cuts affected the kind of things the group wanted to do. The work of the group became focussed on campaigning as various services were threatened.

The SUFYRG played a major role in campaigning against the cuts.

They helped to keep some centres open. They provided a chairperson and group facilitators for a Seminar in February 1997 organised by a range of concerned individuals for 100 service users to **Get A Voice**. They ran a similar seminar for 200 people in 1998, and again for around 100 people in February this year. The Assistant Director of Social Work and two Local Labour Councillors have spoken at these events. The workshops developed questions for the panel to answer, mainly about the cuts as well as possible future directions for day services for people with learning disabilities.

Group Report

The group produced their own report in 1997 and launched it in a Seminar with workshops to discuss the findings. From the immense amount of work that went into this document, it is obvious that there is a great deal of commitment from the SUFYRG members.

Community Literacies help people get a voice: developments arising from this group

The group has gone from strength to strength and, with support from star and volunteers, has:

- presented small group delegations and individual delegates to the City of Edinburgh Council
- met representatives of the Scottish Office
- become part of a new Social Work Strategic Development Group for People with learning disabilities in the City of Edinburgh.

An Edinburgh City Advocacy Facilitators group came together to help the local advocacy groups link together, run big events, co-facilitate workshops and so on. This group produced a Consultation Document, currently being discussed by the local council through their Disabilities Equalities Forum.

All of the 'Get a Voice' seminars for people with learning

disabilities against the cuts came about as a direct result of the work of both the SUFYRG and the ECAFG.

It is clear that without the involvement of Adult Basic Education (ABE) and Community Education, it would not have been possible to help this group of people present their views outside of the Social Work day centre setting. It is essential that the support continues. The group themselves are clear that they need the further involvement of ABE to help them continue their important and innovative work. One of their stated aims was to see other groups like their own within the city to help increase the voice of, and improve services for, people with learning disabilities.

Excitingly, from these initiatives and through the work of the groups involved, ABE, Community Education, Social Work and the Voluntary sector met together with service users to plan, develop and set up a new group for service users in the south of the city. The new group met for the first time in January 1999 and is continuing to flourish. Watch this space for further developments!

Note

* Graphic recording is a system of recording what is spoken about in meetings using accessible symbols and pictographs to support literacies.

B5 International literacy

Further RaPAL references about international literacy

Mace, J. 'The Politics of Literacy Seminar Report – UNESCO Shifts in Policy', *RaPAL Bulletin* No. 3, 1987.

Murphy, S. 'Den Bosch Days. Conference Report From the Netherlands', *RaPAL Bulletin* No. 13, 1990.

Jones, V. 'Adult Literacy in the Soviet Union. Conflicting Purposes and Values', *RaPAL Bulletin* No. 14, 1991.

Clarke, J. 'Review of Alan Rogers' Publication', *RaPAL Journal* No. 50, 2003.

The challenge of literacy in Latin America

David Archer and Alan Murdoch
RaPAL Bulletin No. 4, Autumn 1987

David Archer and Alan Murdoch are research and development workers for the Oxford-based Community Education Direct Research Unit.

In Britain illiteracy* tends to be regarded as an isolated problem. Literacy teaching programmes struggle to identify and make contact with the illiterate population (which is difficult in a society which stigmatises illiteracy) – but only 65,000 out of a recently estimated 7 million, come forward. The programmes often focus on the absence of a set of expected skills within the individual, and the teaching takes place in a semi-confidential environment suitable to this almost medical or pathological viewpoint. The social and political context of illiteracy is rarely considered.

In Latin America literacy programmes have taken a very different turn. The experiences of Cuba and, more recently, Nicaragua, together with the writings and actions of Paolo Freire (author of **Pedagogy of the Oppressed**) have stimulated the growth of a 'popular education' movement, which fuses the teaching of reading and writing with political consciousness raising. Literacy programmes have become political battlegrounds, as the power of education to stabilise or radically change societies is recognised by governments, big business, opposition groups, unions and guerillas.

The scale and background to the struggle against illiteracy is also different in Latin America. The lack of established educational infrastructures (especially in rural areas) and the crippling foreign debts (associated now with enforced austerity, which inhibits expansion) lead to multiple problems. Liberal education, imported from Europe, and once held up as the means to democracy and an end to poverty, has failed in those tasks. In formal literacy programmes there are low enrolment rates and high drop out rates; follow-up work has been poor and the newly literate continue to lose their skills through disuse. The number of illiterates in the continent is growing. Governments are hesitant to confess failures for fear of losing popular support or international aid, so statistics are massaged. Colombia claims that one million people were made literate in 1985, but in fact 800,000 of them dropped out, sometimes even before they started a literacy course.

In the light of such responses to failure, many independent groups have taken the solution into their own hands. Throughout the continent alternative methods and structures, inspired by Freire and the Nicaraguan Crusade, are being applied, and have proved to be effective and adaptable. **Autogestion Educativà**, a self-managing group of Indian women in La Paz, Bolivia, are just one example. For them, motivation has been the most critical element in their literacy work. In their experience motivation is not an abstract quality, but one which can be fostered in very definite ways. First comes the thorny issue of language. All the women spoke Quechua or Aymara as a first language, though as they worked in the market and had to deal with authorities, they also spoke the national language, Spanish. The Bolivian National Literacy Service (SENALEP) had produced literacy workbooks in the indigenous languages, arguing that this was what was wanted, and was most effective (though in fact the workbooks never got used). The Indian women, however, universally agreed to work in Spanish.

> *"We will not learn, and literacy will be no use if it does not empower us,"* explained one woman; *"Spanish is the language of power, and we must adopt it if we are to assert ourselves and seek change."* Everything from land titles to books, newspapers, tax forms, rents and contracts, use Spanish. *"The government have some glorified ideal of preserving our culture, but only we can do that – and only from a position of power. We don't want to preserve our poverty."*

Aufogestion Educativà use only Spanish in their literacy work, but that is only the first step in motivating the women to participate. To someone who has lived without the ability to read and write, whose daily routine is full and does not include those skills, then the learning of them does not seem inherently interesting or essential. The awareness of some uses may be overwhelmed by the difficulty and fears that they anticipate and experience. To help overcome this, Autogestion Educativà have produced a workbook, **Nosotros Podemos** (*We are Able*). It was prepared by the women and has 25 sections, each on a different theme, building out from the individual ('I am a person and nobody should hit me'), to the family ('In my family we all give/share'), to the group ('To participate is to change our lives'), to the community ('Actions are stronger than words'), to the country ('We are fighting for justice'). For each theme there is a photograph (taken of the women by the women) and a sentence with a key word, which act as a focus for dialogue.

The immediate relevancy of the materials to the women is of vital importance, providing a sense of identity and a basis from which to express themselves. However, the materials will be of no use if presented in an imposing or threatening environment. So the literacy sessions take place within the market itself (overcoming any problems of transport and accessibility) during quieter periods of the day. There is no role for a traditional, professional teacher (fortunate also, as there is a shortage of them). Instead it is believed that anyone with the basic skills of reading and writing can coordinate a de-centralised learning process. The group is never more than ten women at one time, so it is an undaunting environment, and one in which newcomers feel able to participate. By being in the market place it has a high profile for women as yet uninvolved.

Groups like Autogestion Educativà have had enormous success. From only a few dozen women a few years ago, there are now 5;000 actively involved, and by the end of 1987 they expect to be 12,000 strong. The power of such a united group will doubtlessly show as they extend to other towns and into the rural area. Health education work has become important to them now as they find a voice to organise, challenge and change their world. The absence of skills is only a part of the submerged existence of the illiterate population. As the women learn the language of power, they also use it.

Hundreds, perhaps thousands, of groups like Autogestion Educativà exist in Latin America. Many struggle against hostile reactions from governments (and the harassment of security forces) – and even more struggle against funding difficulties. Yet increasingly governments are being forced to recognise the efficiency of such groups and the economy of their methods. Some Ministries of Education attempt diluted versions, but disseminating them from above is inappropriate (though possible if given the exceptional circumstances of a country like Nicaragua). It would be more appropriate for international aid agencies (who already fund often highly inefficient government programmes) to direct their resources to initiatives like that of Autogestion Educativà – where a small investment may yield remarkable results. Results indeed which may prompt some literacy workers in Britain to reflect upon how illiteracy has remained out of the political arena, and how it is that it could be treated as such an isolated problem for so long.

An illustration from Adult Literacy materials used in Colombia: this picture is a starting point for discussion in class

Note

* Perhaps in the UK the word 'sub-literate' is more appropriate than 'illiterate' – but in Latin America many people are in the more absolute condition, in that they have often had no previous access to education. The word 'illiterate' may have emotive connotations here, but it is the only suitable and accurate translation of the Spanish word *analfabeto.*

Bilingual-bicultural education in Nicaragua

Jane Freeland
RaPAL Journal No. 5, Spring 1998

Jane Freeland teaches language on the Latin American Studies course at Portsmouth Polytechnic.

In 1980 Nicaragua had a mass literacy crusade which is accepted as the most successful literacy campaign in recent memory. Nicaraguan education has now entered another innovative phase: bilingual bicultural education for the ethnic minorities of its Atlantic Coast. Like the 1980 Literacy Crusade, this project breaks new ground in popular education, of vital interest to ethnic minority and majority groups everywhere.

Several groups live on the Atlantic Coast: about 80,000 Miskito Indians, 7,000 Sumu Indians, 30,000 Afro-Caribbean descendants of slaves and immigrant workers (Creoles), around 800 Creole-speaking Rama Indians and 1,500 Creole-speaking Carib Indians. Bilingual education began in 1980, when the government and the minorities jointly mounted mother-tongue adult literacy programmes in the two surviving Indian languages, Miskito and Sumu, and in English for Creoles and Rama. Despite huge problems, 21% of potential learners became literate in their own languages. In 1981 a Law on Education in Indigenous Languages authorised bilingual education in kindergarten and primary schools for Miskito and Creole children, though growing tension between certain Miskito leaders and the revolution, escalating into counter-revolutionary war, eventually halted progress.

Nevertheless, in 1983 Miskito leaders proposed pilot programmes for bilingual education in kindergarten and first year primary school, for adoption under government reforms of the national curriculum. They ran experimentally in four Miskito-speaking kindergarten and primary schools in 1984. Similar experiments followed in 1985 in six Creole-speaking schools, beside a pilot project in Sumu, for which the law had not originally provided. A Bilingual-Bicultural Education section of the Ministry of Education, locally staffed and run, took charge of developments in late 1984. By 1986, pre-school programmes were starting wherever communities were stable and war conditions permitted, and the Sumu Spanish experiments were extended. Miskito, Sumu and Creole children now begin preschool, in their mother-tongue.

This year, Adult Literacy programmes are being resumed, based on the 1980 literacy manuals revised jointly with the committee. They are using the same methodology but are revising the themes in the literacy manuals. They are doing this in consultation with the local communities. Adult literacy and child literacy complement each other. Parents not convinced of the need for bilingual education for their children are being persuaded of its value as they become involved themselves.

On paper, the plans may look like many standard mother-tongue schemes designed to absorb people into the dominant culture. The reality is more complex. Maintaining and developing their languages is important to the communities. Yet most parents, conditioned to pre-revolutionary educational realities, need convincing that mother-tongue education is good for their children, and that it improves their children's chances in Spanish based education. Creoles, moreover, insist on bilingual education in Spanish and Standard English, to enable them to participate in both international cultures; for them Creole-based education is only acceptable as a transitional stage.

The long-term aim of minority leaders is to increase emphasis on maintenance and development of the minority languages. Already the programmes are bilingual and bicultural: the communities, are closely involved in developing materials based on their culture and oral tradition; the Spanish materials aim to help children critically evaluate the dominant culture. The idea is to increase pride in local languages, and a demand for more than transitional programmes.

But this approach contains a potential trap. Changes in language attitudes rarely happen through the classroom alone, but through transformations in the relationship of minority groups and their languages to the broader society. Educational developments on the Coast have always been inseparable from its political dynamics, both positive and negative. At each stage, political initiatives have stimulated educational ones, and vice-versa. Thus, the 1980 mother-tongue literacy campaigns, the first demand of the Indian representative body, MISURASATA, empowered both pro- and counter-revolutionary leaders; the ensuing conflict damaged the campaigns and threatened future developments. Equally, the campaigns increased the minorities' self-confidence, developing activists committed to working for their communities with the revolution.

In 1985, this work entered a new phase. Peace negotiations began with Miskito counter-revolutionaries, and the *autonomy process* was initiated: two years of intensive dialogue with every community shaped the Autonomy Law finally passed in August (see *Central America Report* No 35, 1987, for a fuller account). The rights it enshrines, to language, culture, religion and territory, mark a milestone for all the Americas. More importantly, the process itself, by making the minorities subjects and not objects of political development, made new demands on their languages.

Last April, I witnessed discussions of the final draft of the law at a multi-ethnic assembly of elected delegates from every community. As at any international conference, the proceedings were interpreted between four languages: Spanish, English, and the two Indian languages shared equal status. The debate, and the consultation preceding it, has dramatically extended the range of the Indian languages in particular; even English, not normally used for political discourse in costeno society, is being stretched. These experiences, which will increase as autonomy comes into effect, are changing people's perceptions of their languages, and with them their attitudes to mother-tongue education.

This exciting project is also daunting. War and economic blockade have drained Nicaragua of human and economic resources; trained and capable people on the Coast are still few, the demands on them many. Creating authentic materials alone is a huge task, requiring research into the learning and teaching habits of the cultures and months of recording and transcribing oral material. Teachers need linguistic and anthropological skills to analyse and use the material for language and other teaching; there is little existing research, all of it by foreigners. Yet people cannot be spared from their work for training. In Creole classrooms, few teachers have a basic English teaching dictionary to consult.

A group has formed, providing technical and material aid to the Bilingual-Bicultural Education Programmes, and organising on-the-spot workshops and seminars in the National University to train community linguists. The group is currently called *Linguists for Nicaragua* – but membership isn't confined to "linguists"! Nicaraguans I spoke to feel there should be a branch in this country. Here are some of the ways British linguists and literacy workers might help.

Technical aid:

- on-the-job collaboration with Nicaraguan teachers and researchers; action-based research in the field.
- action-based research in the field

Material aid:

- financing in-region scholarships (the Swiss fund places in Mexico on appropriate courses in indigenous language analysis);
- funding campus scholarships or placements for Afro-Caribbean teachers to train in this country;
- finance for desk-top publishing of materials.

Help would be channeled through World University Service or Nicaragua Solidarity Campaign.

Critical language awareness in adult literacy classrooms

Guilherme Rios
RaPAL Bulletin No. 43, Winter 2000

Guilherme Rios is a Research Student in the Department of Linguistics and Modern English Language at the University of Lancaster (CAPES/Brazil Scholarship).

Introduction and Method

This short paper provides a critical analysis of discourse within a Community Adult Literacy Programme in a lower social class community near Brasilia. The programme aimed to increase the critical consciousness of teachers and learners and this study was designed to find out how language is used to sustain existing power relations and to contest or overcome them. It was part of a lengthy collaboration between the University of Brasilia's research group on 'Literacy and the Community' and the Paranoa Centre for Culture and Development; and it draws heavily on the theoretical work of Fairclough (1992a; 1995), Street (1984; 1993) and Clark et al (1991).

My aims were:

- to find out the identities which the participants were constructing through language and the ways in which they did this.
- to examine the nature of the linguistic practices in the classroom and their role in reproducing and/or creating socially dominant relations
- to contribute to community based Adult Literacy within a critical and ideological perspective of literacy.

The methodology used was 'Critical Ethnography'. This has developed out of concern about the historical reproduction of social inequalities in conventional ethnography and it aims to uncover broader social processes of control, domestication, power imbalance and the symbolic mechanisms that impose a set of meanings over others. The research strategy was to produce a case study of three different classrooms involved in different levels of literacy work. Each was observed during the second semester of 1996 and the first and second semesters of 1997.

Two types of discourse were analysed: verbal interactions between learners, teacher, co-ordinator and researcher which I shall call *'language in use'*; and *'language as an object for teaching'*, that is, the things that are being taught and supposedly are regarded as language. The methods of data collection were video-camera recordings, field notes, photocopies of participants' texts and interviews.

I organised the data into 'discursive events' (Fairclough 1992b), concrete social situations in which social practice is mediated between the discursive and the social dimensions. And I moved from discursive events to 'discursive literacy practices' (Magalhaes, 1995; Street 1984; and Fairclough 1992b).

Results

I classified these discursive events in three sections. For the first two I have briefly summarised the main observations. For section three I have provided a detailed example of the coexistence of traditional and critical literacy practices in the classroom. All the student names cited in this are fictitious.

1. Language sustaining existing power relations

I found that participants in the adult literacy class drew on language in ways which contributed to maintaining unequal relationships between learners and tutors. This was evident in both

- language in use: how we refer to others; how we address others; and feelings about literacy (for example, that vernacular literacies are not as good as schooled literacy); and
- language as an object for teaching. The way in which writing is treated as separate from the learners and their everyday lives, as something 'out there to be acquired'. Written language is seen as closely related to school and its uses outside school are not highly valued.

2. Participants' identities in the interaction

It was clear that language was used to produce social identities in ways which were:

- disempowering: via the modality used (the use of the speaker's authority) or by the labelling of the named group (e.g. 'person with learning difficulties')
- empowering: via equal initiation of, or turn taking, in conversation; and the negotiation of meanings.

3. Coexistence between traditional and critical discursive literacy practices in the same content

Traditional literacy practices include: a classroom activity silently-led for the learners; the treatment of linguistic units as decontextualised from their concrete situation; and the disempowering way in which people refer to, or address, learners, placing them as social inferiors. These were coexisting in this situation with **critical discursive literacy practices**. By these I mean the moments when learners ask questions about the form or meaning of texts; when language is taught in a way which contextualises words and syllables in the concrete situations in which they are produced; and when people refer to, or address, learners in an empowering way, placing them as someone important in his/her social group and in the social world.

One example of this coexistence . . .

- In the introductory class on the chosen theme, the teacher stuck a text on the blackboard which had been written by the school co-ordinator:

The Centre for Culture and Development of Paranoa (CEDEP) is a civil entity without profitable ends and discrimination of philosophy, colour, sex and politics. The CEDEP develops creativity through critical consciousness.

The teacher started the discussion showing the relationship between the initials and its extended name. A student, Antonio, intervened with a question about the use of the term *'without profitable ends'*. He considered that this would give a 'very strong' negative impression about the outcomes of the programme and its importance in the community. He felt that the quality of the work by the teachers **would profit** the development of the learners.

The learner (critical) can see the coexistence of critical and traditional discursive literacy practices in the choice of the pedagogic material by the co-ordinator (traditional) and the initiative of negotiation of meaning.

- The teacher introduced three grammatical ideas: the article; the sentence; and the noun. The teacher gave definitions, examples and an exercise to explain the points.

Then, following a prior decision to produce a piece of collaborative writing on the theme of 'landless peasants', the teacher asked for a volunteer to start to write some text which would be read out in the next Forum (the general course group meeting to present and evaluate what had been achieved and to plan subsequent themes and activities.) Alvaro walked to the front of the classroom and I accompanied him with a view to coordinating the activity. I offered some explanation about

> what a speech was and tried to decrease their resistance to the idea of building text for this purpose.
>
> While Alvaro was writing on the blackboard and I was providing possible syntactic links and checking that the class approved of the text being constructed, another student, Ana, decided to start writing it out. She read aloud part of the following text in the Forum and Geraldo who had a gift for public speaking, provided an introduction and finished it in the following class.

Coexisting in this situation were traditional views of literacy, in the restricted metalinguistic work done by the teacher, and critical discursive literacy practices in the collaborative process of text production.

The Agrarian Reform is good not only for the country, but also for all of us because the prices decrease and the producers can have a place to live and plant.
The Agrarian Reform also generates more jobs in rural areas, improves the life quality of the Brazillans and increases exportation.
The Agrarian Reform is an excellent action and all Brazillans ought to support it in order to have a 100% better country.

Conclusion

Through the analysis I could see that this coexistence is harmonious in some events and conflictive in others. In the latter, both discourse conventions and the pattern of the social interactions are problematised and can be understood as a struggle between different linguistic and social practices. As a result, new discursive conventions are drawn up, for example, when the learners argue about the previous decision taken by 'empowered' participants in the particular setting. Further, new discursive conventions can lead to change in the local, institutional and societal discursive order (Fairclough, 1992). I think that critical discursive literacy practices in non-formal schools may contribute to reflection on social relations and the empowerment of learners within institutional and political domains. I see this work as a contribution to the emancipation of disenfranchised groups in Brazilian society.

References

Clark et al. (1991) 'Critical language awareness part 2: towards critical alternatives.' *Language and Education. 5.1.*

Fairclough, N. (1989) *Language and power.* London: Longman.

(1992a) (ed) *Critical Language Awareness.* London & New York: Longman.

(1992b) *Discourse and social change.* Cambridge: Polity Press.

(1995) *Critical Discourse Analysis.* London: Longman.

Magalhaes, I. (1995) 'Praticas discursivas deletramento: a construrao daidentidade em relatos de mulheres.' In A. Kleiman (org.) Os *significados do letramento. Campinas:* Mercado de Letras.

Street, B. (1984) *Literacy in theory and practice.* Cambridge: CUP.

(1993) (ed) *Cross-cultural approaches to literacy.* Cambridge: CUP

Growing into Europe: A report of a writing weekend in Belgium, June 1993

Kate Tomlinson

RaPAL Bulletin No. 23, Spring 1994

Kate Tomlinson is a tutor in ABE at Stroud College of Further Education.

For four years students and tutors from an ABE group in Stroud, Gloucestershire have been exchanging letters with a similar group in the Belgian Ardennes. Last year, we participated in a writing weekend at Trier, in Germany, where we also met a group from 's-Hertogenbosch in Holland. Last year, groups from Trier, 's-Hertogenbosch and Southern Belgium were invited to a writing weekend in Gloucestershire. (See RaPAL Bulletin No. 19 *More Ripples in the European Pool.)*

Each year the task of raising the money to make these events happen, and the organisation involved, seems harder. But each year the elation produced in everyone lucky enough to attend, and the benefits in broadening understanding, increasing motivation to write and use skills learned in the classroom in a variety of real situations, makes the next invitation harder to resist. The international chain of events at grassroots level is growing, and this June another link was added when Stroud students were invited back to Belgium to join ABE groups from 's-Hertogenbosch, Trier, 'Lire et Ecrire' (Belgium) and Sabrosa, in Portugal, at an outdoor activities' centre at Herbeumont, in the middle of the forests of the Ardennes.

Aims of the weekend

The aims of the weekend, as laid down by our Belgian hosts, were:

- to motivate participants to write by offering unusual opportunities
- to offer them the opportunity to express their life experiences through the workshop
- to widen experience; most of the participants had never left their own country before
- to enlarge the network of people sharing problems of exclusion and isolation
- to allow them to consider literacy problems as a social rather than an individual problem
- to discover the Ardennes region
- to exchange ideas and methods of work

Six students and two tutors went, all from community groups in the Stroud area. As far as they were concerned, all of the aims of the weekend and more were met. Realising the project through our own efforts had made each of us use new skills and deepen existing ones.

Wordpower

Two of the students, Stephanie and Roger B. were working towards 'Wordpower' certificates at Foundation and Stage 1 respectively. In this situation, it wasn't necessary to contrive situations in order to meet performance criteria. Tasks needed to be done: filling in passports and E111 forms; phoning or writing letters to ask for sponsorship or thank people; lists and messages to plan the journey; reading the programme that arrived from Belgium; reading timetables, planning the journey and organising transport; looking at charts about duty free and foreign exchange; reading signs and security notices at the airport; reassuring each other on the plane; talking to strangers; talking to other groups about our experiences on our return . . . and many others. Roger has just achieved his certificate.

Beyond Wordpower

But other learning was also going on. We got together for a basic French course; we practised and performed our own peculiar version of 'A Policeman's Lot' in plastic helmets; we tried food and drink from five countries, danced, sang in Dutch and Portuguese and, above all, talked and shared experiences despite the language barriers. As Roger B. said:

> *When I first came here i had a problem to communicate with the groups in a different language. When I got on, I got better – a member of the group translated it for me.*

Changing roles, changing people

As the weekend went on, we all began to see each other in new lights and roles. After I had been stopped and searched by security at Heathrow, Roger B. was the one brave enough to ask to visit the cockpit of the aircraft. Stephanie who during the French classes said she "didn't fancy kissing all those strange men" ended up by kissing everyone. Enid (tutor) admitted that she wouldn't have attempted climbing down the precipitous slope to the river in any other circumstances. Roger G. revealed himself to be expert in what appeared to be an ancient Japanese art of plate throwing, and he and Terry, as irrepressibly over-the-top as usual, destroyed a few stereotypes about reserved and uptight English people. I went through the same process of drafting and correction as students when one of the Belgian tutors helped me translate our writing into French. Sally spontaneously kept a diary of events. Keith took photos, talked to as many of the other participants as he could, and despite the language barrier, tried to explain to M. le Depute about G.O.A.L., our county-wide, student-led organisation for which he is local chairman, and through which we raised a large proportion of our funds. Several of us cried when it was time to go.

Research for RaPAL

RaPAL had given us a donation, and asked students to engage in some small-scale research:

a) How did students from other countries use the skills learned in the classroom in their every day lives?
b) How was provision organised in the areas from which they came?

This was above and beyond the demands of the writing workshop, but from the interest it generated, could itself have been the basis for an international weekend.

It was difficult to get the English students together for any length of time before our departure as they all attend different groups in a rural area, and we didn't have a chance to study the lessons learned during RaPAL weekends on this topic. However, I circulated everyone with information about the idea and asked them to think of the questions they might ask students from other countries. Each person developed a questionnaire, some of which were sent on in advance to Belgium, some distributed during the weekend, and some posted off afterwards.

The Problems

The problems we encountered were various:

- The language barrier. It is difficult enough asking the right questions in the right way to English people. We found that the answers we got were superficial and imprecise. Both students asking the questions and those answering them were frustrated at times because they so much wanted to understand one another!
- The weekend already contained a packed programme of events and we had to snatch a moment here or there after meals to do our research. However, everyone seemed keen to persist despite this.

- We needed to have piloted and refined our questionnaires – and probably used multiple choice rather than open questions as these are easier to answer and respond to, particularly in another language. We needed a combined, definitive version rather than five or six different ones.
- Because tutors, on the whole, were the people who spoke a second language, questions and answers had to be filtered through them, which distorted the information. Dialects and accents sometimes also caused a problem here.

We also went armed with pocket tape recorders to record interviews during the weekend and have returned with yards of tape to transcribe. Dealing with the data and trying to turn it into some sort of report will provide us with work for the future.

The Pluses

Both Gloucestershire students and those we questioned were very keen to participate. Everyone was genuinely interested and motivated to find out about each other's lives and circumstances. As Stephanie said at the end:

> *I felt more confident and 1 could have talked to people. I felt 1 could give things a try . . .*

A real feeling emerged of similarities, of 'being in the same boat' as well as a realisation of differences: French for speakers of other languages and ABE happen within the same groups in Belgium; in Portugal there is a complex and much broader range of subjects encompassed by the term Basic Education; at Trier students have to pay a minimal sum for their classes. Roger G. said:

> *I have learned I can communicate with the other groups even though they speak a different language. They're all a friendly lot and they all muck in. I discovered that when you are here you are in a different world – picking up different items. I felt a different person. 1 felt all relaxed and in a big family all laughing together.*

The Dutch, whom we didn't find time to interview properly, were interested enough to ask us to post on questionnaires to them. Maybe this will be the start of a more organised and focused attempt by students of different nationalities to find out about each other.

The Writing Workshops

The writing workshops gave us the opportunity to experience some of the imaginative learning methods we had heard about from Michele, one of the Belgian organisers, in her letters. She and Catherine, the other organiser, had recently attended a training course on structuring the kind of workshop we experienced. They had sent us information in advance explaining the stages we would go through. We had had our reservations: students looked doubtful and tutors expressed their own inhibitions as they wondered how these down-to-earth students with such practical literacy goals would feel about writing about the legendary Nutons of the Ardennes forests, or a story beginning: "I am a book and this is my story . . ."

In the event it proved to be a very positive experience. It is true that like any writer, we floundered at times. We decided to work as a group, and I was amazed at how quickly these people who were not accustomed to working together, got into their stride.

The first task was to create a portrait of the Nuton and a Calendar of his feast days based on the elements of Fire, Earth, Water and Air. The initial stimulus was a walk during which we collected objects and took photos of whatever made us think of these elements, and explored these in writing, using all our senses. Roger B. said of this:

> *I liked writing about the objects we found. And putting it into writing. We all worked as a group. We made a lot of ideas. My object was a piece of coal. And I described it.*

Ideas came so thick and fast that we had difficulty in getting them down, and in this atmosphere, students were automatically selecting and rejecting ideas, modifying and enlarging on them. Everyone contributed and we all felt a sense of ownership over the writing we produced. Roger G. said:

> *It was quite good. All the ideas that we had we wrote down – we put it on paper and we put it into proper paragraphs and it come out lovely. I liked the group with all the ideas that they had. They put their heads together for the story and it seemed to come quicker. It seemed to help us all. We had some ideas and then we had some better ideas which seemed to click – so we chose the ones we liked the best.*

The second workshop took place at Redu, a village full of second-hand book shops that is twinned with Hay-on-Wye, and books were the stimulus for the writing.

> *I enjoyed all the writing – every bit of it. In the village we didn't get much time to look around. But as soon as we got back in the hall I felt confident and ready to go.*

We chose to work in the same way, and the end product was the history of a book within whose covers all life was contained. We were very proud of it, and Sally read it aloud in front of everyone at the official reception attended by the Bourgemestre, the Deputé and the regional president of 'Lire et Ecrire's. She explained:

> *In the writing events we all contributed as a group and we found we all had imagination to write a story. When we got back to the place where we were staying I found I had the confidence to read our story in front of the Bourgemestre . . . and about fifty other people. We all have got pen friends over there now.*

We have been told that our story will be published in the "Journal d'Alpha", "Lire et Ecrire's" newsletter.

We celebrated that evening with a buffet of specialities of food and drink, from the five participating countries, followed by a barn dance. The following day we wrote evaluations on cardboard cut-outs of clothes we had packed in our luggage. Roger G. summed up the weekend like this:

> *I like sleeping at the Centre. It made me think of barracks in the army*

– I liked that. Sometimes the talking took too long, e.g. at the official reception . . . We were all pleased with our writing and the ideas that came to us. Saturday night with the group was great fun, everyone dancing . . .

And we haven't stopped talking or working on the weekend yet.

One world – One vision?

Linda Kirkham and Chris Wild
RaPAL Bulletin No. 14, Spring 1991

Linda and Chris are participants in the University of Leicester Certificate in Adult Basic Education

International Literacy Year held particular significance for people doing the certificate course. We were offered a module entitled "International Literacy", which involved a study of five major modern literacy movements. These were the Nicaraguan "Crusade" from 1979, India's continuing efforts of recent years, the Kenyan programmes following independence in 1963 and some of the current programmes of Europe and North America. Through a "distance-learning" model we each received a package of materials from or about the country in question. This ranged from teaching materials and accounts of field workers; journal articles by interested parties and well-known names in literacy circles; to reports, research, statements and statistics from governments, policy makers and other involved agencies. From this, and anything else we could glean, we made our own analysis and followed up any aspects that particularly interested us as individuals and as A.B.E. practitioners.

At the same time, we were constantly trying to relate the international experience of literacy to our own practice and philosophy. From the developing world to the western bloc, through history and across the political spectrum, we gradually drew together common threads. Ideas and issues began to emerge which we could relate to our own experience in the U.K. as well as other countries. Amongst the most pertinent of these common threads were:

Women and Literacy

Globally and historically women have had less access to educational and vocational opportunities than men. Cultural and social views of female roles have often meant that their potential in these fields has been ignored and underrated. In some countries it is clear that literacy programmes can represent the first chance for women to create their own opportunities – for example the take up in Kenya, where 77% of learners are women, and in Nicaragua, where over half of the young tutors were women. In both countries there has been a corresponding strengthening of women's organisations through, or from contact with, the literacy movement. There are also many examples of women being able to increase their earning potential through literacy related work schemes, as in India and Kenya. Here in the U.K., recent impressions suggest that literacy programmes are attracting more women and, in our experience, offer a starting point to further opportunities.

Economy

One of the early motivating forces for initiating literacy programmes was the idea that mass literacy would lead to economic "take-off" – indeed this idea took hold so strongly in the 1960s that a formula of percentage literacy to economic performance was devised. But this idea does not reflect the true complexity of the relationship -whilst there is evidence of the development of writing being intrinsically involved, with early commercial developments, (the need to record transactions), we also know that, in Britain most obviously, industrialisation preceded the development of mass education and literacy.

The simplistic connection often made between a country's economy and its people's literacy levels can engender harmful attitudes. A document from Canada highlighted this by drawing attention to "Canada's illiteracy bill", which set out to show that illiteracy was draining the economy through "lost taxes" "jail for frustrated illiterates" "inflated consumer prices to cover mistakes" "dwindling revenues for publishers". "reduced international competitiveness" "wages lowered by illiteracy" and so on. It seems that the implied correlation between literacy and economic well-being may sometimes be used to side-step other social injustices such as bad housing, discrimination and low standards of living, all of which cannot be combatted by literacy alone.

Statistics and statements

These never seem to reflect the reality of the situation – how are literacy statistics compiled? Whose definitions are used to assess "illiteracy"? (In different countries numbers are collected and literacy defined in different ways; yet these figures are added together to give us a global figure which is often quoted as a reliable statistic). Further, in whose language and to whose standards are people expected to be literate and for what purpose – who is blamed if they fall short? (In one document the lack of participation of the pastoral peoples in the Kenyan programmes were not in question). Where these questions are not carefully addressed and reflected in a programme there can be harmful consequences – just as when the "brigadistas" (tutors) of Nicaragua attempted to teach in Spanish to the peoples of the Atlantic Coast. These people had several different languages and had not been directly involved in the Socialist Revolution – subsequently much hostility was created and fuelled by the language policy and teaching materials of the literacy crusade.

Relevance

Looking at the levels of take-up of campaigns, their drop out rate and other evaluative evidence we saw that the less successful campaigns had the common feature of being "pre-packaged" and, therefore, often lacking in relevance to the participants' realities. There seemed to have been little attempt to question, along with potential students, how the efforts they would have to make to acquire, or improve, their literacy skills would fit into or enhance their lives either individually or collectively. There were often statements, made by policy-makers about the need for content and materials to reflect local diversity and individual concerns – but, in practice, the same primers and the same basic pedagogy were used nationwide. In contrast, movements of Latin America hold the process of consultation and "dialogue" with potential students as a principal foundation of literacy programmes.

We came across several reports of literacy workers in different settings who were quietly dropping the "official" curriculum and working with their students on issues and with materials that were directly relevant to their concerns. But as we focussed our international study on the meaning and impact of literacy as experienced by the literacy student, we were often frustrated by the lack of first hand accounts from the students themselves – about their motivation for learning, their experiences in doing this and the impact it may have on their lives economically or otherwise. This crucial element is often missing from the research and reports of eminent bodies and policy makers. We felt that this omission is a central issue for everyone: for the literacy tutor, paid by the "power groups" to deliver literacy within a pre-defined framework; and for the literacy student and also for the adult who cannot, or does not want to, tap into education as it is presented and whose voice is not heard.

In the process of doing this study many of us have found it fruitful to discuss these issues with our own students in Leicestershire. When this particular question of the relevance of literacy and how it is presented was discussed in a West Leicestershire town group it prompted many expressions of what literacy meant to the people there. This is a summary of their views:

- literacy is more than reading and writing – it's about developing confidence and feeling good in oneself;
- it is about tackling immediate problems such as spelling and writing cheques in public;
- it allows you to develop functional skills such as reading menus,reading newspapers and writing letters;
- to improve job prospects;
- to be able to help your children with their education;

- being able to read and write allows you to develop "privacy";
- are people being **kept** "illiterate" to maintain the status quo?
- being literate means avoiding embarrassment when filling out forms in public;
- literacy puts your life into perspective – it helps you to understand events that shape your life – it's about meaning and understanding rather than about ticks and crosses;
- literacy gives you something to aim for – developing confidence to study and improve your life chances;
- literacy enables you to challenge people's attitudes, to evolve and develop yourself.

This list clearly shows the individual differences in motivation and needs in one small locality. We conclude that we would find this rich diversity in requirements for literacy whoever we asked, whether the people are from the same street, village, city or nation. We hope that the insights gained from international Literacy Year will help to improve and develop literacy programmes everywhere – so that the real benefits and meanings of literacy can emerge from the myths about literacy which maintain inequality.

B6 Family literacy

Further RaPAL references about family literacy

Bird, V. and Pahl, K. 'Parent Literacy in a Community Setting', *RaPAL Bulletin* No. 24, 1994.

Palmer, J. and Rhodes, K. 'Measuring Success in Family Literacy: Seeing the Wood and the Trees', *RaPAL Bulletin* No. 24, 1994.

Keen, J. 'Family Literacy in Lothian. Connect – Community Learning Programme with and for Parents', Living Literacies Conference, *RaPAL Bulletin* No. 28/29, 1995.

Exploring family literacy

David Barton
RaPAL Bulletin, No. 24, Summer 1994

David Barton teaches in the Linguistics Department at Lancaster University

The current emphasis on family literacy in Britain provides many possibilities for innovative programmes and effective literacy work. However, the phrase means different things to different people. It is one of those ideas which we all approve of, even if we are not quite clear about what we mean by it. The general feeling of "it's a good thing" is contained in both the word *family* and the word *literacy*. When the words are combined and appear in a newspaper article or in a photograph of a mother sharing a book with a contented four year old on her knee any critical comment seems churlish. However, the narrowness of many images of family literacy are the basis of the oversimplifications which the media and the political proponents of family literacy get carried away by.

Here I want to draw attention to two areas which can contribute to an understanding of family literacy. Firstly, research on the reading and writing which people do in their everyday lives can provide insights. Secondly, since there has been a decade of experience in the United States with family literacy programmes, it is worth examining what North American educators reflecting on the work there have to say.

In Britain and the States there have been a small number of studies of people's everyday reading and writing. In Lancaster we are completing a study of the reading and writing done in the home and the community, documenting the literacy events which people participate in and the literacy practices they draw upon in going about their day-to-day lives. By studying people's actual lives there are several points which can be made about family literacy. I will list them as six points here although inevitably they interact. For further details, see the references at the end of the article.

1. *Literacy is more than book reading.* A wide variety of literacy goes on in the home. There are many different literacies beyond book reading. Literacy at home is tied in with daily activities and combines many sorts of reading and writing as well as drawing upon spoken language, numeracy and much more. It is this range of practices which children are exposed to and participate in.

2. *Family is more than mum.* There is a wide variety of households, including single parents, grandparents and other relatives, in the home or nearby. Related to this there is a wide variety of support for literacy: it is not just associated with mothers. Often when adults recall who were the significant people in their childhood's, in terms of education, it is not their parents whom they mention, but other relatives or family friends. (See Sarah Padmore's study of "Guiding lights" listed in the references at the end.)

3. *Home life is* different from school life. Home literacy practices are often different from school practices. It is important to find out what actually goes on in the home, rather than make assumptions about it. There are many literacies and schooled literacy is just one form of literacy. Family literacy can sometimes be an invasion of school and its practices into the home.

4. *Family literacy is lifelong.* Participating in and learning about literacy is not only important at the age of five. Households are significant from infancy right up through the school years. The teen years are of special significance (and in inner city America, at least, a sizeable minority of school age children live in households without any adults). Older adults in their sixties take on new literacies – and may learn these from their children.

5. *Everyone participates in literacy activities.* Parents with problems reading and writing nevertheless engage in a wide range of literacy activities. They keep diaries, write poetry, take phone messages, send letters. They are not empty people, living in barren homes, waiting to be saved and filled up by literacy.

> *intervention, the danger is that they will turn away the very people they are designed to assist, people who are already overburdened with the challenges of everyday life. Thus, the research suggests that we need to stand the "from the school to the family" model on its head, allowing what happens in families and communities to inform schooling (rather than vice versa).*

Elsa Auerbach sketches out a broader notion of family literacy. It includes parents and children interacting around literacy tasks, both school-based ones, such as homework, and ones derived from the home, such as cooking and going on outings. Equally important in her approach are other aspects of family literacy work:

1. Parents working independently from their children on reading and writing.
2. Using literacy to address family and community problems, increasing the social significance of literacy in their lives.
3. Parents addressing child-rearing concerns through family literacy class.
4. Supporting the development of home language and culture.
5. Interacting with the school system.

In contrasting a school transmission model and a socio-contextual model of family literacy she accepts that most projects fall somewhere along a continuum between "a prescriptive interventionist model and a participatory, empowering one." She concludes that "as educators, we need to reflect on where our own practice puts us on this continuum and how the family literacy movement can become a vehicle for promoting change, rather than a bandwagon which impedes it."

References and further reading:

Elsa Auerbach, *Which way for family literacy: intervention or empowerment,* in a book edited by Leslie Morrow, to be published by the International Reading Association.

David Barton & Sarah Padmore, *Roles, networks and values in everyday writing,* in D. Barton & R. Ivanič (eds.), Writing in the Community. Sage Publications, 1991.

Susan Benton, *Networks of communication between home and school,* in M. Hamilton, D. Barton & R. Ivanič (eds.), Worlds of Literacy. Multilingual Matters, 1994.

Margaret Clark, Young Fluent Readers. Heinemann, 1976.

Sarah Padmore, *Guiding Lights,* in M. Hamilton, D. Barton & R. Ivanič (eds.), Worlds of Literacy. Multilingual Matters, 1994.

Denny Taylor, Family Literacy. Heinemann, 1983.

Denny Taylor, *Family Literacy: resisting deficit models,* TESOL Quarterly 27(3), Autumn 1993, pp.550–553.

Initiatives in Southwark with parents and children

Foufou Savitzky and Helen Sunderland

RaPAL Bulletin No. 24, Summer 1994

Foufou Savitzky is the PACT Co-ordinator and Helen Sunderland is a PACT trainer and worker for the Langauage and Literacy Unit, Southwark, London.

The Language and Literacy Unit has been given the brief to develop PACT (Parents and Children Together) work in the London Borough of Southwark. The work is funded by the Inspectorate, through the GEST budget and, in the London Docklands Area, by the London Docklands Development Corporation. As the work develops, we are looking to other sources of funding to expand the projects further.

We were excited at the idea of being able to combine our experiences as trained school teachers, adult education workers and parents to develop a variety of new strategies for involving parents in their children's education. The London Borough of Southwark is a typical inner city area, with wide differences between schools. School populations vary; some are almost exclusively white working class, others have children from a very wide range of linguistic and cultural backgrounds, including a large number of refugee children and children of West African descent, and yet others are predominantly Bengali.

The acronym PACT can be understood in two ways: *Parents and Children Together* and *Parents, Children and Teachers*. As the work started, people were also beginning to talk about the concept of 'Family Literacy'. Our aims encompass all of these. Our primary aim is the improvement of literacy standards among children of primary and secondary age. We achieve this by the greater involvement of parents in their child's education and the development of closer links between teachers and parents. Teachers gain an increased awareness of what parents are able to contribute to the school and of possible literacy difficulties of some parents. Parents gain an understanding of the literacy process and methods used in schools. Both parents and teachers develop a greater understanding of the advantages of bilingualism in children. We also aim to be a first point of contact for parents who want information, advice and help on a wide variety of issues concerning their own, or their child's, education.

Because this was a relatively new area of work we decided to approach it in a variety of ways. We began by consulting with teachers in schools, and making suggestions as to the type of work we could offer. Different schools opted for what they considered the most relevant and feasible for their situations. However, once we had started making contact with parents, different priorities sometimes emerged. Below, we give examples of the different approaches we have used, and describe some of our experiences.

Parents and Children Making Books Together

One of our most popular projects has been with parents and children working together at making their own books along a variety of themes. These books reflect the linguistic and cultural backgrounds of the children, which are often not represented in mainstream publishing.

Initially, the schools send letters to parents inviting them to bring photographs which they can use to make books for their children. Parents are also approached individually by teachers, and this generally proves far more successful than the letters. In some schools, only one parent came to the first session, but the quality of the book produced (we use a laptop computer to produce the high quality printout, and the illustrations are provided through photographs and drawings by the children) and the experience of sharing the book with the whole class (i.e. the child 'reading' the book to the class) so inspired other children that they harassed their parents to come in and make books themselves. Even parents who had always avoided any involvement with the school were persuaded by their children to come in. These included one woman who came in to whisper that she wouldn't be able to make a book because she could not read and write herself. She has since begun making a book, with one of the project workers as scribe, and is looking for an appropriate literacy class for herself. Since then she has come to every single meeting the school has had for parents. Several parents take this opportunity to raise issues concerning their child's progress with literacy or their own concerns with reading, writing and spoken English. We are able to give advice or to refer people on to appropriate agencies.

This initiative has proved to have very positive effects on children's attitudes to literacy as well as on parental involvement, as the following comments show (the original spelling has been left in each case):

> *"I liked it (making a book) because I liked to make it for mum. I read it at home with my dad and sister."*

> *"We both enjoyed makeing the book very much. My son was proud to be able to show of the book he made with mummy at school which I think has encouraged him to read books a lot more."*

> *"I think that this group is an exalant idea. I enjoyed it and so did Nicola. As she dose have a few problems with her reading im sure that this will encorage her as it is somthing that is of intrest to her."*

Many children are choosing to spend time making books, both at home and at school, and involving siblings and their parents. A teacher has commented,

> *"parents have been bringing in books from home to share with the class, that they've made and brought and children have cultivated a love of books and are constantly asking their parents to buy books".*

One child, in particular, who appeared to have very little experience of books, and was reluctant to even try reading or writing, was one of the first to make a book with his mother and is now making rapid progress in class. Parents are ordering copies of books to give as presents to absent relatives. This project has drawn in some parents who do not usually have any formal contact with the school. Teaching staff are delighted with the success of the project and are closely monitoring it to see what long term effect it has on children.

Training for Parents

As parents ourselves, and in our contact with parents in Southwark schools, we realised parents' need for training and information on how to support the work schools are doing with children. We offer training to parents in the areas of reading, choosing books, writing, spelling, bi/multilingualism and maths. To date we have worked in ten different Southwark schools. These meetings are generally well attended and parents come away with practical strategies for helping their children as well as a basic theoretical understanding of how children learn in these areas. The following comments are typical of those made by parents following sessions:

> *"I talk to my children in my language but they will only answer in English. I found out that I should continue to do so – the children will be learning new words and it will help them with English."*
>
> *"My children do not like speaking my language. Now I know it's important to explain the advantages to them!"*
>
> *"It gave me a chance to talk about my child's problems."*

The sessions also provide a forum for parents to discuss other issues concerning their own or their child's education (for instance, parents who are concerned that their child may be dyslexic).

Most of this work has taken place in primary schools. However, two secondary schools have been involved. In one, at a meeting targeted at the parents of children receiving extra support with reading and writing, we were inspired by the huge attendance and commitment on the part of parents, children and staff who came out on a bleak January evening, in the middle of an OFSTED inspection, to discuss reading.

In the same school, we will be running a six week intensive course for parents. This will provide them with basic training to enable them to work with secondary age children on their literacy skills in the classroom. The aims of the course are to:

- allow parents to reflect on the issues they are likely to encounter
- gain an understanding of the theory of literacy learning
- be able to apply the theory to the individual young person
- select and look critically at available reading materials

Working with parents already attending English for Speakers of Other Languages courses at Southwark Adult Education Institute, we found that they had very little information about how the system of education in England works, or how they were expected to relate to schools and teachers. We developed a course intended to give parents the confidence to participate in their child's education. The areas we covered included: *how children are taught reading, writing and maths; choosing books for children; the National Curriculum; testing; progression through the system; expectations of parental involvement; communicating with teachers; practical issues such as school outings, uniform, dinners, etc.* Throughout the course parents were encouraged to draw on their own experiences of education and compare different systems and approaches.

Several interesting issues arose out of this course. For instance, teachers may have preconceptions about cultural and religious matters which they use to exclude children from certain school activities. An example was given to us of a Muslim child excluded from taking part in a nativity play even though her mother would have been happy for her to be included. The Parents' Charter is intended to give parents the right to information. Many parents, however, cannot make use of this right. For example, a parent on the course had not been aware that part of her daughter's school report included her SAT results and, in fact, had not known that her daughter had been tested. Parents were universally diffident about approaching teachers and were unaware that they could be involved in the work of the school.

Counselling For Parents

Although we always encourage parents to bring us their concerns about their child's education, in one school we made ourselves available solely for this purpose. Among the issues we dealt with were those of:

- how a refugee mother could explain to her six year old child why she had "abandoned" him back home (having arrived in Britain a few months previously he was now presenting severe behavioural problems in class)
- how to maintain the culture and languages of the home country in children who were either born here or arrived at a very young age
- how to help children to grow up to be bilingual, including finding a community language school
- how to play an active role in children's education, both at home and in the school
- how to help children with reading, when the parent is not sure of the English pronunciation of words
- changing teachers' perception of parents to see them as valuable resources (for example, one parent had been translating popular children's books into her language for her child and then throwing away the translations; her skills are now being used with children in the classroom, and the translations saved)
- communication between parent and teacher (for example, when a project worker was able to interpret an interview, it was the first opportunity for both parties to express their concerns and share essential information).
- how a mother whose child has problem eyesight, but who believes him to have other language problems, can get him assessed.
- how a mother who cannot read and write can help her child and get help with her own literacy.

Training for Teachers

All training sessions for parents have been open to teaching staff. In many cases teachers have indicated that they have learnt from these sessions and have asked for similar training to be made available to their whole staff group. We have provided training specifically for teachers on:

- the teaching of spelling
- using Caribbean literature in the classroom
- the bi/multilingual child

We have found that the most successful work has taken place in schools with a real commitment to parental involvement. This is extremely exciting and rewarding work to be involved in, and we feel that it is showing tangible rewards in terms of parental rights. We hope the work continues and expands and feel there is a lot of scope for further development.

B7 Workplace

Further RaPAL references about the workplace

Halliday, E. 'Literacy: The Community's Business', *RaPAL Bulletin* No. 15, 1991.

Bootle, R. and Rowley, J. 'Trade Union Perspectives on the BAXI Workbase Programme', *RaPAL Bulletin* No. 17, 1992.

Frank, F. 'Workplace Basic skills – Some of What's Happening Around the World', *RaPAL Bulletin* No. 17, 1992.

Nieduszynska, S. *et al.* 'Workplace Basic Education: Trade Union Perspectives on the Baxi Workbase Programme', *RaPAL Bulletin* No. 17, 1992.

Tobias, K. 'The Confidence Gap: Helping Adults Deal with Interviews', *RaPAL Bulletin* No. 17, 1992.

Frank, F. 'Workplace ABE and Community ABE. Two Different Kettles of Fish', *RaPAL Journal* No. 51 2003.

Frank, F. and Rodrigues, E. '"We are the governor's dogs": Students' voices and policy-making. Developing a new framework for literacy practitioners in empowering learners to engage with workplace change', Occasional Paper. *RaPAL Journal* No. 51, 2003.

Love, literacy and labour

Jane Mace

RaPAL Bulletin No. 17, Spring 1992

Jane Mace is a lecturer in adult community and continuing education at Goldsmiths College, London and has been involved with workplace basic education programmes for a number of years.

> *Local government will have to come to terms with instability which will effect almost every area of the human resources activity . . . It is a strange fact that the growing realisation of the importance of human resources management has, in many authorities, been accompanied by a diminution in the importance and scale of the personnel function. (1)*

> *It was only with the development of capitalist society, in which the capacity to labour is exchanged on market like any other commodity, that the personal emotion for the mass of the people can appear as separate from work. (2)*

There was once a group of women who met early one morning in a classroom. They met twice a week for ten weeks. During those meetings, they read, talked, wrote, and talked some more. At the end of the ten weeks, they went back to what they usually did at that time of day: Hoovering, washing down toilets, and buffing corridors – and in one case, cleaning the classroom in which they had been meeting. At that time of the day, they were cleaners, working for a college. (At other times of the day they were grandmothers, story-tellers, consumers, local residents, people with diverse histories, dreams and realities).

Later in the day, in the rooms they cleaned, others worked; and over the years, the piles of paper exhorting these people to market what they had previously merely taught, grew higher. Every day, these paperworkers struggled to read and write important policy papers about national vocational qualifications, customer care, accreditation, marketing, and cost effectiveness. And month after month, the waste-bins that the cleaners had to empty grew fuller.

One of the women in the original group, however, went on with the writing she began in that classroom. She wrote poems. She wrote letters. And she wrote hymns. All of this, of course, went unnoticed by the paperworkers whose offices and classrooms she and the other women cleaned every day.

The course this woman had attended was a workplace training course. I remember it with a kind of wistfulness. It was 1984, and my first experience of recognising that the places where people worked was where community education could happen, too. It was also my first experience of teaching at 7 in the morning. In the years since then I have learned a little more about such work: ducking and diving with nifty bids for paid educational leave courses with the local authority where I work (Lewisham Borough Council, in South London). Half-day courses, two-day courses, even a couple more ten-week (four hours a week) courses; inventing names that somehow both please the funders and attract the students. ('Report-writing' 'communication', 'improve your confidence in spelling', and most recently, 'paperwork made easy'). Always trying to insert something of the naughty and joyful bits of literacy work into the courses argued for as providing a means for 'better communication in the workplace'.

There are three things that bother me now, recalling that group of women and looking at the talk we are in the midst of now about workplace training. One is about the growing primacy of the employer's motives; the second concerns ideas about 'change' (more accurately referred to in the first quotation at the beginning as 'instability'); and third is something Sheila Rowbotham (1983) was writing about which I summarise, from the second quotation (from her) as the separation of love and work.

Employer motivation

Let us not knock employers. They are only human, and some of them still have equal opportunities policies. But their interests in workplace training are bound to be more and more circumscribed by other, often more pressing interests: they have to balance the books.

What we have to do, as adult literacy educators, is persuade them that they can argue for this 'workplace training' to their auditors. We are having to learn evermore tricky disguises for what we do; and in the process, all that trust that we rely on with our students becomes more and more fragile. In arguing for the benefits of literacy, we are used to more homely language: no tests, we say: nothing like school; you will be in good company; everyone at their own pace. But this kind of stuff will not stand up to the accountant's questions. We are in the territory of training; and training will cost their employer money: not only in training fees, but also in lost production – or, (if they accept the arguments for cover) in the costs of covering those released for the training.

Motives such as good industrial relations, a 'feel good' factor, humanitarian principles, or even equal opportunities policies have to be tempered with those of profit. Training professional staff is one thing: it is an investment. But training the low paid, part-time, shift working staff is more risky. Put money into the first lot, and there is some hope (not always borne out by reality) that they will go back to the job willing and able to take more responsibility and attract more customers; but put money into this other lot, and before you know it they might get ideas above their station and leave (also not borne out by reality, as the story I told at the beginning suggests).

ALBSU's Basic Skills at Work Initiative will this year be reporting on a national survey of employers' perceptions of the literacy demands of specific occupations. It should make interesting reading. The argument presumably is that the findings will convince more employers to release more staff to have time off work for education that will develop their literacy. What literacy research has told us, time after time, however, is that what workers think they need and want in terms of literacy education may have little to do with the profit or patronage of their employers. What is more, workers often have a great many more literacy competences than employers give them credit for. To put it another way, employers do not always know what their workers know; yet the funds for workplace literacy increasingly depends on the assumption that they do.

Change

You know about change: I know about it. It is all around us. Changing needs; changing society; rapidly changing markets; technological change. We have got to train for it, adapt to it, keep up with it, not get left behind. But what happened to people as agents for change? Where in this concept of change, is the literacy student who makes their own changes? What about the change people are living with already? (It is worth noting, in passing, that women undergoing the 'change of life' are precisely in the age bracket where least training opportunities are on offer).

Stress at work is the result of change without meaning. Being restructured, reorganised, and reviewed is now an everyday common or garden feature of worklife.

Organisations once designed to provide education and health are now selling modules and packages. Those who work for them, with literacy interests deeper or wider than the prescribed 'needs' of the organisation, had better look elsewhere if they want literacy education that meets those interests. Workplace literacy has to make them fit the changes others have in mind for them. In this context, literacy educators have their work cut out to convince, not the employers, but the workers and their elected union representatives that workplace training offers any possibility for them to make their own changes. Little wonder the main currency of literacy work – trust – is now being debased. The 'gaffer's angle' is in the way. (3)

Love and work

Mary Wolfe, when she worked for Workbase, was the first person to point out to me a fundamental misconception of workplace training with manual staff: namely, that workers are doing a good job as long as their managers think they are. Actually, a large number of manual workers of my acquaintance have a higher standard of work that they do every day than that expected of them. They care about their work. Permission from managers and bosses to let them go to a course is not always the generous gift that those employers may think it is; for what it puts at risk is their own high standards of work. It may be ok for those paperworkers to leave their desks for half a day a week to go on a course; but unless there is someone to replace the cleaner or care assistant or maintenance worker when they go on a course, *and someone who works to their standard,* something far more precious than their employers training programme is at stake: their own job satisfaction.

Despite low pay, despite insecure conditions, despite having to get up when others are still drowsing in warm beds, despite health and safety hazards, countless thousands of people in manual grade jobs put love into work. No-one pays them extra to do this: and capitalism prefers it if they leave emotion out of it. The thing in the market economy of the 1990s that literacy educators have to engage with is an education job with employers. We have to remind them that literacy beyond the instrumental uses they have thought of also has use and value in a workplace culture. Hymns, songs, poems, letters – creative literacy, in short – may not cut the wages bill; but they certainly might do something about repairing the broken bridge between love and labour.

Notes

(1) Rob Pinkham, 'Centre Stage' in Local Government Chronicle 4 (October 1991), p.16

(2) Sheila Rowbotham, **Dreams and dilemma: collected writings,** Virago, 1983, p.187

(3) 'What was the Gaffer's angle – giving us time off with no obvious advantage to him?' was the reaction recalled by a group of Sheffield council workers to the first publicity about a workplace training course; from Cathy Burke et al, 'Take Ten', in Jane Mace and Martin Yarnit (eds.), **Time off to learn; paid educational leave and low paid workers** Methuen, 1987, p.108

Inspecting the consequences of virtual and virtuous realities of workplace literacy

Geraldine Castleton

RAPAL Bulletin No. 39, Summer 1999

Geraldine Castleton is a lecturer in the School of Language & Literacy Education, Faculty of Education, at the Queensland University of Technology, Brisbane.

Workplace literacy is a new and highly contested site of activity. The field has evolved so quickly in response to social, economic and political imperatives, that the term "workplace literacy" has achieved a level of commonsense acceptability so that its constructions have largely gone unchallenged. Many of these commonsense views, now grounded in discourses around workplace literacy, are in need of exploration and explication. This is because discourses readily become institutionalised and understood as natural or commonsense ways of knowing, understanding and talking about the world. In this way discourses can be seen as not only *reflecting* social reality but also as producing these social realities.

This paper reports the findings of recent research (Castleton, 1997) into how workplace literacy is socially described. Among the ideas that provided a theoretical framework for this project is the work of Dorothy Smith (1987, 1990) who has argued that institutional texts establish 'relations of ruling' that, in turn create a 'virtual reality' brought about through distinctive patterns of reading and writing these texts (Smith, 1990:62). In addition, it was informed by Lena Jayyusi's (1988, 1991) interest in the moral foundations of social order that become "reconstituted and reestablished as 'grounds' for accountable, rational, intelligible actions, inferences and judgment within discourse" (Jayyusi, 1991:236).

Reading *Words at Work*

Within the Australian context many of the "commonsense" views on workplace literacy are evidenced in the 1991 federal government report, *Words at Work: Literacy Needs in the Workplace (1991).* A number of recommendations from this report were taken up in the policy document, *Australia's Language: The Australian Language and Literacy Policy* (1991). This resulted in extensive government funding of workplace language and literacy programs that became part of the federal government's agenda of reform in Australian industry and the education and training sectors, in order to produce a better skilled, adaptable and competitive workforce.

The document can also be described as an Australian counterpart of what Darrah (1992) labelled "future workplace skills literature", and included in Gee and Lankshear's (1995) category of "fast capitalist" texts. Emanating in the first instance from the United States but taken up in many western nations, this literature, that presents what appears to be a rational account of how work must be done in 'new times' has quickly become institutionalised. Much of this literature also characteristically constructs literacy as a functional, employment skill closely tied to a nation's economic progress and productivity. According to Smith (1987, 1990), it is within such institutional texts that particular `relations of ruling' are established that create a 'virtual reality' (Smith, 1990:62) for and about those who are constituted and positioned within the discourses presented.

New workers for "new times"

These institutional texts make a strong case for workers to be innovative, flexible and highly skilled. For example Hammer & Champy (1993) describe a 're-engineered' workplace marked by new kinds of work arrangements, including 'process' teams in which workers share joint responsibility for performing and knowing about the whole production process. This literature relies heavily on our culturally shared understandings and acceptance of the accounts offered of how work is done. These understandings include particular sets of norms, rights, capabilities, social and moral responsibilities for those who belong to the category of 'worker' (Castleton, 1997:174). Discourses established within and by these texts present a particular version of knowledge about the place of literacy in the workplace, that sees workers, in very particular ways, contributing to the nations' inability to compete effectively in the international marketplace. What is assembled is a picture of 'workers' who are clearly operating outside the parameters of what is means to be 'good workers', including possessing adequate levels of functional literacy skills. (Castleton, 1997:181).

Talk of words at work

The project also drew on interviews with a number of people, including industry representatives, government bureaucrats, trade union representatives and workplace literacy teachers. These people were selected because of the various ways in which they were involved in the field of workplace literacy and could be expected to have well developed views and opinions on this issue.

The talk of these key stakeholders was examined to interpret the various links they made between the salient features of 'work', 'workers' and 'literacy'. This analysis showed that their views typically conformed with those presented in the institutional texts including the government document and "fast capitalist" texts. There was general agreement on a world of work in which there is a problem that has been created by some "workers," and that this problem has dire consequences for society as a whole (Castleton, 1997:207–208).

(E)merging discourses at work

The outcomes of this project demonstrate how knowledge of workplace literacy has been socially constructed out of a range of prevailing discourses on work and on literacy. Out of this fuse of ideas a particular version of "workplace literacy" has received a preferred reading and hearing. However, the findings support the call by Hull (1993:44) to "amend, qualify and fundamentally challenge the popular discourse on literacy and work", as there are significant omissions and silences in these contemporary understandings that do not allow for a proper account of workers, for how 'literacy is made' at work, and for its place in the everyday lives of all people.

Grounded as it is in other discourses that are themselves incomplete and flawed, the dominant discourse on workplace literacy not only offers limited understandings of the inter-relationships existing between literacy and work, but also works to constitute and strengthen particular power relations within work settings and beyond. These findings reinforce Foucault's (1980) contention that power and knowledge are interrelated and inseparable so that any field of knowledge constitutes at the same time certain power relations. Within the increasingly segregated workforce found in today's workplace such formulations define some as winners and others as losers.

Informants to the study, for example, typically talked of how workers' literacy practices "used to be okay", but now, under new working arrangements, these same workers' skills were framed as "not okay". The very skills that enabled them to be described as the 'good' workers, in traditional ways of doing work, including the particular ways in which they communicated with one another in individual work contexts, now were used to separate them out as the 'poor' workers, that is those with limited literacy skills. Workers from non English speaking backgrounds, once valued as the 'good' workers because they "put in a hard day's work without complaining", were now determined to be not acceptable because they could not, for example, abide by the written instructions and procedures that now dominated workplaces. In fact, one informant described how his company used to seek out non-English speaking background employees, but now had instituted selection procedures including literacy testing, that effectively excluded these workers. He defended this position, stating that his company could no longer "take the risk" of having employees with poor language and literacy skills. Native speakers with poor literacy skills fared no better, with a number of interviewees recounting how traditional practices, that involved workers relying on each other to meet 'on the job' literacy tasks, were no longer acceptable in workplaces marked by an increasing need for reporting and accountability procedures. These accounts reinforced the message presented in the official government document, as well as the accounts offered in "fast capitalist" texts.

Contemporary discourses of work privilege a particular account of work that creates new social realities and identities in workplaces. These new discourses on work name those experiences considered to be important to contemporary society. They value the ways in which some people's experiences are talked or written about, while denying credibility and validity to the experiences of others The strong emphasis given to workers' skills within these discourses displays their heavy reliance on the functional literacy discourse in their descriptions of workers' literacy competence, and on binary oppositional categories like skilled" and "unskilled".

According to Street (1995:125), the common practice of relying on the functional discourse to explain literacy at work 'disguises and effectively naturalizes the ideological role of literacy in contemporary society', and allows for certain conceptualizations of worker identity that carry with them particular moral implications. Literacy from this mechanistic and technicist sense becomes a representation of many of contemporary society's most serious problems: workers with limited literacy skills can be held responsible for a nation's poor economic performance. From this situated perspective, blame is located in individuals and explanations are framed in terms of workers' ethnic backgrounds, attitudes to work and their socio-economic circumstance, not in institutional justifications of fiscal difficulties, organisational mis-management or market declines. Literacy takes on a signification that far outweighs what features of social life can be adequately and appropriately explained in terms of its actual role in people's lives.

Alternative discourses on literacy at work

More meaningful formulations of literacy than those offered in the functional literacy discourse need to be applied to the context of work to fully appreciate the role literacy plays both for workers and for work.

Accounts of work offered by the informants in this current project demonstrated the rich interplay of communicative practices that do exist in workplaces, though they are not often given the legitimacy they deserve. Informants talked of workers typically relying on other workmates, particularly in those tasks that involve some form of literacy, as they go about their jobs. This process is a fundamental part of social life both within and beyond workplaces that, rather than be ignored, must be accommodated in our understandings of work and of literacy, that include recognising its plurality. There are examples in the literature of networks operating in workplaces that give a far more accurate representation of how work is actually achieved through the sharing of knowledge and skills rather than by individual performance (Gowen, 1992, 1994; Darrah, 1993; Hull, 1993, 1997; Prinsloo & Breier, 1996; Castleton, 1999). These accounts allow for more positive ways of recognising workers' attributes and. can capture more effectively the collective and communal nature of work and of literacy, as well as account for how people may move in and out of particular roles as work is accomplished.

Understandings of workplaces as "communities of practice", characterised by rich social networks, would assist in removing the present concern for the individual skill level of workers. Recent work undertaken in Australia (Pearson et al, 1996) and the United States (Hull, 1997) shows that a deficit model, grounded in the functional literacy discourse, still predominantly drives workplace literacy programs in those countries. While there is some evidence that such deficit approaches may result in improved performance in the short term, Gowen (1994:134) has warned, that they do little to help organisations "restructure themselves into more humane and democratic workplaces". Approaches to workplace literacy therefore may need to focus less on what we all have been socialised to think of in terms of traditional education and more on addressing the role and value of literacy in the lives of the workers to whom these measures are addressed.

The future of workplace literacy

A workplace literacy pedagogy must emerge that allows for and sustains the voices and experiences of workers as they interrogate workplace texts, social relations and practices to determine how they can take on more active roles in determining their own futures. Critical readings of commonly found workplace texts and accompanying practices, such as "Vision Statements" and "Core Values" for example, may create opportunities for workers to put forward alternative ways of achieving increased productivity. Gee and colleagues (Gee et al, 1996:158) have argued for a socio-cultural approach to literacy and to new-capitalist business that requires learning for performance to include the acquisition of tacit knowledge through immersion in communities of practice. However they further maintain that such an approach must "go beyond simple immersion to gain the ability both to reflect on one's tacit knowledge and to critique the communities within which one has achieved it" (Gee, et al, 1996:168). Such an approach would extend the catchcry of "fast capitalist" texts of worker emancipation to give workers an authentic and authoritative say in their own destinies. Rather than working to establish the cultural conformity implicit in new discourses of work, workplace literacy programs need to attend more closely to the benefits of recognising and acknowledging the diversity of practices that do exist in workplaces that can be built upon, rather than ignored and devalued.

All those involved in the field are therefore urged to process, and avoid perhaps becoming unintentionally implicated in education projects not of their choosing. Workplace literacy "work" must be seen as more than a "virtuous" response to workers' purported needs, and must fulfill real purposes for all stakeholders, providing for all 'the means of grasping the social relations organizing the worlds of their experience' (Smith, 1987:153). Otherwise it runs the risk of becoming, in Foucault's (1980) terms, a disciplinary discourse that sets out to legitimate and sustain certain forms of power/ knowledge.

Bibliography

Australia's Language: *The Australian Language and Literacy Policy (1991)* Canberra: Department of Employment, Education & Training. Castleton, G. *(1997). Accounting for Policy and Practice in Workplace Literacy. (e)merging discourses at work* Griffith University: Unpublished Ph D thesis.

Castleton, G. *(1999) Understanding work and literacy: (e) merging discourses at work* Melbourne: Language Australia.

Darrah, C. *(1992)* 'Workplace Skills in Context.' *Human Organization Vol. 5,* No. *3. pp. 264–273.*

Foucault, M. *(1980) Power/Knowledge: Selected Interviews and Other Writings 1972–1977.*

Gee, J., & Lankshear, C. *(1995)* `The New Work Order: critical language awareness and 'fast capitalism' texts.' *Discourse.' studies in the cultural politics of education. Vol. 16,* No. 1. pp. *5–19.*

Gee, J., Hull, G. & Lankshear, C. *(1996) The New Work Order.* Sydney: Allen & Unwin.

Gordon, G. (ed) New York: Pantheon Books.

Gowen, S. *(1992) The Politics of Workplace Literacy. A Case Study.* New York: Columbia College Press. Gowen, S. *(1994)* 'I'm No Fool: Reconsidering American Workers and their Literacies.' In O'Connor, P. (ed) *Thinking Work Volume l: Theoretical Perspectives on Workers' Literacies.* Sydney: Adult Literacy Basic Skills Action Coalition.

Hammer, M. & Champy, J. *(1993) Reengineering the Corporation: A Manifesto for Business Revolution.* London: Nicholas Breatey Publishing. Hull, G. *(1993)* 'Hearing Other Voices: A Critical Assessment of Popular Views on Literacy and Work.' *Harvard Educational Review. Vol. 63,* No. 1. pp. *20–49.*

Hull G. *(1997)* (ed) *Changing Work, Changing Workers. Critical Perspectives on Language ,Literacy and* Skills. Albany: State University of New York Press.

Jayyusi, L. *(1984) Categorisation and the Moral Order.* London: Routledge.

Jayyusi, L. *(1991)* 'Values and moral judgement: communicative praxis as a moral order.' In Button, G. (Ed) *Ethnomethodology and the human sciences.* Cambridge: Cambridge University Press.

Pearson, G., Bean, R., Duffy, J., Manidis, M., Walkenberg, T. & Wyse, L. *(1996) More Than Money Can Say. The impact of ESL and literacy training in the Australian workplace.* Australia's Language & Literacy Policy, Canberra: DEETYA.

Prinsloo, M. & Breier, M. (eds) *The Social Uses of Literacy.' Theory and Practice in Contemporary South Africa.* Bertsham, SA & Amsterdam: John Benjamins.

Smith, D. *(1987) The everyday world as problematic. A Feminist* Sociology! Toronto: University of Toronto Press.

Smith, D. *(1990) The Conceptual Practices of Power: A Feminist Sociology of Knowledge.* Northeastern University Press.

Street, B. *(1995) Social Literacies: Critical Approaches to Literacy in Development, Ethnography and Education.* London: Longman. *Words at Work: Literacy Needs in the Workplace (1991)* Canberra: Australian Government Printing Service.

Literacy and the capitalist workplace

Rob Peutrell

RaPAL Bulletin No. 43, Winter 2000

Rob Peutrell is an ESOL and literacy tutor, and union activist, in a Further Education College in Nottingham

Introduction

Attention has increasingly been drawn to the necessity of analysing the competing discourses which prevail in the field of 'workplace literacy'. In particular, there is debate about the limitations of approaches grounded in functional/deficit literacy and managerial ideologies; and about the need to identify those which are rooted in workers' own experience, which may be used to create **'more humane and democratic'** workplaces ((Castleton, Rapal Bulletin 39, 1999) The latter perspective posits a view of the workplace as a 'discursively constituted reality', which is open to reconstitution through the introduction of 'different' discourses. Furthermore, it assumes that it is possible to achieve a community of interest between managers and workers in the workplace, however difficult this may be in practice (Holland 1998 p.42). In this paper I shall consider the issue of whether this alternative 'stakeholder' perspective actually moves beyond traditional workplace literacy to offer workers **'an authentic and authoritative say in their own destinies'** (Castleton, 1999).

Economic Realities?

The starting point for this discussion has to be our understanding of the workplace. The difficulty here is not the emphasis workplace literacy theorists put on discourse but their dissolution of workplace reality *into* discourse. We do need to make sense of capitalism's changing discourses but must anchor them into a capitalist *reality* constituted through the process of capital accumulation and class struggle.

The process of capital accumulation is what defines the capitalist workplace as capitalist and it has enormous implications for workplace relationships and for the scope of workplace reform. Within capitalist enterprise, management has the task of holding down relative wage costs whilst increasing productivity; and asserting their 'right to manage' whilst engaging workers' co-operation. This complex set of management responsibilities results in an equally complex range of management strategies. At the local level, there may be some scope for pursuing different approaches but at global level, dominant forms are constructed and reconstructed over time. Hence we can compare the corporatist 1960s and 1970s with the neo-liberal 1980s and 1990s and the current attempt to articulate a 'third way' stakeholder capitalism (Giddens 2000). However, the apparent contradictions between 'competing' management approaches should not mask their common goal; both the seemingly democratic approaches of the stakeholder enterprise and the overtly authoritarian practices of the sweatshop are strategies which aim to facilitate capital accumulation.

These imperatives of capital accumulation conflict with the workers' need for autonomy – manifest in demands for more money, job control, time, pleasure and so on. These needs are asserted in a variety of counter-discourses and practices, ranging from covert absenteeism to overt trade union activism. Whilst in the short term negotiated compromises may be reached, they have always proven unstable, vulnerable to the vagaries of the competitive economy and the unremitting pressures of struggle.

The analysis of the workplace as a site for accumulation and struggle calls into question both the idea of a **'humane and**

democratic' *capitalist* workplace and the anticipation of a workplace literacy able to **'fulfil *real* needs for *all stakeholders*'** (Castleton, 1999). Perhaps we need to say what we actually mean by the word 'democracy' in this context. Do we mean *capitalist workplace democracy,* in which an always contingent 'democracy' remains ultimately subordinate to the requirements of capital accumulation and, therefore, managerial power? Or do we mean a *workers' workplace democracy – an* insubordinate project aimed at ending that subordination.

Literacy and Struggle

Consistent with the perspective on the workplace outlined above is Freire's stark challenge that we must choose between 'domesticating' and 'liberating' literacy education – between the literacy of the 'oppressor' and that of the 'oppressed' (Freire 1985). Post-modernists would ridicule this Freirean 'choice' for its crude, 'discursively incorrect' *binary positioning* but the point I am making is that work-place conflict represents something more fundamental than competing, but ultimately negotiable, discourses rooted in different, but equally legitimate, experiences.

From this perspective, what is needed is a literacy pedagogy that connects not with *workers in work,* but with *workers in struggle.* Struggles can be seen as learning processes through which participants can find their own individual and collective 'voices'. Through struggle workers can become autonomous, self-active subjects able to articulate their own needs and values (Cleaver 1992). Such struggles have resulted in counter-cultural and resistant discursive practices in which literacy has played a rich and essential part. There is a wealth of material – including workplace graffiti, workshop/office chat, jokes, internet sites, minutes of workplace meetings, poetry, fiction etc. which illustrate the various ways in which workers have communicated their resistance, orally and through diverse print and electronic technologies.

It is important not to view workers' struggles as, at best, limited to a kind of subterranean survival culture; and the literacy worker's role as one of 'giving' them more of a say in their working lives. In that case 'democracy' or 'empowerment' would be achieved as a result of outside, professional intervention, rather than as a feasible outcome of workers' own struggle.

Without doubt, funding to provide literacy support for workers **in struggle** would be a scarce commodity. Nonetheless, there are two contributions literacy practitioners can immediately make. First, we can develop our critique of functionalist/deficit literacy and also resist our co-option into this *structurally* situated discourse, including an apparently progressive stakeholder view of literacy: one that denies the class antagonisms which inevitably structure relations between managers and workers. Second, there is the solidarity which can help sustain workplace (and other) 'communities of resistance'. Literacy practitioners may have tools and knowledge to offer, such as helping people to produce/publish their own material; (co-) researching counter-cultural/community literacy practices (Cameron 1992); and articulating a critical sensibility to help nurture a democracy of grassroots voices and experiences.

Literacy Professionals

Finally, I think we need to put ourselves into the discussion on workplace literacy. In particular we need to tease out the ways in which our position and identity as educational `professionals' relate to the theories and perspectives we adopt in respect of the workplace and literacy.

The idea that social relationships can be reconstituted through the disinterested intervention of professionals introducing 'better' discourses is not new. Rather, it has been one of the legitimating myths of professional power throughout the 20th century. It is one way in which professionals lay claim to social territory, status and resources. However, in the context of the workplace, this claim both obscures the structural, class origins of workplace conflict; and the way in which the apparently 'humanistic' and `democratic' can in practice constitute a more sophisticated and less transparent regime of control.

Perhaps the current changes in adult education will cause all of us to reassess who and what we are. Like workers everywhere, we have been the subject of 'fast capitalist' ideology. Our deprofessionalisation, and casualisation parallels craft workers' 'deskilling' and the rise of the ubiquitous 'McJob'. Our common experience with other workers might provide the basis for a different workplace literacy pedagogy: one rooted in an ethic of solidarity and a commitment to the potential of workers ***through their own self-activity*** to take control of their working lives. It would start by acknowledging the limits for radical practice in existing education and training.

References

Cameron, D. *(1992) Researching Language,* London: Routledge.

Castleto, G. *(1999)* 'Inspecting the consequences of virtual and virtuous realities of workplace literacy' *RaPAL 39*. London: Avanti Press.

Cleaver, H. *(1979) Reading Capital Politically* Sussex: Harvester Press.

Cleaver, H. *(1992)* 'The Inversion of the Class Perspective', in Bonefield W, Gunn R and Psychomedis K (Eds.) *1992 Open Marxism vol. 2* Pluto Press London. (Also, check out Cleaver's web page)

Freire, P. *(1985) The Politics of Education,* Bergin and Garvey.

Giddens, A. *(2000) The Third Way and its Critics,* Cambridge: Polity.

Holland, C *(1998) Literacy and the New Work Order* Leicester: NIACE.

Footnote

The author's union is NATFHE, the National Association of Teachers in Further and Higher Education.

Built-in or bolted-on: Literacy practices in the civil construction industry

Jean Searle and Ann Kelly
RaPAL Bulletin No. 47, Winter 2001

Jean Searle is the Director of the Queensland Centre of ALNARC (Adult Literacy and Numeracy Australian Research Consortium). She is a Senior Lecturer in Adult Literacy and Communication for Adult and Vocational Education at Griffith University.

Ann Kelly has worked as a teacher, researcher and administrator within the adult literacy field for over 15 years. Currently she is completing a doctorate on the oral interactive competencies of clerical workers.

Introduction

The Queensland Centre of ALNARC (Adult Literacy and Numeracy Australian Research Consortium) has conducted two research projects within the civil construction industry. The first of these focused on an examination of the effects of the inclusion of literacy and numeracy in training packages[1] on the quality of learning and work outcomes. The second project examined how staff employed at different levels within one company viewed training, in particular literacy training. In this paper, data from both projects are drawn on to examine representations of 'literacy' and 'numeracy' and to discuss the perceived advantages and disadvantages of a taking a built-in (integrated) or a bolted-on (end-on) approach to literacy training in the workplace.

The workplace

The work sites at the centre of the two studies were both part of the Pacific Motorway project – the construction of approximately 100 kilometres of motorway. Overall, the project involved five companies and 1760 workers. The two companies selected for this study were major employers of these workers, had their own training sites, and demonstrated a commitment to training using the civil construction training package.

During the period of the first ALNARC study (Kelly & Searle, 2000) a range of training programmes was conducted on the two sites. These programmes were provided by company trainers, external civil construction training consultants, and TAFE[2] institute literacy and numeracy teachers. In some cases, these programmes were based on materials that had been commissioned by the motorway companies as an alternative to those developed by the Australian National Training Authority (ANTA, 2000) to serve the civil construction industry generally.

Integration of literacy and numeracy competencies within the civil construction standards: built-in or bolted-on?

In developing the industry standards, two approaches were taken with respect to the integration of underpinning literacy and numeracy skills: built-in, as implicit, underpinning skills, and 'bolted-on' explicit skills.

1. Built in – implicit underpinning skills

While literacy as explicit speaking, reading or writing skills is not addressed at AQF certificate level 1, a number of literate practices are covered in the level 1 unit of competency:[3] "*Carry out interactive workplace communications*". Specifically, in order to be assessed as competent, workers must demonstrate they can "*receive and convey information"*, *"carry out face-to-face routine communication"*, *"work with others"* and *"participate in simple on-site meeting processes"*. Data in the following table were collated from our most recent study and these indicate some of the on-site practices in which all workers must engage.

Table 1: Workplace activities

Activity	When	Who	What
Inductions	When hiring new employees	Work health & safety officer	Induction to company WH&S
Work Activity Briefing (WAB)	Commencement of new job	Project officer, engineer, WH&S office and crew	Site plans, training crewneeds, equipment
Pre-start meetings	Every morning (sometimes evenings)	Leading hand and whole crew	Objectives for the day Problem solving Discussion of previous day
Pre-start checklists	Start of shift	Individual	Equipment checks Safety checks
Task specific briefings	Start of shift	Individual	Task objectives Problem solving
Job Safety Analyses (JSA)	Commencement of new job or task	Whole crew	Analysis of safety procedures Environmental issues
Toolbox meetings	Once a fortnight	Leading hand or Foreman to crew	Job issues Safety issues

The training co-ordinator for the Motorway Project stated,

> *"In effect a labourer has to be able to understand what the language is, how to apply that language, talk in that language to the engineer as well as other workers, and understand what it's about. And of course, the other part of the competency requirement is to be able to complete the associated paperwork. Every day he [sic] has to fill out a series of activities of what he's done for the day: a time sheet so to*

speak. So, a time sheet, in actual fact, is a literacy component".

At no time did we observe any explicit training in these organisational literacies. The assumption is that workers are 'socialised' into these literacy practices.

Observations of on-site training were conducted to identify the strategies used by the company trainer (male) to integrate the underpinning literacy and/or numeracy skills. The observations were in relation to training in the erection of scaffolding. Strategies adopted by the trainer to ensure that the underpinning literacy and numeracy competencies were achieved by the participants included:

1. A close integration of material aspects (for example, the individual scaffold components) with their representations which were illustrated on the whiteboard as they were discussed;
2. A logical organisation of concepts and relationships was developed along with the oral presentation, that is, a kind of 'top level structure' was developed to reiterate the critical points that were being made;
3. The ANTA non-endorsed learning materials on scaffolding were provided as a resource for independent use by the participants;
4. Assessment items related directly to the oral presentation.

In addition to these strategies, the company employed other *built-in* approaches to ensure that the literacy and numeracy competencies that are integral to "technical" competencies were acquired. These included:

1. The development of graphic posters to illustrate the main points in relation to on-site 'toolbox' meetings, where everyone gathered around to hear about a particular procedure or use of equipment. These posters were displayed in the crib rooms as well as being available for individual workers.
2. The commissioning of learning materials from a private consultant which were specific to skills required on the motorway project and took into consideration the

literacy/numeracy levels of the workers. Use was made of colour photographs, labelled diagrams, graphics and cartoons. Further, plans are in place to develop mini-cassette recordings of instructional material.

2. Bolted on-explicitly stated skills

In order to develop the required literacy and numeracy skills that underpinned a number of technical" competencies, a workplace English language and literacy (and numeracy) programme (WELL) was implemented. The programme was conducted in a training room on-site by two female language and literacy teachers who travelled to the site. One of the sessions observed was based on helping students comprehend learning materials which focused on the safe usage of oxyacetylene equipment. The strategies utilised by the teachers included:

- Use of company training materials to contextualise the skill development;

- A series of language activities based on this text, for example:
  - accessing prior knowledge of technical terms;
  - labelling a diagram with 'common usage' terms and comparison with the technical terms;
  - reading for meaning (short passage from training materials and in groups answer questions);
  - discussion of difficult words and spelling patterns;
  - completing a retrieval chart;
  - matching definitions, and so on.

The contextualisation of training

During interviews, with trainers and workers, a consistent theme became the necessity for trainers to be experienced in civil construction. One trainer, when asked whether he perceived any differences between teachers and trainers, claimed that *"trainers are teachers with experience"*: For him, experience 'on the job' rather than years of learning theory was the major defining feature of an excellent trainer. Interviews with workers supported this belief although there was some notion that basic skills needed to be taught. For example, it was suggested that a worker who was having difficulty in his 'doggers'[4] course would have benefitted from a short revision of subtraction and multiplication applications. On the other hand, one of the trainers recounted that a worker had discontinued attending the numeracy program because he wanted to learn how to 'screen' sand in the preparation of concrete and the numeracy teacher did not understand this technique.

Although the teachers were using texts from the industry training materials, they were aware that they were 'outsiders' contracted to 'fix' a perceived worker deficit. They were concerned that 'on *the part of the trainers, there was no sense of integrating"* and further that there were *"problems with the logistics . . . of timetabling, of people turning up for classes, of people who volunteered initially and then didn't follow that through . . . [and] none of the supervisory staff really had much of a clue about the classes so there was no culture, really"*. They suggested that a partial solution to the integration problem would be to employ an enterprise-based literacy/numeracy teacher who would work across the various motorway sites and be available to deliver training as required.

A question of culture and pedagogy

So, the question at the heart of this is: Is there a dichotomy between the culture of the teacher/educator (often female) with a background in school literacy discourses, and the culture of the workplace trainer (usually male) in the construction industry, with an industry background and a basic certificate in workplace training and the assessment of competencies? Each would have their own Discourse (Gee, 1996: 127), their ways of being in the world – the "saying (writing)-doing-being valuing-believing combinations". Each would teach or train what and how they know best. The teachers demonstrated excellent practice in terms of schooled literacies, using the language of educators, but this is markedly different from the language and practices of the workplace. It's a question of 'fit'.

More than ever before, teachers and trainers are being called upon to interpret sets of competencies and related performance criteria in ways that are, firstly, responsive to industry and workplace environments and, secondly, legislatively, financially and morally accountable. From our study we offer the following implications for consideration by teachers and trainers.

1. It is imperative that there is an understanding of; the discourses that operate within the specific industries and sites in which teachers and trainers are operating.
2. Enterprise-based teachers are more likely to understand the culture of the workplace and the specific literacy and numeracy practices of that; environment, than other literacy/numeracy teachers who only train workers at a site on a limited basis.
3. In working with training packages, while it is important that teachers and trainers become familiar with their components, it is equally important that these endorsed competencies are seen as being situated within workplace practices. It is the competent performance of these practices that is fore-grounded in training packages, not competence in relation to sets of abstract skills and understandings.
4. Similarly the knowledge, including literacy and numeracy knowledge, that underpins competencies, should be perceived as embedded within those competencies and not as additions or pre-requisites to the competencies.
5. The move for increased employee accountability has led to increased textualisation in the workplace. Therefore there is a need for appropriate training so that all workers are informed and also competent in engaging in these new literate and numerate practices.
6. The successful development of a learning culture within the workplace requires the implementation of effective and varied communication mechanisms at all levels of the company

operation. However, workers may need training in order to actively participate in these communication processes.

Notes

1 The term 'training package' refers to a set of industry competency standards and assessments (in this case the Civil Construction package) which relate to the Australian Qualification Framework (AQF).
2 TAFE = Technical and Further Education institutions.
3 At the time of the research, the Civil Construction training package was developed to Certificate Level 3.
4 'doggers' – the job of load securing for lifting with cranes.

References

Australian National Training Authority (2000). *The packages: civil construction training package.*

Gee, J. P. (1996), *Social linguistics and literacies; Ideologies in discourses.* Second Edition. London: Taylor & Francis

Kelly, A. and Searle, J. (2000). Literacy on the *Motorway: An examination of the effects of the inclusion of literacy and numeracy in industry standards in Training Packages on the quality of learning and work outcomes.* ALNARC Report to DETYA: Melbourne: Language Australia.

B8 Prisons

Further RaPAL references about prisons

McGahan, H. 'A Day in the Life of a Prison Tutor', *RaPAL Bulletin*, No. 21, 1993.

Wilson, A. 'A Creative Story about Prison Writing', *RaPAL Bulletin* No. 21, 1993.

Wilson, A. 'Three Days and a Breakfast – Translating Time in the Literacy Lives of Prisoners', *RaPAL Bulletin* No. 37, 1998.

Kirman, J. 'Naughty Literacies – Nice Words', *RaPAL Bulletin* No. 38, 1999.

The illiteracy myth: A comparative study of prisoner literacy abilities

Stephen Black, Rosemary Rouse and Rosie Wickert

RaPAL Bulletin No. 15, Summer 1991

Background

There is a popular perception [in Australia] that 'illiteracy' is a characteristic of prisoner populations, and that this illiteracy rate is several times greater for prisoners than the general population. The current Minister for Corrective Services [in New South Wales] certainly supports this perception, as too does the Department of Corrective Services generally (e.g. in their Annual reports) and several prison educators who have made reference to illiteracy in prisons. However, this popular perception hasn't been tested. There hasn't been a satisfactory assessment of the literacy abilities of prisoners in NSW (or elsewhere in Australia) and in consequence it hasn't been possible to compare prisoner and general population literacy abilities. Until now that is, with the recent publication of the first national Australian adult literacy survey (No *Single Measure* – Rosie Wickert, 1989). Our study applied the same national survey questionnaire to two prisoner samples (100 male prisoners at Silverwater and 100 female prisoners at Mulawa prison).

In undertaking this study we acknowledged the concept of literacy made clear in Wickert's study, that literacy is relative "to social and cultural norms, to time and place, to purpose and intent", and that literacy has to be seen as "the application of specific skills in specific contests". Our intention was to provide a "profile" of prisoner literacy abilities, not to determine an 'illiteracy' rate, which in fact wouldn't have been possible within the concept of literacy applied here.

Three dimensions of literacy were looked at: Document (e.g. filling in forms, writing cheques), prose (e.g. understanding newspaper reports) and quantitative literacy (e.g. mathematical skills involved in keeping financial records or understanding schedules). Within each of these dimensions a number of literacy items were used, scaled in difficulty according to the American population norms in the study on which Wickert based her survey (i.e. *Literacy, Profiles of America's Young Adults,* by Kirsch and Jungeblut 1986).

Comparative data are presented in the research under three sets of interview samples: complete interview samples, and those for the 18-24 year old group and the 25–34 year old group. This was in recognition that most prisoners are in the younger age groups, making for more appropriate comparisons with the national sample.

Aims

As indicated, the main aim of the study was to make an assessment of the literacy abilities of two prisoner samples and to compare the results with those of the recent national adult literacy survey. A secondary aim was to determine if there were any major differences in the literacy abilities of male and female prisoners that might justify different educational provision in prison.

Outcomes

Overall, on all three literacy dimensions, the prisoner samples did perform less well than the national sample. On document items the differences were not that marked, but they were greater on the prose and quantitative dimensions. This tells us little though; of far more interest are the findings on specific literacy items.

The percentage of correct responses for prisoners on some

document items such as completing a job application form was poor, indicating possibly that for prisoners this is a literacy item they are not very experienced at undertaking. However, some quite remarkable differences were found on the literacy item requiring the identification of specific dosage instructions on a pharmaceutical packet. These figures are the reverse of the job application form item. Prisoners, and in particular young female prisoners did exceptionally well in this group.

In other document items such as writing details on a cheque, the prisoner sample figures were again down on the national sample, but on identifying information from a paint chart, it was the Silverwater male prisoner sample which performed so much better than the other two samples, indicating possible prior experience with this type of activity.

On prose literacy items which included identifying and using information from newspaper articles, prisoners performed poorly, with few correct responses, especially in the more advanced prose items.

On quantitative literacy items at a basic level, differences between the three samples were not marked, though interestingly, the Mulawa sample performed very well on totalling the entries on a deposit slip. On keeping a running total in an account book and using airline schedules, again the prisoner figures were often quite markedly lower then the national sample, but there was little difference found on calculating change from A$5 based on menu charges.

Discussion and conclusion

On the basis of these findings it would be quite inappropriate to continue to refer to prisoners as 'illiterate', or in fact to regard the literacy abilities of prisoners as being much worse than those found in the adult general community. The percentage of correct responses were generally lower, though it needs also to be recognised that on some literacy items their literacy abilities were superior. Clearly literacy abilities have been shown to be related to the need to use particular literacy items, or prior experience in undertaking them.

These are significant findings as they make the aforementioned popular perceptions about the poor literacy abilities of prisoners quite inaccurate. It also makes quite tenuous any perceived casual link between literacy ability and criminal activity.

Differences between male and female prisoners generally were not great, and in cases where there were quite large differences, these could be partly explained by sex stereotyping (e.g. on the paint chart item and the dosage instructions). These differences would not merit different educational provision for male and female prisoners.

Despite the overall finding that prisoner literacy abilities are not that bad compared with the general population, this cannot justify less of a focus on literacy programs in prisons. For example, there can be little comfort in the findings that only half the prisoners could keep a running total in a bank account, fill in a job seeker form or a cheque to pay a bill. Very few prisoners could cope with more advanced prose items. The findings provide greater direction for prison education programs. More emphasis on maths would be justified, and also a greater focus on jobseeking skills, especially for the younger prisoners. More advanced literacy work to enable prisoners to cope with prose literacy, essential for undertaking further education and training courses, would also be justified on the basis of these findings.

Literacies within prison settings: A fourth space?

Margaret Herrington and Tom Joseph
RaPAL Bulletin No. 43, Winter 2000

Tom Joseph is currently a postgraduate student at Nottingham Trent University and Margaret Herrington directs the Study Support Centre at the University of Nottingham.

Introduction

Debates about literacy in prison invariably focus on the evidence about the 'low' literacy standards of prisoners. Yet prisoners have traditionally used various forms of literacy for a range of purposes during incarceration: to maintain their sense of identity; to maintain communication and relationships with the outside world; to account for crimes they have committed; and to challenge miscarriages of justice etc.

In a recent publication Anita Wilson has attempted a new conceptualisation of literacy activity in prisons (in Barton, Hamilton & Ivanič 2000, ch. 4). Her extended research with prisoners revealed new evidence about the kind of literacy practices to be found in this 'situation'. Her analysis led to the suggestion that when prisoners leave their outside world and go into the prison system, they create a metaphorical 'third space', between the two worlds, within which they adapt both the internal official documents and the forms of literacy from outside, to use in their everyday lives in prison. Wilson shows how literacies are generated in this 'third space' to subvert authority, maintain sanity, and amass currency and clearly believes that the situation of prison shapes the literacies to be found there. In this article we shall briefly review this concept and also explore her claim that 'the longer a man remains in Prison, the stronger he seeks to retain his ties with his outside worlds.' (Wilson, ibid. p.68)

We, the authors, have worked together in a student/learning support tutor capacity in higher education for several years. During our sessions, Tom, who had been a long-term prisoner in top security gaols, revealed his remarkable journey from experienced criminal to poet, from criminal identity and activity to upright citizen. He had described himself as 'semi-literate' when he entered prison and yet, on leaving, he read English Literature at a major British university. His experience of literacies in prison meant that he was well placed to consider the value of Wilson's hypothesis. Margaret had first developed an interest in the literacy concerns of young offenders whilst teaching in a probation day centre in the early 1980s and had worked since then as a literacy educator in the post-16 sector.

Interpreting prison literacies

This experience led us to consider that Wilson's idea of 'the third space' was useful but may need further development. Our first question was about Wilson's illustrative examples of literacies in the third space and with her interpretation of those. The examples themselves did not resonate with T's experience and the interpretations were not shared. For example, when Wilson (p62) read a strange message from a young prisoner addressed to the governor, Mr B * * * * * *'s, which read:

> *"Dear Sir*
> *I am a transvestite like dressing up*
> *in women's clothes see you soon you big sexy get*
> *So much love*
> *to all the world*
> *I'm a transvestite*
> *Queen v Lee*
> *Show the world*

Show the judge this boy is
A transit van"

she suggested that this note followed the conventions of letter writing on the outside but challenged authority by "the informal and unconventional content of . . . [the] correspondence". However, she also formed the opinion that the author was a "psychiatrically disturbed youth".

Tom agreed that the inmate's stance in the writing-up and posting of the message was one of anti-authority, but he interpreted it as a macho expression of defiance, one in which the writer trades in his masculinity in order to establish his perceived effeminate take on "THEM" (judges in wigs, stockings and robes). Having done so, he ends the note by reclaiming his masculine self with the implicit rhyming slang for, "man", i.e. transit van. Tom thus disagreed with Wilson's interpretation, arguing that it involved nothing more than a play on words, common among young men like Lee, who laugh in the face of the authority. His experience of prison from the inside allowed him to deconstruct the literacy with his contextual knowledge of language and led him to quite different conclusions. For us, this suggested that a number of interpretations of third space literacies were possible – including those of other prisoners and prisoner researchers – and that those of the single 'outside' researcher should be seen alongside them. It also encouraged a more general question about the balance of power between researcher and researched within ethnographic methodology.

Prison Education: A Fourth Space?

Despite our concerns with Wilson's examples it was clear that in proposing a 'third space', she had unlocked the potential to shed light on the, hitherto, largely unacknowledged world of informal literacy practices in prison. Her diagram reprinted below shows the third space as one in which the literacies from outside and the official Literacy inside are made into new forms, unique to the prison situation.

Figure 1: The third-space. (reprinted from A Wilson, 2000, page 63)

However, our second question concerned the position of Prison Education in relation to this framework of a 'third space'. The Education classes are provided as part of the. official Prison system and yet are a contested part of it. Traditionally, education provision has had to struggle for its place as part of the rehabilitative processes within UK prisons. Yet the classes are sites/spaces/microclimates (Knights 1991) within which formal and informal literacies develop. We would like to argue that these should be considered as a 'fourth space' which can generate very complex interactions between Literacy outside and inside the prison. The examples described below illustrate the point.

* * * * * * * *

Example 1. Tom writes:

During September 1986 while serving time in HMP Liverpool I was categorized as a category A standard escape risk. In 1987 I received a 12-month sentence for assaulting a prison officer and a further charge of threatening to kill him was ordered to lie on the file. This sentence ran concurrently with a 2-year sentence imposed for firearm offences. The authorities at HMP Liverpool were unhappy with the outcome and they made it clear to me that I was in for a rough ride. They allocated me to HMP Frankland, one of the two POA [Prison Officers'

Association] militant strongholds in the Durham area, where a review of my category status was carried out on an annual basis. There was no appeals mechanism in place, and prison security reports were "classified information". It was impossible to know whether or not the review process was being handled fairly. The continued A classification had a detrimental impact on any prospects of early release as category A's are ineligible for parole. In April 1993 my case went before the category A committee where a decision was made to decategorize me to B status. Yet in May 1993 I received the obligatory knock-back to the first parole application.

In September 1993 I enrolled on an Access course at a local college and in November I sent off the UCAS application which listed my chosen universities. During that same month a visiting tutor was refused entry by security officers and over the next three months a small group of prison officers made a concerted effort to destroy my future release plan by repeatedly abusing the process of the Sentence Planning Scheme. Their aim was to provoke me into assaulting one of them or alternatively to frustrate and distract me to the extent that I would have to abandon my studies. My response to the sabotage was to complain but my complaints fell on deaf ears. Out of sheer desperation I produced a written analysis and report highlighting the abuses and submitted the finished product as an Access Course assignment.

I began by identifying the legal obligations with regard to Sentence Planning Scheme (Criminal Justice Act 1991) and then itemised the ways in which these had been compromised. I wrote in the third person but used my own experiences to highlight the types of abuse of process, for example: "the entering of erroneous information in the initial profile document, which proved highly prejudicial to the inmate's chances of parole." I wrote in the official report format, using the appropriate jargon, clarifying the terms of reference, the sources of evidence, the findings, recommendations and annexes.

I had explained to the Education Coordinator what I was doing and mentioned that I did not intend to send a copy to the governor. I compiled and completed a very thorough report which earned me a good mark as an assignment.

When I eventually left the prison I passed my report on to another prisoner to use as he wished.

* * * * * * * *

In this case the prison education system was both containing the challenge and rewarding T for it. This seemed to be an example of a prisoner using a formal literacy format from outside the prison (a formal report) within an official Prison context (education classes) to challenge the ways in which Prison Literacy practices were being used to subdue a prisoner, and to gain accreditation from an accrediting body outside the prison (and eventually to pass it back into the prisoners' domain). Officialdom had created the space in which this occurred. It contained formal structures and programmes and also, intrinsically, the potential to articulate challenges. It seemed to suggest a 'fourth space' which is shown diagrammatically below

This shows a slightly different representation to Wilson's description. It shows the 'outside' world, surrounding both the Prison and the third space, and including the educational institutions which provided ACCESS Course staff. It then shows the official Literacy of the Prison alongside the third space in which prisoners develop their own literacies (mixing outside and internal Literacy as in Wilson's 'third space'). Prison education is shown as a 'fourth space' straddling the official and unofficial worlds. One half contains the

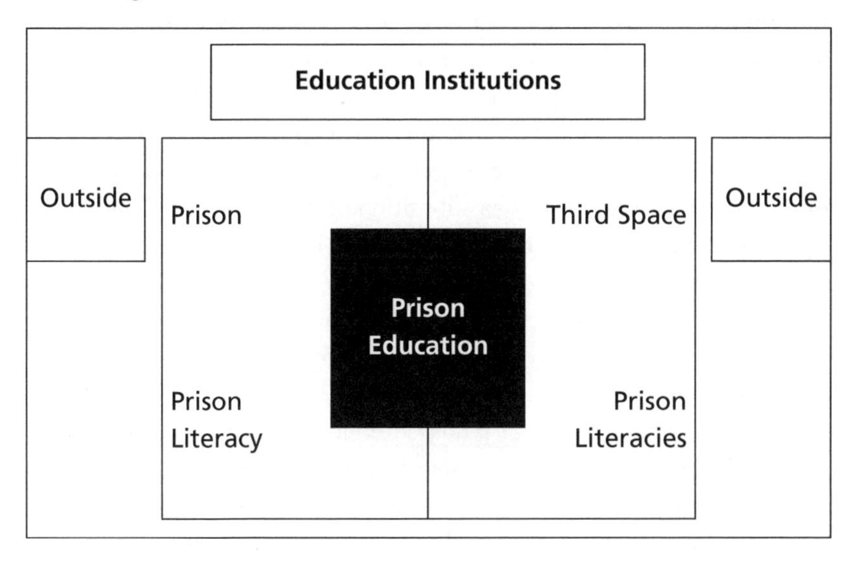

Figure 2: Prison Education: A fourth space?

formal programmes and the other shows the space in prison educational settings in which informal literacies develop.

This can be seen as an attempt to define a particular part of both the official and the 'third space' more sharply but we think it may be worthy of a separate term because the educational setting brings a more powerful dimension into this literacy activity; one which offers huge possibilities for real challenge and change in prisoners' lives. In T's case this 'fourth space' was not just a means of 'prisonisation' inside prison but provided his key to a different life outside.

* * * * * * * *

In the second example, further complex interactions are evident.

Tom writes . . .

In 1989, having passed three GCSE examinations, I enrolled on an advanced level English Literature course. It was during this period that a tutor suggested that I write a poem – something I had refrained from attempting because I thought erudition a prerequisite for writing poetry. However, he gave me an anthology of prison poetry to read and showed me that this was not the case.

My poetic awakening came quite suddenly. There was a gestation period in which ideas slowly began to grow in my mind, followed by a burst of creativity. I began to experiment with words more and more and each new poem was submitted to my tutors for their criticism. If I agreed with their suggestions, I would make the necessary amendments.

Eventually I began to envisage the compilation of an anthology – an intellectual project in which my ideas, thoughts and feelings from past to present were to be made tributary.

The criticisms of my poetry helped develop a critical faculty I previously lacked. This process also generated an extended letter, a piece of literary criticism which I submitted as a piece of discursive writing for my Access course and which got a good mark. When the admissions officer at Nottingham University requested recently written work, I sent an essay and a batch of poems from my anthology and was offered a place on the strength of them.

FROM MADNESS TO QUIETUS

A fellow prisoner's sudden departure-
Stranded here where my vision is restricted,
From cell to cell I heard it being answered:
That sincere voice of valediction.
Then a scream to pierce my calm composure
That felt like a hurtful foreign splinter!
He had walked from his confine into conflict
Then I heard the keepers calling for assistance.
An angry mob chorus were chanting the consensus:
The dirty bastards have kicked him senseless
But then a con's lone voice appeals for silence,
No (it shrills) he attacked the screws first!

Evening's closing in on the sinking autumnal sun;
From the sky's furnace a pall of black cloud.
I hear Labi Siffre crooning It must be love:
Imponderable effects from incongruous voice
At a time so tragic, have warped all logical thought.
The incident no longer a matter for grave concern,
Only now this quiet descending air, slowly, gently,
Makes its return like disturbed dust resettling to earth.

Later, I hear a train hurtling its way north,
Its approach growing louder from the south,
The intensity of its thunderous roar at midnight
Mirrors the chaos of the day's madness-
Now, pitch passing – volume dying off into the distance,
The fading clamour seems to represent the ageing process.

All calm restored to the farthest depths of silence
I feel I know how peaceful the release of death is.

HMP Frankland 1991

The correspondence, critique and poetry were composed on a typewriter in my single cell. The space was vital, especially in the quiet, early hours and was a 'microclimate' replicating aspects of the education block environment, within which it became possible to explore and develop my literary potential. Further, as a candidate for parole release, I was invited to take part in a formal Literacy practice. I was provided with a piece of paper entitled, `Prisoner's Representations', on which I was required to write the reasons why I thought parole should be granted. I wrote . . .

> *"I would ask you to grant parole in this case and by doing so you will enable me to join a university. Please find attached offers from Teeside University and Nottingham University. I hope to receive an offer from Exeter University in the near future. This is the only real chance I have had of escaping from my criminal past – I know that I will lead a law abiding life at university (and beyond) but just need the opportunity to prove this to everyone who has helped me to get this far."*

I was granted parole and released from HMP Frankland its August 1994 and went to Nottingham to study English.

* * * * * * * *

In this case, T used his relationship with one part of the outside world to create material which helped him to access another. His access to the powerful literacies outside helped him to deal with those inside. In these examples we can see that T was always using literacies to move his own intellectual project forward and to re-write his life script [Knights, 1991].

In terms of Wilson's observation that:

> *"the longer a man remains in prison, the stronger he seeks to retain his ties with his outside worlds (p. 68),*

T was a long-term prisoner who consciously avoided contact with his previous outside world in order to create a new lifescript. He used the new outside worlds made available to him, to create a route to a different world.

Conclusion

In the current policy climate of claims about low literacy levels in prison, the insights from New Literacy Studies provide a broader operational framework within which to view prison literacy and a more accurate account of what actually happens in a prison. Wilson's work on the 'third space' has, in particular, provided an important springboard for further analysis. The examples discussed here show one prisoner engaged in a complex process of literacy development. The inside and outside worlds and the formal and informal literacies are consciously intertwined and prison education as 'a fourth space' is a creative site at the heart of the process; a gateway to a new 'outside' world. For experienced literacy workers, this acknowledgement of the significance of prison education is not new but it appears to be something which has to be repeated again and again.

References

Knights, B. (1991) 'Writing Relations in a Men's Prison'. *Free Associations*, vol.2, part 1, no.21, p65–85.

Wilson, A (2000) 'There Is No Escape From Third-Space Theory. Borderland Discourse and the in-between literacies of prisons', in Barton. D, Hamilton. M. & Ivanič, R. (ed) *Situated Literacies. Reading and Writing in Context*. London: Routledge.

Third space theory: A response by Anita Wilson

RaPAL Bulletin No. 44, Spring 2001

I believe a writer should always feel privileged if their work acts at a forum for further debate and I found the comments put forward by Tom Joseph and Margaret Herrington on my work to be both imaginative and provocative. I feel it is my duty to respond, partly to defend the 'third space' but also to add further to the debate!

Before detailing specific points I would like to point out how delighted I was to read a piece of writing which was both collaborative and balanced. In Situated Literacies, the constraints of time and space made it difficult for me to put over my corresponding view that all voices should be heard equally. I do believe strongly however that attention to proper and fair representation in my research is reflected in the willingness with which prisoners continue to support my work. I would like to make the following comments which are, of course, open to further critique!

As a general observation I see prison education as one dimension of the 'third space' rather than an 'extra space'. To place prison education at the centre – as Tom and Margaret propose – presumes that prison education mediates all literacy-related activities and practices that go on in prisons. This assumption does not take into account the fact that many prisoners engage in literacy-oriented endeavour without ever being part of the education department and fails to recognise that many prisoners view education departments as something more than academic spaces. To retain prison education as one dimension of my 'third space' in fact acknowledges the fine line it treads between two institutions – education (outside) and prison (inside). Its timetables operate between pedagogical and carceral parameters and it constructs its participants as both learners and prisoners for example.

It is interesting that the writers feel I make 'assumptions' about 'Mr. B's' letter and then make some of their own! I would prefer to see both as interpretations based on personal experience. Mine – from my third space position as outsider-insider – are generated from being situated in the actual event, based on a sound knowledge of the young man in question and focused on the relevance of day to day literacy-oriented activities in the face of considerable adversity. Tom's – from his third space position as insider-outsider – are based on his own experiences of prisons and his desire (and ability) to put an 'academic' reading onto a non-academic event. To clarify the issue, the entry from my journal notes that this young man did indeed have serious mental difficulties. My point is that despite his distress, he still managed to illustrate an awareness of generic convention and to show a remarkable inventiveness in his use of literacy. Also, within the cultural 'situatedness' of the young offender institution in which this event took place, the term 'transit van' is aligned to 'transvestite' rather than rhyming slang from the adult prison world.

One final point – in relation to the rebuttal of my observation that *'the longer a man stays in prison the stronger he seeks to retain his ties with his outside worlds'.* My reading of 'outside worlds' is intended to be understood broadly and not confined to personal worlds which are historically linked to crime. Long-term male prisoners who collaborate with my work and who are unlikely to be part of any 'outside world' for many years tell me that they construct a physical and metaphorical space where they grow plants, keep pets, and
keep up-to-date with world news in order to keep 'institutionalisation' at a distance. Only some tell me they draw upon education as one way of situating themselves in a reality other than that of prison.

Since I 'launched' the third space theory it has travelled

extensively. Recently it appeared in the USA as the basis of a paper on Ethnography in Education. It has made an on-line appearance via South Africa as part of a global debate on literacy. It has travelled with me to Canada, Scotland and Europe. It is now circulating courtesy of RaPAL. Prisoners – who were central to its inception – are delighted and so am I!

Response to Anita Wilson by Tom Joseph and Margaret Herrington

RaPAL Bulletin No. 44, Spring 2001

In her original article, Anita Wilson failed to acknowledge the significance of Prison Education within her metaphorical 'third space' between prison and the outside world. Yet in her response above she argues that Prison Education should be seen as one dimension within the 'third space' she has identified.

We would like to make three points by way of rejoinder: First, we do not presume that Prison Education mediates all literacy related activities in prison and it is difficult to see how that conclusion could have been drawn from our illustration and explanation. We argued for a separate 'fourth space' precisely because we did not think that it was included in, or included, the 'third space';

Second, any exploration of the literacies which develop in prison needs to include reference to Prison Education. It seemed to us that this could not be defined as within the 'third space' entirely, because it involves a formal set of educational activities, organised by the prison authorities, with a set of literacy/educational objectives and a number of far reaching outcomes in literacy, learning and personal power. It does not primarily involve prisoners creating literacy from an informal mix of prison literacy procedures and forms of literacy from outside, with the intention of creating currency, posing small challenges to the prison system and protecting themselves from prisonisation. There could be some overlap but the difference in objectives, outcomes and in the relationship with the prison authorities, suggested the need for a different term, a 'fourth space'. If Anita Wilson wishes to subsume everything within the third space, then she would need to create a robust classification of quite different mechanisms of literacy creation within it. Surely there is a difference in kind between the reappropriation of a canteen form and the creation of poetry in response to a formal curriculum?

Third, whilst Anita Wilson understandably appears to see the 'third space' as a creative centre of literacy practices which gives a kind of power to the prisoners in resisting prisonisation, for Tom this was not real power which was going to make any difference outside prison. He regarded the informal literacy examples from the third space, quoted in the original chapter, as relatively unimportant in terms of personal power. For him the important shift in his access to literacies was to move beyond the 'third space' towards the higher realms of the education system inside the prison and thence to the outside. He thus felt very strongly that this access to real, sustained personal power should be described separately.

We feel it important for this debate to move forward by taking proper account of the full complexities of prison literacies and not to subsume everything within a relatively undifferentiated 'third space'. We continue to find the term 'fourth space' helpful.

B9 Literacy and gender

Further RaPAL reference about literacy and gender

Solity, J. 'Women's Literacy', *RaPAL Bulletin* No. 1, 1986.

Snatched time: Mothers and literacy

Jane Mace
RaPAL Bulletin No. 35, Spring 1998

Jane Mace is a founder member of RaPAL and editor of Literacy Language and Community Publishing: Essays in Adult Education (Multilingual Matters, 1995). She now works at the Centre for Continuing Education and Development, South Bank University, London.

One hot June morning in 1913, a crowd of women met in Newcastle. They were working-class women who had travelled there from all over the country. The occasion was a conference of the Women's Co-operative Guild, founded twenty years earlier. One by one, women came up to the platform. They called for reform of the divorce laws, a minimum wage, shorter working hours, improved maternity welfare, sanitation, and education; they demanded the right for women to vote. Many of the speakers were mothers, with little spare energy for any kind of reading, let alone writing. Yet they conjured up the image, for one witness, of women with `the indiscriminate greed of the hungry appetite' for all that reading could mean.

That witness was neither a mother nor working class. The appetite she thought about, however, was one for which she had a particularly keen sympathy. She was Virginia Woolf. As she watched and listened that day in Newcastle, Woolf felt frustration at the continued powerlessness of pre-suffrage women:

In all that audience, among all those women who worked, who bore

> *children, who scrubbed and cooked and bargained, there was not a single woman with a vote. The thought was irritating* ***and depressing*** *in the extreme (Woolf in Llewellyn Davis [1931] 1990: xxi).*

But far greater than the despair she recalls is the sense of hope she felt, inspired not only by the 'vitality' of the women she heard speak that day, but the 'inborn energy' she found, some months later, reading the written memoirs sent in by some of the Guild members to its Secretary, Margaret Llewellyn Davies. What she found was the determination of women, against all the pressures and hardships of their lives, to reach out for other possibilities via literacy. 'They read at meals; they read before going to the mill', she wrote. These women (as she saw it) read,

> *. . . with the indiscriminate greed of the hungry appetite, that crams itself with toffee and beef and tarts and vinegar and champagne all in one gulp. Naturally such reading led to argument. The younger generation had the audacity to say that Queen Victoria was no better than an honest charwoman who had brought up her children respectably. They had the temerity to doubt whether to sew straight stitches into men's hat brims should be the sole aim and end of a woman's life. (ibid: xxxvi)*

For us, reading that piece by Virginia Woolf over half a century later, there is something exciting here, not only about an important moment in women's history, but also in the ideas about argument and appetite. Women's literacy and collective experience will 'naturally', according to Woolf, lead to 'argument'. In everyday language, this word has two contrasting meanings: one associated with speech, the other with writing. The first (as in: 'Stop arguing, you two!') is that of loud and emotional disagreement. The second is measured, rational, set out with carefully presented evidence and at the heart of academic endeavour. Women's reading, at that moment in history, evidently led to both kinds of argument: the noisy and the measured. For it had also led many of them, not only to get troublesome in public meetings, but also to set something of their life experiences in writing and send these into the Guild offices. These uses of literacy, undertaken in conditions of poverty and hardship, represented and expressed women's demands for change: demands made with 'audacity' and 'temerity'. 'Thus it came about', Woolf concludes,

> *. . . that Mrs. Robson and Mrs. Potter and Mrs. Wright at Newcastle in 1913 were asking not only for baths and wages and electric light, but also for adult suffrage and the taxation of land values and divorce law reform. (ibid: xxxiii)*

And thus, too, did it come about – in 1928 that women finally achieved the suffrage.

The present-day context in which I have been thinking about this is a policy framework which requires the mother to nourish her children's literacy appetites before her own, and which continues to insist on a view of literacy as about schooling, rather than about a whole range of social and cultural experiences. For 'family literacy', since its inception in the UK in 1993, despite consistent use of the plural 'parents', has had an explicit agenda to 'teach the mother, reach the child' (McLeod 1993) as a means to resolve the perceived 'crisis' in national standards of basic skills in schools.

This context has led me on a search for an alternative agenda which might show mothers as other kinds of readers and writers too. The mothers I have been learning about were bringing up children between the 1890s and 1930s – a time when they were neither expected nor invited to participate in their children's schooling, and when few, particularly the large mass of working-class mothers, had the time to. I was looking for data on how reading and writing mattered, or did not matter, in their lives.[1]

Hungry Reading

One of the things I found, among others, was a picture of children whose 'avid reading', far from encouraging their mothers, worried or even maddened them. Literacy posed a health hazard. With limited lighting (gas, paraffin or candles) reading could harm eyesight: and there are many accounts of mothers prohibiting or discouraging their children from reading because it might be 'bad for their eyes'.

For one young girl such activity had cost her her job:

> *My grandmother had been a housemaid, dismissed for 'wasting candles' by reading at night in her attic bedroom (P1906)*

There was also the widespread belief that public library books carried germs of contagious diseases. Possibly even worse, for some, was the contagion of 'ideas' which books might give rise to. But children's reading worried, annoyed or infuriated their mothers in other ways, too. Women who, like those observed by Virginia Woolf, had all too little opportunity to exercise such literacy as they had themselves, saw their children's reading – especially of their daughters – as a distraction from the chores to be done in the home.

Nevertheless, many accounts suggest passionate reading by children of the period; in the fields, in the bedroom, by the fire in the kitchen. One woman, born in 1890 had, as a child, been

> *. . . an avid reader, always with a book in her hand, often shutting herself in the toilet to read (R1025).*

Another (G1041), now a retired librarian, described herself as 'reading so ravenously' as a child that her mother became alarmed and appalled at my turning out a bookworm, she would not let my sister learn to read until she was 9.' Such 'avid', 'ravenous' or (sometimes) 'voracious' readers (in the same metaphoric world as Virginia Woolf's 'indiscriminate greed of the hungry appetite') are nearly always portrayed as either children or older people. Very rare indeed is the suggestion that the mother herself might have been such a figure: for concentrated literacy, like concentrated activity of any kind (including watching television) is an anti-social activity, and the mother at home must be constantly available.

The sight of children absorbed in these activities is in theory a delight to parents: in practice, for Eleanor, living in a remote Lancashire village, it felt like the last straw:

> *She felt it was "a waste of time when there's work to be done", and my mother remembers her throwing books on the fire, or lighting the fire with their pages, with the remark that "It's all they're good for".*

According to her granddaughter, Eleanor's 'antipathy' to books co-existed with an enthusiasm for other kinds of literacy. She enjoyed what she saw to be 'real life' reading: whether letters from her family in the village twelve miles away or the newspaper, which, she said

> *had same point to it. It was about real life and not the fanciful stuff of books (S481).*

For the mother who could neither read nor write, however much she wanted to encourage her child to achieve more than she had been able, the child's love of reading could simply be infuriating, as Joan recalled of her mother, Eliza:

> *When you read, it used to annoy her. She tried not to get annoyed, but you know how you get lost in a book? and she would take a book off me sometimes and she'd say "I've spoken to you three times" and I'd say, "I'm sorry mother I haven't heard you." She found it very difficult to understand this fascination that my father and I had for books* (interview: Stocksbridge, April 1996).

This, like other accounts showed fathers as more sympathetic to reading than the mother. Against the demands of water to be boiled, sheets to be mangled, vegetables to be peeled, clothes to be mended and children of all ages to see to, and in cramped accommodation, for the mother at home her children's reading must indeed have often seemed 'a waste of time'. Once children had grown and gone, with the greater freedom of a home with no children in it, some of these same women were able at last to indulge as great an appetite for reading as the children they once scolded. The 'fascination for books' that Joan's mother Eliza could not understand as a younger woman was evidently one she discovered herself years later; for, as Joan also told me, Eliza, at the age of 78, set about learning to read.

For more privileged women like Virginia Woolf, with leisure and light, 'baths and money', reading – even avid reading – was far more of a possibility than it could be for Eleanor or Eliza. If their children's reading sometimes seemed a 'waste of time', reading anything for themselves was an impossible luxury. So the evidence that some, like the women at that meeting in June 1913, succeeded in snatching time for this luxury in spite of all the odds stacked against them must stand as an inspiration. One account, from the Women's Co-operative

Guild campaign for state benefits for mothers is particularly so. It begins with these words:

> *I was married at twenty-eight in utter ignorance of the things that most vitally affect a wife and mother.*

During the years of 'weariness and hopelessness' with an 'utter monotony of life', with five childbirths and a constant struggle to survive, the writer says, 'I could give no time to mental culture'. The best she could do was to read as she scrubbed:

> *I bought Stead's penny editions of literary masters and used to put them on a shelf in front of me [on] washing-day, fastened back their pages with a clothes-peg and learned pages of Whittier, Lowell, and Longfellow, as I mechanically rubbed the dirty clothes. (Anon, 1984)*

As the history of women's education suggests, such determination may indeed lead to argument'. The argument that I believe needs developing is the one which says: the first priority for mothers in a literate society is to nourish and develop enormous appetites for their own literacy. For that argument to be articulated as boldly as it deserves, the women who inspired Virginia Woolf that day continue to inspire us:

> *they touched nothing lightly. They gripped papers and pencils as if they were brooms. (ibid: xxiv).*

Note

1 Funded by a grant from the Nuffield Foundation, a 'directive', using my research question, was sent in November 1995 to the volunteer correspondents of The Mass Observation Archive at the University of Sussex. This entailed another literacy activity: the writing of some 250 people round the country, drawing on first-hand memories and recollections offered by their relatives. In line with the Archive's policy, correspondents are identified by number, not name.

References

Anon (1984) *The toll of motherhood* c. 1900, in: Horowitz Murray, J. **Story minded women and other lost voices from 19th century England** (London, Penguin)

Llewelyn Davies, M (1931, repr. 1990) **Life as we have known it: letters from Co-operative Women.** (London, Virago)

McLeod, D (1993) *Literacy drive puts accent on family.* **The Guardian,** 16 June: 4.

Stitches in time

Tricia Hegarty
RaPAL Bulletin No. 35, Spring 1998

Tricia Hegarty is project manager and group tutor of The Next Step Project – an adult literacy initiative by The Verbal Arts Centre which is based in the Northwest of Northern Ireland.

The Verbal Arts Centre was established in May 1992, to support lifelong creative, imaginative learning and practice – whether personal or group-centred and in both formal and informal settings. We have developed work in three broad areas: within health settings, in education and library contexts and in a community outreach programme. We work primarily through building partnerships with others to complement existing provision. It was while working part-time with the Centre and part-time as an ABE tutor with the local F.E. College that I first saw the possibility of developing a project which would combine elements of both areas of my work. **The Next Step Project** complements existing ABE provision by providing a range of creative learning opportunities that support the leap from everyday functional uses of literacy to uses of literacy in imaginative expression.

I was invited to present a workshop at the RaPAL Study Conference by Jane Mace, external evaluator of The Next Step Project, who had visited Northern Ireland in March and met with learners participating in the project. At the time of Jane's visit we were putting the final touches to an exhibition of writing and artwork by women from both sides of the community for International Women's Day, March 8th. This work had come out of a Next Step Project writing development activity which I had designed and called "Stitches in

Time". This combines talking and writing, sewing and art activities around the theme of significant clothing and culminates in a one day creative workshop. "Stitches in Time" is an activity which is representative of the aims and methodology of the Next Step Project and so it seemed a good way of describing the work being done and the thinking behind it to the delegates at the Study Conference.

The 'Stitches in Time' Group

My workshop took place on the Saturday of the RaPAL Study Conference from 11.30 am to 1.00 pm and was attended by twenty participants. Having been a literacy worker for just two years this was the first time I had been called upon to talk about my work to a gathering such as this so I began to feel quite nervous as everyone took their places in the semi-circle of chairs which faced me and I was glad to see the reassuring smile of Judy Pringle of Belfast's Adult Basic Skills Resource Centre, the other half of Northern Ireland's representation at the conference!

I began by explaining the background to The Next Step Project and the work being done. I described how we stimulate people's interest

in writing through asking them firstly to talk to each other about their own lives and experiences. Once people see that others are interested in listening to them they feel more confident about attempting to record these stories themselves in writing and combining the writing with drawing, painting and sewing takes away the pressure of open and blank white sheets of paper which daunts beginner writers.

Busy Stitchers

In "Stitches in Time" we wrote on the artwork, choosing the most significant words and lines from the stories and using these decoratively. To show how this had been done I brought along photographs and writings and one of the smaller pieces of artwork from the exhibition. I also read out a story written by a woman who took part in the project.

The workshop group discussed the concept of improving oral and written communication competencies through creative activities and agreed that there should be more of this kind of work in ABE groups. We talked about how visual representations made by individuals themselves help towards an understanding of their complex life experiences – including their own "learning".

To give participants a direct experience of the methodology behind a writing development activity such as "Stitches in Time" I asked everyone to spend the next fifteen minutes on a piece of writing about the significance of clothing, or one particular item of clothing, in their lives. I then invited people to share what they had written with the group, just as I would in a Next Step Project writing class – and, just as with Next Step learners, there was the same awkward pause while people considered whether they would be brave enough to be the first to speak up! In sharing our stories we found that a common feature of several was the idea of clothing as a kind of language which people adopt to suit different situations and audiences. Several people had written about the clothes they wear to work and what these say about their attitude to what they do and how they wish others to see them in their role of tutor, organiser, head of department, activist and so on. This led to a discussion about how not being literate in a given situation might feel like not having suitable clothing – and how being competent in a range of literacies can be compared to having an extensive wardrobe, with clothes suitable for all occasions.

Writing around this given theme had brought to the surface shared feelings and stimulated thinking about differing attitudes and experiences. People learned something about each other and hopefully a little more about themselves also. The workshop itself was a literacy event, designed to evidence once again the empowering potential of creative writing in adult learning, through which the Next Step Project works.

Response to my workshop was very positive, with participants expressing how much they had enjoyed taking part and commending the work of the Next Step Project. Those involved in ABE tutoring said that they would try the "Stitches in Time" activity with their learners. At a group discussion which took place just before we left the conference I talked again about how the use of professional writers, playwrights, poets and storytellers (which is a cornerstone of the Verbal Arts Centre's work) can enrich the learning experience of the adult returner and inspire first time writing. The subtitle of the Next Step Project is "Wordpower through the verbal arts", chosen in order to highlight the concept of wordpower, now ingrained in association with the units, elements and performance criteria of assessment frameworks for the adult learner's progress towards literacy, and to realign it in its wider context, a context which includes creative power as well as those skills associated with what are often described as "basic" or "everyday" literacy functions. I hope that my Conference workshop demonstrated that The Verbal Arts Centre, The Next Step Project and writing development activities such as "Stitches in Time" are helping to ensure that the paradigm for research and practice in adult literacy always includes imaginative and creative uses of Wordpower, access to which the Next Step Project aims and works to improve.

Note

Verbal Arts Centre is based at Cathedral Old School, London Street, Derry BT48 6RQ, Northern Ireland.

Gender-specific images in reading and writing

Fie van Dijk

RaPAL Bulletin No. 23, Spring 1994

Fie van Dijk is a Dutch adult educator and researcher based at the University of Amsterdam

The start was quite simple: now and then I got postcards with love from friends. I don't remember when it startled me that a certain pattern showed up in the pictures of reading and writing. And that pattern is that women are reading and writing in quite different ways from men.

After this discovery I decided to collect postcards with images of reading and writing. Friends told each other and when I became seriously ill this brought me extra post. At a rough estimate I now possess more than one thousand cards. Now I get duplicates of many Dutch cards, so I presume that my Dutch collection is nearly complete, except for some very old cards.

Commercial = Mainstream

The Lancaster Literacy Research Group published an inspiring article about *Photographing Literacy Practices* (Changing English 1 (1) 1993). They consider reading and writing not as sheer technical skills, but as cultural practices, embedded in social action. They try to catch these social contexts by photographing literacy events in different countries and use the pictures as a tool to examine literacy practices. But they found that the photographs could not be fully 'decoded' without access to background information from the photographers.

My collection of postcards is different: a postcard is a commercial product, meant to be sold. For sure, every image has a background story, but it doesn't need to be revealed. In a way the postcard is autonomous.

The commercial aspect also brings about the fact that the images nearly always reflect *mainstream notions,* particularly concerning role relationships and the representation of men and women. Thus, the cards form a tool for analysing gender-specific aspects of the traditional view on reading and writing. The postcards also generate questions about literacy events which are not represented.

Here is an example. Many cards portray Jewish men reading the Tora, a peaceful and dignified picture. It would not come to one's mind to pose questions. But I possess a *newspaper* picture with the caption: *'An unusual everyday happening at the Wailing Wall in Jerusalem: American women claim equal rights'.* Without this explanation the picture could only be interpreted by insiders. But now I realise I have no single card of women reading the Tora. Women must keep their hands off the Tora. Religion is for men's sake. Are not the books of the Bible written by men? Are women allowed to be a priest in many churches?

As another example, most postcards depict women in subordinate positions. They look up (very often in a provocative posture) at the man who gives them orders, they type their texts even literally blindfold, and above all they read the letters of their lovers; they never study. They are always white. But on postcards sold by liberation movements or literacy campaigns in the Third World women instruct men or read academic books. These cards are not commercial, but are aimed to support liberation and literacy organisations. That is why alternative images of women can be represented.

A card with a man, lonely in his study, reading thick books, does not generate questions. But when one has seen thirty cards with learned men, one wonders where the learned women are - not to speak about black women (and men). There are no cards with learned or important women. The only exception is a card with Margaret Mead and one with the Dutch queen at a worktable overloaded with papers.

Reading versus Writing

My collection consists of three or four times more cards about reading than about writing. I can imagine why: a reader can read in more different postures than a writer does, and could, therefore, be more picturesque to photograph or to paint.

But perhaps it has also to do with what Kenneth Levine mentions in his book The Social Context of Literacy (1986), that there is a tremendous literature on *reading*, but far less on *writing*. He supposes that this is due to the fact that schools start with reading, with the assumption that writing can be acquired through practice providing that reading has been mastered (p. 23). He remarks that

> *writing is a powerful and distinctive medium of self-reflection, and it has a great potential for recording and conveying innovation, dissent and criticism. It offers what is often the only channel of access to the decision-makers in bureaucratic organisations, and is virtually a necessary condition for effective participation in the organised political process in parliamentary democracies. (p. 42).*

This critical definition of writing is not represented on the postcards, and this fact provides information about mainstream ideas on reading and writing.

There are more women reading than men. That could have something to do with the picturesque postures that reading calls forth. But it could also stem from traditional views on literacy, that define *writing* as *active* and *reading* as a *passive* activity. Not long ago *active* was associated with *masculine*, and *passive* with *feminine*.

The character of the images

A great many of the postcards are reproductions from old manuscripts or engravings. Other cards are photographs, sometimes presented as works of art, sometimes specially arranged for the card. There are also cards with drawings, intended for children.

Sex plays a role on many cards, and I wonder whether this is connected with reading and writing or with the function of postcards.

Even Ed Schilders, a Dutch columnist, has never understood the combination of *sex and books* (De Volkskrant, 4.7.1992). He made a collection of clippings and pictures about the subject. To give some examples: a woodcut from 1518 is called *'The learned man at work'* and shows a man fondling a book while in the background his wife is fondled by a lover. An ex-libris label shows a man in his library looking hot-blooded (his pipe falling out of his mouth) at a scarcely clothed girl (high heels, stocking suspenders) who wants to deliver him some books from a staircase. Schilders' story shows that it is not only on postcards that reading and writing are associated with sex.

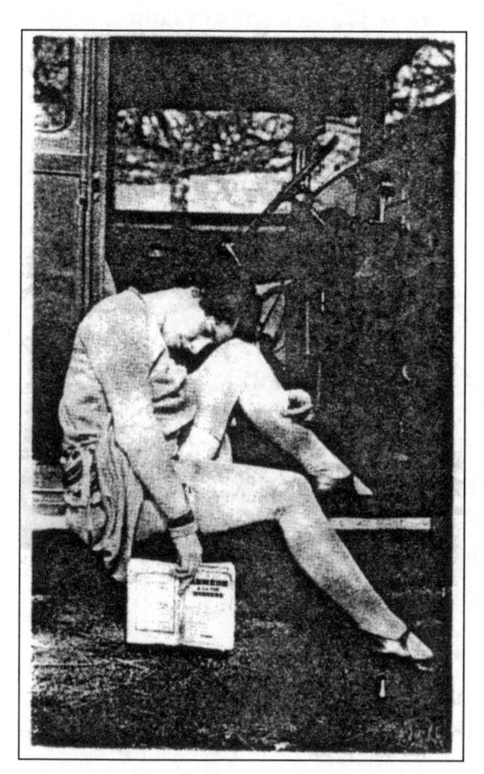

For the present I assume that most pictures are made by men, who paint the world from a man's point of view: men are learned, competent and independent, women are mothering, not very smart, sexy and not to be trusted.

What is read and written?

Men, especially, read newspapers. And Playboy. Little boys read Playboy already in the baby-carriage, and a priest reading Playboy seems to be extremely funny. One postcard even pictures an Aboriginal, with a bone through his nose and a boomerang within reach, enjoying Playboy. Even male animals read newspapers and Playboy, Play-fox, etc: cocks, pigs, apes and elephants. Men also study thick books, or are absorbed in thinking behind typewriters (I've only one card with a word processor, with a nun of all people).

Women read and write letters. From her posture and her ecstatic look one concludes that the letter comes from the woman's lover, or is written to him, sometimes covered with tears or accompanied with crumpled up drafts. The oldest cards in my collection (from the time of the First World War) show photographs of ladies, dressed up romantically, who read a letter from, or write a letter to, a soldier at the front; he floats in a cloud above her head, aiming a gun in a trench. Sometimes a woman bows, sobbing, above her typewriter. As a matter of fact, a man never cries. Many cards show a woman typing work for her boss, even in the back seat of a sports car. One woman is so helpless that she pushes a bottle five times

bigger than herself with a sign *HELP!* into the sea.

Newspapers are hardly read by women. They read gossip papers and needlework books. When a woman is portrayed with a man, she reads the gossip and he the financial comment. On one card a woman is knitting, with a newspaper in her knitting bag on the chair-back. In reproductions of old paintings women often read the bible. This seems to be an indication of virtue; even the virgin Mary is often portrayed reading a holy book.

Women also read aloud, to ill people or to children. Even little mouse women read to little mouse kids. And girls to their younger brothers and sisters. Boys never read to anyone. Only one card shows a grandfather reading to his grandson. If a couple is shown together, the man reads the newspaper, the woman pours tea, does needlework or is bored.

Where and how is reading and writing done?

There is no doubt: men control the public domain. They read their newspapers on the street and in the pub, their thick books in study rooms and libraries, with pipe, cigarettes and coffee, and surrounded by globes, computers and robots. They are always dressed up correctly, even with a hat. Even male pigs wear trousers and read newspapers, while females walk about naked, with their handbags.

The private domain is the domain of women. They lie gracefully on sofas, in bedrooms, in gardens, surrounded by roses, holding a novel loosely in one hand. Often they are naked, or in a sexy outfit. The legs, especially, draw attention, often provided with stocking suspenders and bare thighs. There are a few cards with naked men, but they obviously belong to the gay circuit and are, like the women, sex objects.

If the woman is not portrayed as a whore, then she looks like a Madonna or mother: modestly reading the bible or surrounded by children. Girls in pastel colours are sweetly reading or writing. But some cards radiate hatred and ridicule of women: a woman with curlers, smoking heavily, reading *HOT ROMANCES;* an extraordinarily fat woman reading while standing in a sort of massage apparatus (on the back of the card: *'Reading your Weight* away'); ugly viragos reading a letter: feminism is ridiculous.

Marriage is not held in great respect. The traditional breakfast image is not lacking: he reads his paper and she pours tea and wants to talk with him; he is sitting on one side of the stove and reads his paper, she is sitting on the other side, knitting, or fumbling with her handbag. *'Don't marry, be happy!'* is printed above this picture. On a painted card a man writes with the hand of his murdered wife (the knife still in her back) the lines: *'Ich soll Erich treu sein. Ich soll Erich treu sein'* (I will remain true to Erich). Another card shows a man lying in a halved bed (the saw still lies on the floor), reading the letter of his ex. Even the painting above the bed is halved. The only words left on it are: *HOME, SWEE.*

I don't see any difference in the images of women and men represented between the oldest card that I possess, stamped 1903, and the newest ones. One of the latest painted cards shows a pavement with two men and four women. One man is reading a book, the other one a newspaper, it's a card from Finland (La-Uusim). The waitress serves, one woman is fumbling with her - guess what? - bag, another is just being beautiful, and the fourth one is pregnant, walking along the street with a child ahead. And our society may be more multicultural, but there is no one from Surinam, Turkey or Morocco reading or writing, not to speak about black women reading or writing. The only exceptions are, of course, the cards from the Third World.

Captions

A card with a man puzzling over his papers has a caption: *'The brain is a wonderful organ'*. A girl in sweet colours is reading a book in misty light. Comments on the card: 'A *good book is like a wise boyfriend'*. Some other examples:

- A little boy with pipe, dog and newspaper. Caption: *'Now I'm big', this little chap is thinking, 'Because I have father's pipe and I have father's newspaper'.*

- An old man writes with a goose-quill a letter a young girl dictates to him: *'Der Schwieriege Liebesbrief (The difficult loveletter).*
- Beautifully dressed ladies (much lace and roses) read letters, accompanied with texts like:

DER LIEBESBRIEF. (The Love Letter)

So wird's gut sein, denke ich.
Wenn mein Brief ihn hat erreicht,
Wird er wissen, was ich meine,
Denn erraten kann er's leicht'

(The loveletter. It will be alright, I think. When he receives my letter, he'll know what I mean, because he can easily guess).

or:

'Grand soldat, gue j'aime, pour to recompenser je t'envoye cette fleur, et mon pluc doux baiser'.
(Great soldier whom I love, to reward you I send you this flower, and my sweetest kiss).

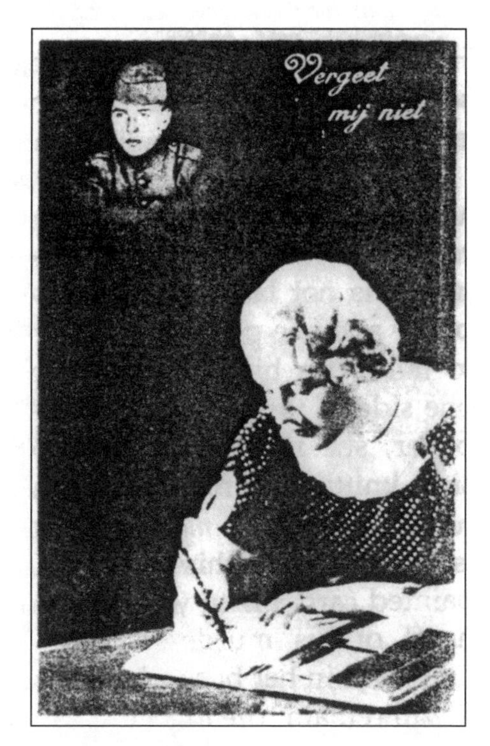

- A woman is reading the newspaper while her husband is scrubbing clothes in a washing tub: *'The New Woman - Wash Day'*. It is clear, the man is a softy.
- A man is reading the newspaper. This wife says: *'Darling, I'm going out tonight . . . for about a year, o.k.?'* The man answers: *'Yes dear,*

don't forget to clean the smell of cat piss out of the carpet'.

These sorts of captions confirm what most of the pictures already show: that you are a man when you read a newspaper, that women above all read and write letters, and that a reading man doesn't listen to his wife. They don't add much to the pictures.

Conclusion

In the preface to her book **Een beeld van een vrouw** *(An image of a woman),* a visualisation of femininity in a post-modern culture, Rosi Braidotti (1993: 12) postulates

> *In the visual culture of today one has to learn to look, 'read', and interpret. Many of us have to learn to be visually 'literate', as we not so long ago had to learn to handle the computer. This new 'literacy' creates the instruments that enables us to interact and to converse with the post-modern culture.*

Women with little formal education have nothing to do with any interaction with the post-modern culture, but the 'reading' of postcards (a popular cultural form of communication) and the questioning of the representations of women and femininity give them, as well as professionals, knowledge and understanding, not only about the gender-specific aspects of reading and writing, but also concerning the position of women in general and their own social position - without the need to use suspect terms like 'feminism' or 'emancipation'.

In a lecture entitled De future of reading, held in Amsterdam on 23.2.1990 (published in **Vriji Nederland** 10.3.1990) the critic of English culture, George Steiner, said

> *The luxury of private space, of closing your door in order that nobody else can enter – and don't forget that reading may be tremendous selfish; we do have a wonderful example in Montaigne: his wife and children were not allowed to disturb him. When Montaigne was reading, nobody was allowed to disturb him.*

Well, isn't it time for culture critics to realize that their wonderful examples are wonderful only for men?

The representation of women and men reading and writing is not only a matter of academic discourse, but reflects a bitter reality. Thus, in Allende's Chile, everywhere in the slums of Santiago, libraries arose where women's groups debated on subjects that were important to them. After the arrival of Pinochet the libraries vanished and the women had to start knitting again. Everywhere in the world women have less access to education than men. Statistics show that of all 154 million new illiterates, between 1960 and 1985, 133 million are women (Nelly Stromquist, 1990). To a lot of men it is threatening if their wife learns to read and write.

There is as yet little knowledge about the gender-specific aspects of reading and writing. What we need is research into the needs and expectations of women who want to learn to read and write, and into the power relations that are involved in learning to read and write. Perhaps the reading of images such as photographs and postcards can be a starting point for this.

References

Braidotti, R. (1993) **Een beeld van een vrouw**. Kampen.

Lancaster Literacy Research Group, (1993) *Photographing Literacy Practices,* **Changing English 1 (1).**

Levine, K. (1986) **The Social Context of Literacy**. London

Schilders, E. in **De Volkskrant** (4.7.92)

Steiner, G. (1990) *De future of reading* (lecture, 23.2.1990) in **Vriji Nederland** (10.3.1990).

Stromquist, N. (1990) *Women and illiteracy: the interplay of gender subordination and poverty.* **Comparative Education Review** , 34 (1).

Part 3

Learning about dyslexia through research and practice: A longitudinal research and practice case study in professional development

Margaret Herrington

Introduction

So far we have considered the research in/and practice evident in the RaPAL publications. We have identified research and practice narratives in the collection and have given a flavour of the themes that RaPAL members have explored. In this section, I shall explore a longitudinal research and practice narrative about dyslexia, which exposes the interplay of research, development and practice over a period of 20 years.

But why do this? And of what value will this be to newcomers in the field? First of all, an analytical reflection over time can demonstrate the importance of seeing research and practice as an organic way of working rather than as an occasional activity, undertaken for a further degree or within a defined research project. A longitudinal time frame provides a rich description of interconnectedness which encourages new staff to feel more at ease with longer periods of development and to be more aware of how long it can take for deep expertise to develop. The narrative below shows some of the hinterland behind my own writing about dyslexia.

A second reason is that professional development narratives in relation to dyslexia tend to be oversimplified. Training sessions often focus on a top-down delivery of a body of knowledge about dyslexia with little reference to the theoretical integration of this knowledge with that of teaching and learning in general. My experience suggests that it is much more productive to encourage staff to see themselves and their students as actively constructing their own narratives about dyslexia, asking and answering their own questions in the context of their own personal and professional lives. The narrative below, with its fits and starts, gaps in knowledge, delays and occasional leaps, challenges oversimplification and provides an example of a professional route with which staff can compare their own experience.

Third, these narratives make an important and often unacknowledged contribution to the generation of knowledge about dyslexia. For example, the dyslexic learners' descriptions, in their own words and on their own terms, only occasionally appear in the research literature and their significance as research is rarely recognised. Literacy practitioners have a unique opportunity to collect this data and to connect it with the existing research literature. The narrative gives a flavour of this kind of knowledge generation.

Finally, at the time of writing, the policy discourse focuses on how best to respond to dyslexia as a *disability*. Under the Disability Discrimination Act Part IV, dyslexia is so designated and staff in educational establishments have to ensure that they do not discriminate against dyslexic learners. This poses three challenges for literacy staff: the conceptual challenge of recognising the ways in which particular literacy concepts and practices 'construct' the disability; the challenge of finding a way of thinking about dyslexia and non-dyslexia in basic skills classes (what are the criteria and where are the boundaries?); and finally, a more practical challenge of asking whether their general practice as adult literacy educators requires any adjustment in relation to dyslexic learners. In reviewing my own experience and constructing a written narrative, I shall show how my thinking and practice developed in these respects.

Longitudinal narrative as a research method

We have referred earlier to narrative as a method of enquiry (Clandinin and Connelly, 2000). A variant of this is autobiographical work, and this has been used by professional educators to make meaning of their own development (Schmidt, 1997). Several UK literacy workers have used elements and types of this approach with regard to both themselves and their students (Clarke, 1996; Ivanič, 1998; Ivanič, Aitchison and Weldon, 1996; Mace, 1992).

In general, autobiography offers a powerful mechanism for revealing the varying complexity of professional development. It offers a means of reviving 'lost' or 'forgotten' knowledge, making the implicit and hidden parts of lived experience explicit and visible and so adding to the historical record. Further, it provides an opportunity to create knowledge: a new 'gestalt' narrative (Abbs, 1976) which is interpretive, connecting, synthesising, imaginative and fictive (Graham, 1991). This research method frees the researcher to engage in a multi-dimensional journey, exploring and problematising earlier professional selves, contexts and emotions. In so doing it releases material that is otherwise not acknowledged (especially important in the case of marginalised groups). However, it is complex in its requirement to work with several selves (prior and current) at the same time and to negotiate the tensions between the historiographic and the poetic.

This approach has been challenged on a number of grounds. For example, the nature of the knowledge that autobiography produces can be seen as too personal, about one life only and limited by self-reporting. In the case of the new interpretation, if it is fictive then in what sense is this knowledge? Graham (1991) has addressed the issue of under-theorisation which lies at the heart of these concerns by returning to epistemology and in particular to constructivism. For him, and from a constructivist perspective, 'knowledge, like the self, is provisional, changing and socially constructed'. Autobiography involves 'reflective self consciousness and . . . the active construction and reconstruction of personal experience and so produces knowledge'. He goes so far as to say that autobiographical processes are engaged in all knowledge generation and should be acknowledged as such (Graham, 1991).

Other challenges are about autobiography *in use*. Hidden assumptions have been identified: the uncritical construction of private narratives without reference to their capacity to distract from necessary social change (Brain, 1988); the relationship between the development of autobiographical study and the individualist western culture; and also the tendency to use chronology in an over-simple way and to present the story of lived practice as a picture of ongoing progress and improvement (Usher, 1998), with the discontinuities almost airbrushed away (Huberman, 1995). There is also a more fundamental challenge to narrativity from philosophers who believe that it does not adequately deal with the issues of truth (what is 'true' in a multi-narrative context?) or of representing those who do not tell their stories but who just live their lives (Strawson, 2004).

Despite such reservations, this approach offers a chance to articulate individual stories and to discuss the impact of time within them. The narrative constructed here is a kind of analytical testimony in which I have acknowledged the difficulties in retelling the lived story: the unwitting as well as the witting selectiveness and the ongoing question of whether the testimony is really about what I thought at the time or what I now think I thought then. Inevitably, too, in any autobiographical analysis there is representation of the voices of others, and I am fully aware of the responsibility this brings and of the subtleties of the processes involved.

Working with these complexities, the following narrative shows how I made sense of dyslexia: how I worked in the spaces between research discourses and the individual experience of dyslexia. I wanted to describe the integrative rather than accumulative nature of the processes of learning about dyslexia, the unevenness and uncertainties involved (not just a spiral of successive action research or reflective practice cycles), and the identifiable transfers of knowledge and practice from one context to another – with the transformation involved.

Specific method

But how to construct it? The simplest approach seemed to be a thematic analysis, detailing my learning about particular issues. However, I wanted to give a feel of how particular phases felt in general and wanted to show the unevenness of the process. I decided, therefore, to use four successive, professional episodes over a period of 20 years, from 1982 to 2002, as the broad organising structure for producing 'experiential artefacts, to draw out generalisable messages' (Clandinin and Connelly, 2000).

The four phases chosen include two periods of work in basic education and two in higher education:

1. Distance learning in Leicestershire, 1982–1990.
2. Leicester University/Leicestershire Adult Basic Education Service, research project, 1987–1989.
3. HEFCE project on supporting students with special needs at the University of Leicester, 1992–1994.
4. The University of Nottingham Learning Support Unit/Study Support Centre, 1995–2002.

These phases involved the managing and developing of practice as well as individual research and development projects. Figure 3.1 shows how research was undertaken within specific projects and within everyday practice, and often at the same time. The model here is not one of sequential research and then practice, or vice versa. The interesting mix of research in practice within more broadly based development projects demonstrated that this was not a matter of either/or.

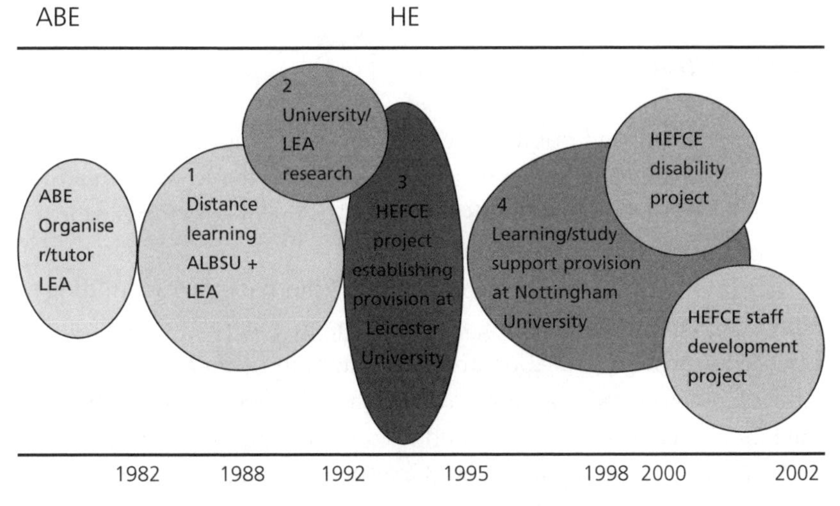

Figure 3.1 Four phases 1982–2002: building research into the infrastructure of professional practice

I reviewed each period for key memories, learning and turning points as I tried to work through the main questions that any literacy, numeracy and ESOL practitioner even now has to address:

- What is dyslexia? How can we conceptualise it? How can we make sense of the different narratives about it?
- How are people dyslexic? How is dyslexia identified? Can we make sense of existing methods for doing so? Who should do this?
- How does this knowledge about dyslexia affect how we view all our students?

- What should our responses be as teachers? How should our methods change, if at all?

The narrative that follows draws on archive material (sessional notes, reports). It is not a comprehensive professional history; rather, I have used the archive material to fill in the detail of what I have remembered as being important.

Finally, I looked back over the whole period, re-questioning the new gestalt.

Starting point

Selecting the starting point for such a narrative is not easy. In the event, I selected 1982, seven years after I had started to work in adult basic education, because in this year I took a major step forward in understanding something about dyslexia.

It is important to say at the outset that I did have many building blocks in place about teaching and learning in general. I was a trained teacher for post-16 education and had taught in further and higher education before moving into adult literacy. The experience within the literacy scheme had sharpened my understanding about learning and failure to learn in general, and about failure to develop literacy skills in particular. By 1982 it had already given me an array of methods that could be used to assist adult learning, both generally in terms of learner-centredness and 'voice' and specifically in terms of language experience, creating text for reading and writing, teaching phonics, and using visual, auditory and kinaesthetic methods of teaching spelling. I used these in a variety of community contexts: the home, community centres, libraries and at a probation day centre.

Though Freire had written his seminal work about literacy and power by this time (Freire, 1979), my theoretical understanding about the broader cultural and political implications of literacy was still quite limited. My undergraduate studies in Economics and Modern History had left me with a human capital perspective of education, and not a very critical one at that. Though my teacher training had included some work on the sociology, philosophy and history of education (Bernstein and others) and the foothills of sociolinguistics, I generally saw literacy quite simply as desirable and empowering, both personally and economically for most people. I was interested in encouraging students to focus on what they wanted to learn, to have their say about this and to publish their writing (and I had started to investigate student motivation and drop-out for a research degree), but I do not think that, in 1982, my work was driven by the much deeper critical theoretical perspectives which I was to develop later. And though I later learned about colleagues elsewhere in literacy education who had had a far more advanced radical critique by this time, I was not untypical in mainstream work.

In relation to dyslexia, all I had at this point was a feeling that the discourse about dyslexia, which appeared to involve deficits in brain function and which I had heard of from colleagues in passing, did not fit with any of my actual students. I did know about students who had failed to progress and who tutors described as having 'plateaued', but I did not relate such intractable difficulties to dyslexia. I simply saw them as difficulties that were extremely hard but not impossible to get round, even though I could not always see how to do so at the time.

Main phases

The first two phases of work described below involved a complete shaking up of this position in terms of my understanding about dyslexia itself, and about the role of the learners as co-researchers. This was against a background of deepening theoretical understanding about literacy in general. I would characterise this as experiencing a shifting underlying terrain in relation to the theory and practice of literacy and literacy education at the same time as finding routes in uncharted surface territories.

1. Distance learning in Leicestershire, 1982–1990

The first major step forward occurred when I helped to establish a distance learning scheme in Leicestershire. This was a new form of provision for 'hard-to-reach students' in rural areas. Alongside mainstream group and volunteer provision, a new individual and group distance model was devised (ALBSU Special Development Project, 1982–1983; Herrington, Report, 1985) which involved professional tutors visiting students in their homes on a monthly basis and using distance methods in between. Systematic records were kept from the project and findings were disseminated via staff training to colleagues in the main scheme. It was an example of a development project, ostensibly designed to develop a new system of delivery, which also involved research in practice with each learner.

When this was subsequently expanded to cover the whole of Leicester and Leicestershire, it attracted a wide range of people in terms of employment and level of difficulty, compared with mainstream group-based provision. The methods we developed with the learners over this period enhanced our general repertoire and, in particular, shifted our ideas of time and space in relation to literacy learning. Freed from the constraints of the classroom, adult learners chose a variety of different times and locations for learning (Herrington, 1993).

The catalyst for developing my understanding about dyslexia was this change of teaching and learning context. For the first time I was in a position in which I was not present when the students were doing much of their literacy work. When we had our monthly session, much of the time was spent in discussing process as well as content: metacognitive awareness was both intrinsic and explicit.

For example, I became aware for the first time of how students who experienced severe difficulties with reading were actually using tape to assist their reading. A young man who worked as a carpenter wanted to improve his reading. He found technical manuals too difficult and, encouraged by his wife, wanted to improve generally. Assessment revealed him to be a middle-level reader (according to assessment practices at that time), and our discussions led to a decision that I would place texts chosen by the student onto tape and send them to him by post. At our meetings the student would read the text to me but would also describe how he had used the tape to learn the text. He reported that he would listen all through for the gist and then, to my surprise, he would listen to, read and re-read very small bits of text. He would intersperse this with physical activities (such as working in his garage) and re-read the text again. He did this over and over until he had it in his memory, and would then move onto the next bit. It appeared that intractable difficulties could be shifted if the student really took control of the process.

Though I did not describe this student as dyslexic at the time, his degree of difficulty in developing access to text was highly problematic for him. His work alerted me to the extent to which I had been relatively unaware of process and even relatively incurious about it. There was nothing in principle that would have stopped me finding this out within classroom practice. And though I considered that his micro management of the process of overlearning was so effective that I went on to suggest it to other students, I certainly did not understand the full implications of his interspersing of tactile and kinaesthetic activity with the learning of text.

These discussions about process revealed further unusual descriptions and unexpected subtleties in the accounts of how literacy and numeracy were experienced. One particular student, in a highly responsible position, had ostensibly come to get help with spelling. During my period of working with him, he alerted me to the idea of a cluster of characteristics that I had not really put together before:

- His difficulties in making sense of the moving squiggles on the page during early schooling and the way in which his frustration expressed itself in violence.
- A particular teacher noticing that he was answering all the questions in a science class and providing him with individual tuition.
- Accounts of his reading as an adult showing tracking difficulties and the need to keep reading each day in order to retain his ability; his description of this as 'a way of keeping a path clear'; his experience of occasional, temporary, visual blanks when reading.
- His enjoyment of reading particular kinds of novels.
- In speech he occasionally had to restart his sentences because

his language became jumbled: he seemed to want to say several things and they seemed to conflate as he tried to speak.
- That the root of some of his spelling mistakes involved what was comfortable for him to look at rather than a memory lapse per se.
- That he experienced some days when he could not write coherently, and it still felt extremely frustrating; he could not control when this occurred.
- That when writing 'from the inside', non-formal or set writing, his work was imaginative and fluent. He wrote excellent stories for his son and a fictional account of his schooldays.
- He was good at maths.

From here on I really understood the importance of an investigative stance; of understanding the more subtle mixes of characteristics and the particular sources of frustration, rather than focusing on more general levels of competence and broad brush methods.

Students also described more extreme and intractable degrees of difficulty with memory for words in both reading and writing. I encountered students whose visual and auditory memories for text were so weak that none of the main methods for teaching reading were effective. Our discussions often focused on how they got round the difficulty in everyday life and on their problem-solving strengths. I nevertheless sometimes felt that I had reached my own limits in knowledge about what to do, even after seven years in the field. Everything was experimental. Investigations could lead to dead ends as well as leaps forward. This was particularly true in the case of an ESOL student with limited spoken English.

> S lived with her mother . . . Her spoken English was very difficult to understand and I realised that one reason for this was that the English words were being used with the intonation pattern of her mother tongue. She had no literacy skills in her mother tongue or in English. My feeling was that S should be attending ESOL classes because her spoken English seemed a greater priority. However, she had tried these, not enjoyed them and refused to go. I tried therefore to teach her some language and literacy,

using language experience, easy readers, functional words and phonics. She enjoyed the sessions and the distance tapes I prepared but I realised that her memory was very weak in relation to both sounds and shapes in words. I did not feel that she was making the kind of progress that individual attention can often secure and felt somewhat at a loss. Most of the learning outcomes did not involve literacy and seemed non-measurable: according to her mother she looked forward to the sessions and so experienced a positive learning experience. She enjoyed the fun in the sessions and she used the tapes but I did not feel that any of her learning was really secure and I did not really know what to do about it. I was conscious that there were complexities to unravel but there seemed to be no additional expertise to call upon.

Clearly, without some level of spoken English or access to a suitable interpreter, I did not feel that I could make much progress in understanding the real problems this student had in accessing text.

When the students were willing to act as researchers of themselves and as co-researchers, the knowledge outcomes for students, tutors and the field as a whole were remarkable. One student, V, for example, kept a diary to try and capture her strong variations in performance and met with me to discuss her experience.

MH's notes from a meeting with V about V's 'Bad Day'

1. Weekend. Considerable amount of writing. Flowing well and then 'blocks' started to occur over this three-day period (despite feeling generally relaxed, fine and wonderful). Feeling of 'blocks' in thinking what she wanted to say. Then in attempting to think, more errors started to occur. Feeling of everything having 'gone', 'totally'. Her verbal fluency is also affected. To the outsider she'd seem OK but a friend or trained observer would notice.

   In fact V can write:

Today was a bad day and I thought I had better get some of my thoughts down on paper. The simpler words are coming easier but I am struggling with anything I really want to say. I want to say how it feel my mind just seems to have closed down any words that for me I understand I am not writing them down this is why I have simplified this short note.

Though she cannot 'see' it unless she puts her finger on every word.

There definetly seems as though there is a vail over part of my brain my thought pattern is not at all the same. As I look at the paper it is just a jumble of words. I shall have to read it back to see if I have made sence of what I have written.

I'm still trying to explain what it feels like, the one thing I can say is that I just don't want anyone to look at anything I've written. I know that I am capable of explaining to greater depth of how it feels the frustrations is so irritating it has to be felt to be understood

I'm sure that even if I was copying words from a book I would still have difficulty in following the words. The more I look at this writing, the more confusing it becomes there are just lots of words on a piece of paper all mixed up, and with no real meaning.

I will write again tonight to see if I can get anything else down.

'Still feeling that visually I'm not seeing the words' and she was unsure that they were spelled correctly.

2. Tried looking at three words.
   She V can read this. Letters not clear – slight shimmer of movement

– has to read it letter by letter to see what is there.

Sincerely Though she spells this consistently well on good days, she doesn't know today if this is OK. Could identify where greatest confusion would lie.

Museum Wants to add an a before the u. Letter by letter as above.

3. Looked at words on different coloured paper: yellow clearer, still scanning letter by letter, still don't know if spelt correctly; blue – not as clear as yellow; white? – worst possible combination is black on white paper.

Important to note that 'something' had happened about which V is very conscious. It represents such a drop from how she normally operates that it throws her badly. Because it is unpredictable, it is always a shock. Actual performance in this state is not bad in an 'objective' sense – it is the fact that V can no longer express herself as herself that is so upsetting for her. Although what is on paper is not bad, it isn't her.

The description was so powerful that I was forced to ask myself questions about how literacy could be learned given this kind of cognitive moving feast, and how the learner could derive confidence given such dramatic uncertainties about connecting words inside her head with her written output and about being able to see the words on the page. Though she had a general uncertainty about spelling she could be a fluent writer, but I could see that for her it was the lack of control and the unpredictability of these processes that caused most difficulty. It was when her whole reading and spelling systems felt as if they had suddenly started to run on reduced power that the most acute distress was experienced. Methods of teaching that did not take this on board would have been ineffective.

These explorations fitted well with the general approach to assessment in ABE at the time, which was largely critical of any formal testing with adults (and for a range of very good reasons). No one thought it odd to be exploring with students in this way and clearly, in relation to the above student, any test outcomes would have been affected by how she was experiencing text on the day of the test.

However, it was during this period that I first encountered formal tests for dyslexia. I was asked to accompany a student to the Assessment Centre at Aston University and so observed the educational psychologist's full test procedure for testing dyslexia. I read up about formal testing and encountered, for the first time, the significance of 'unexpected discrepancy' – between IQ and expected performance in literacy tasks – in assessing dyslexia.

I had read about the controversies surrounding IQ testing but had never actually seen intelligence being tested before. The student was extremely uncomfortable with the timed tasks, and I felt very uncomfortable with his discomfort. I was also aware that this method had very little to do with the adult's own descriptions of literacy processes and was all about the completion of tasks set by others. It may have been attempting to record 'abilities', but there were clear limitations in this respect and there was little in the process that empowered the student.

The impact on the student of being told he was dyslexic was dramatic: strong emotions flooded out. His personal and educational history had been very difficult (he had been sent to a special school) and he had suffered enormous and continuing loss of self-esteem as a result of his literacy difficulties. He was terribly shocked by the verdict and was then disappointed that there was no immediate 'cure'. I explained again the basis of the test and the ideas about discrepancy, in an attempt to contextualise this kind of testing and to expose its limitations.

At the time I felt that there was some logic to the idea of a specific rather than general learning difficulty and, on balance, it was his feeling of recognition that he was highly intelligent but with a specific literacy learning difficulty that made me feel it had been worthwhile (despite my doubts about the testing and despite the lack of clarity about how methods of teaching should change). However, the outcome for the student was one of feeling that it was all too late for him and that he was more concerned now that schoolchildren should not have similar experiences to him. At the time, the world of formal testing for dyslexia felt like a parallel universe, characterised by its own certainties but with little to do with the everyday experience of literacy teaching.

A further 'prod' in the direction of dyslexia came from a student who had acquired difficulties with reading and writing as a result of a stroke. Acquired dyslexia was new territory for me and I followed the student's descriptions carefully. She recounted that she had been told by a consultant that if she did not retrieve her literacy within a certain time period she would not get it back at all. Yet here she was several years later still progressing. I could not read up about this at the time, given the pressure of student numbers, but I did log this as another example of students' own experiences challenging existing knowledge. In terms of the discussion in Part 1, I lodged it in my brain as something that I would have to follow up at some point because the specialists' story had been unconvincing. I felt aware again of the gap between disciplines and their ideas and practice in relation to this field.

What did I make of this?

Distance learning contexts had allowed me to take a major step in building my sense of what investigative practice about the nature of literacy learning could be, by allowing a new level of research in practice discussion with students. As a professional educator, I felt renewed by this.

With regard to dyslexia, I began to wonder if dyslexia was about degree and/or kind and/or intractability of literacy difficulty. I was encountering students with extreme difficulties but I also started to pay more attention to intractability and failure to progress. I was also starting to learn more about the good day/bad day variations in performance and the implications of these both for assessment and for explaining the accusations of 'lazy, not trying' from school teachers which so many literacy students had experienced. This was new knowledge for me. I had previously tended to think that students were largely consistent in their level of performance, at whatever level.

I was also happier with the term 'specific learning difficulties' rather than dyslexia because it seemed to offer a way of viewing the heterogeneity of the students I was responsible for. I felt very unsure about the validity of the idea that the term dyslexia could only be applied to those with average or above average IQ. I could see that the difficulties might be easier to spot, but most of the characteristics seemed to be present across the ability range and so I began to wonder if all adult literacy students were dyslexic to some degree.

Finally, I remember feeling rather out on a limb, cautiously wending my way. I was a trained teacher with some knowledge about the psychology of education. I was also an experienced adult educator who had seen the dyslexic students blossom using my developing investigative methods. Yet I felt something of an outsider in relation to professional psychologists. So, I started to read the specialist literature as a means of gaining understanding and confidence. This occurred alongside a major leap in understanding about theories of literacy from an encounter with Brian Street's (1988) work on autonomous and ideological models of literacy at an early RaPAL conference. However, it would be quite some time before I really sorted out how to view what appeared to be in-person processing difficulties within a deeper social theoretical framework about literacy.

2. Leicester University/Leicestershire Adult Basic Education Service, research project, 1987–1989

This project has already been described in some detail (Herrington, 1995, 2001). It was established because of the frustration of one student who felt that his basic education tutors simply did not know enough about his kind of difficulties. An action research project was designed which brought together a group of similarly frustrated students with two tutors and two coordinators to research these difficulties. The methods used included intensive tuition (weekly, over two years) and its evaluation, literature searches, monthly exploratory discussions which occasionally involved outside speakers, and visits to experts.

It would be difficult to overestimate the staff development impact of this work for all the staff involved. My role was as organiser, coordinator and facilitator, helping the group to draw its main findings together in an open-ended way as we went along. My own knowledge was developing on a number of fronts at the same time: directly from students; from the literature that I was now working through more quickly; from experts; and from the tutors who were working directly with students each week. But we were all gaining, not just by producing a list of findings and recommendations but because we were all engaged in the roles of researchers, teachers and learners.

The first, strongest memory was of the pent-up anger among some members of the group about lack of recognition of dyslexia within the educational establishment. This was not helped by the generally hostile tenor of the education debates about dyslexia at that time. From the learners' perspective, it was difficult to understand educators who appeared not to be interested in why they could not acquire literacy skills or in the learners' views about this. Part of the work was to unravel the possible reasons for this situation, for example learning that those professional groups opposing the use of dyslexic labels had a variety of intellectual, financial and social equity reasons for this. Also the definitions of dyslexia were confusing, the formal testing methods were open to question, the costs of testing were considerable, the disadvantages of labelling were understood, and there was the inequity of separating off one group with literacy difficulties from others in order to give them preferential treatment.

In the light of this, the group acknowledged the strong emotions but tried to start from first principles in relation to definition and identification. We did not start with an existing definition of dyslexia (see Klein, 1986) but chose to concentrate on trying to describe members' experiences and on working from there. One of the earliest discussions was about terminology. The students wanted to use the term dyslexia because it meant difficulty with words. No one worried about the label dyslexic – they were pleased to have a name for their experience and it was infinitely preferable to the other labels they had experienced. They did not like SpLDs (specific learning difficulties) and were not impressed by my arguments for its use. We became aware of the different definitions in the literature, and also of the fact that educational psychologists as a professional group were far from united (BPS, 1988). My earlier idea of single positions held by particular disciplines proved not to be the case. This reinforced my concern that we should describe as clearly as we could the knowledge we had within this group as well as continuing to read more widely.

The descriptions revealed different clusters of the following characteristics: various kinds of visual disturbance on the page; tracking difficulties; delays in reading from lack of phonic knowledge; acute memory difficulties in relation to spelling; lack of connection between brain and hand for writing; some difficulties with maths; and good day/bad day variations. The emotional factors unfolded each week, in every session, reflecting past experience, current frustrations in general and frustrations with their present learning of literacy in particular. Individual members dealt with frustration in very different ways but all decried the continuing ignorance and misconceptions in society in general.

Given the emerging sense of plurality and heterogeneity within a core of common characteristics, we felt we should review methods of assessment. The students reviewed some of the informal methods, for example the Bangor test produced by Professor Tim Miles. This test, with its eight indicators, raised questions about underlying processing problems. Students, however, wanted to know what left/right confusions/delays, remembering numbers, multiplication tables and pronunciation had to do with their literacy difficulties. For the students the problems were easy to describe; they did not need other indicators, especially when they did not understand the basis of them.

Alongside the gathering of individual data, some tutors and a couple of students started to delve further into the extensive research literature (biology/physiology/neurology, cognitive psychology, education). The students were interested in reading about dyslexia and were annoyed to find so little accessible written material. I was particularly interested in learning about the main research agendas and methodologies, and quickly realised how little was actually published about adults. Scientific and clinical methodologies seemed to hold sway and I remember trying to relate the 'experiments' reported as research to what students were telling us directly about their literacy and memory. I was particularly grateful for writers such as Tim Miles (1983) at Bangor University, who worked directly with pupils and parents and with academic research and who seemed to be trying to bridge the gaps. Nevertheless, it felt as if the emerging dominant paradigm was one of individual impairment and that very different kinds of discourse and evidence had shaped particular ideas about the nature of the deficits involved. There was also a sense of most research being of a top-down nature, undertaken by non-dyslexic people about dyslexic people and transmitted to practitioners via publications, conferences and pressure group activities.

In the project we set the neurological/psychological claims alongside the unexpected descriptions and the physicality of them (blockages, veil over parts of the brain, breaks in connections). I was concerned that what appeared in the literature as quite a crude mapping of the structure and functions of the brain should not be interpreted in too determinist a way, but I did feel that the physicality of the descriptions was important and that I had not paid sufficient attention to this before. Further, the literature on *acquired* dyslexia was useful for students who had experienced actual damage as a result of illness or accident. One member had lost literacy skills as a result of loss of blood supply to the brain during a heart attack. The literature encouraged us to consider developmental and acquired dyslexia as two quite different conditions, or rather difficulties with literacy stemming from two quite different causes. I was not sure that such a crude polarisation was entirely convincing. Given the presence in many basic skills classes of adults with these kinds of experiences, it clearly is important to know about both.

Alongside the general drive to understand dyslexia, we examined implications for teaching methods. In general, the tutors did not feel they were doing anything new: they had all the core methods at their disposal and were interested in devising effective methods with students. In contrast, the students who were quite clear about what they needed from tutors felt that this was a new approach, and so it was clear that their previous tutors had not been investigative in this way. Tutors felt that an exploratory approach was vital for the more intractable or puzzling difficulties. There were no other answers available for adults in the literature.

One student, an intelligent and articulate man, experienced enormous frustration because he could not write his name. This proved embarrassing on a weekly basis as he had to sign for social security benefits. The exploration in the group revealed first that he could use a keyboard (we obtained a PC and he wrote a 40,000 word autobiography), and thus recognising letters and pointing to them was not problematic; and second, that if he created reverse letters in the air using gross motor movements, he could then write his name on paper. This moved me once and for all away from the standard approach to teaching handwriting: focusing entirely on letter formation, joins, grip. I began to think about inefficient connections and blockages in this area, too, and to think more broadly how such connections could be remade. I realised that IT was not an optional extra but an essential prerequisite for some students.

Above all, this work provided a lived experience of involving students as co-researchers (cf. RaPAL Doing Research, 1989). In designing the project I had been committed to making the dyslexia research agenda accessible to students – their questions seemed to me to be as important as mine, if not more important. The reality proved very complex. Students found it hard to see tutors as fellow researchers: they expected tutors to be the experts. Though this perception changed, students continued to have unrealistic expectations about what could be achieved in the time (their only chance!). Further, we had not taken full account of the competing narratives about our positions within the group or of the power differences among and between students and tutors. The most marked contrast was between those who wanted to use the research for social action and change immediately, and those who wanted to gather evidence in a much more cautious way and take time to interpret and evaluate it. Finding ways in which all could exercise power proved to be a multi-layered and not always consensual experience.

What did I make of this?

The move from being an investigative practitioner on my own terms to being part of a group with shared responsibilities for outcomes proved challenging. I had wanted to find a way for dyslexic learners to have their say, and had found it, in all its complexity. The experience of undertaking a research project alongside my mainstream practice allowed an ongoing dissemination and enrichment, with distance students attending some project meetings and tutors feeding their findings from the dyslexic students back to other literacy students. Although some power could not be shared, the project offered a first-time opportunity for both tutors and students to identify what could be done together. It thus laid the foundation for all my future work with student groups.

With regard to dyslexia, I was now more aware of heterogeneity within the 'syndrome'. If phonological deficits were at the heart of dyslexia (Snowling, 1987) – and I was prepared to believe that phonological matters were important – they did not seem to explain all the characteristics the students described. I was trying to connect our evidence of the spikiness of profiles with the Miles distinction between 'lumpers' and 'splitters' (one basic syndrome or a multiplicity of different difficulties). The impact of personality and personal circumstances in mediating the effects of dyslexia also seemed important. Hence I was less interested in the available screening and testing tools and more interested in individual accounts which brought the in-person and contextual factors together.

I still felt some uncertainties about the relationship between dyslexia and other literacy problems. Was dyslexia really just a case in point? We knew about the opprobrium, stigma and negative stereotyping with which adults with literacy difficulties had to contend. Dyslexic students often seemed to be echoing these and hence were in no way distinctive to many of us. On the other hand, surprising new insights had emerged from their descriptions which encouraged me to investigate further with all my students. This case in point appeared to have something to say to them and to me. In terms of the underlying questions, I felt at this point that the term dyslexia could sensibly be used for any student who experienced severe and/or intractable difficulties within this range of skills and practices.

Other uncertainties stemmed from the research literature. Miles felt that there was more agreement among researchers than I had gathered from my reading at this point (source: a personal interview during the project), but I found it quite difficult to piece together the different kinds of discourse and evidence into anything like a coherent whole. It seemed straightforward to those who operated within their particular discipline framework, asking defined questions. For anyone trying to piece it all together, the extent of agreement was far from clear.

Notwithstanding this, we had started out wanting to know if basic skills methods needed any adaptation. The conclusion was that good basic skills methods needed a re-articulation to include a sharper investigative focus and the key findings from dyslexic learners. I attempted to do this within a new ABE teacher training course at Leicester University (1988–1990), and in the process started to try to locate dyslexia within a 'critical literacy' context. This new development would prove even more necessary when I moved into higher education.

3. HEFCE project on supporting students with special needs at the University of Leicester, 1992–1994

When I became involved in this project, I had already started to support dyslexic higher education students both in the community and during a part-time secondment to Leicester University. The project was designed to establish a centre for students with special needs, including dyslexic students. With two colleagues I established a system of support which involved pre-entry advice, screening and formal testing, individual tutoring, assisting with access to Disabled Students' Allowances and making recommendations for examinations. This 'managerialist' response was an attempt to produce a coherent response within a complex higher education institution, and once again the new context offered me some important new insights about dyslexia.

Support of students in higher education required a deconstruction of academic literacy practices and conventions, and it was very clear that the volume, speed and nature of the information processing required could overload dyslexic students. Most of the work involved helping them to fit in so that they could get their degrees, though I always tried to show that the practices were social conventions designed by non-dyslexic people in positions of academic and institutional power (a mix of academic socialisation and a critical literacies approach). I always encouraged the students to focus on the quality of their thinking.

The HE students began to describe their different ways of thinking especially when they were having extreme difficulties with academic writing. We offered many strategies, and students worked with these: analysing the nature of argument and its various versions; creating visual and kinaesthetic structures and scaffolds for 'blanks' in written structures; setting cue questions; and modelling writing. For most students these methods worked reasonably well, involving both analysis and creativity.

For one student nothing worked, and he appeared to be so disorganised that he would never produce written work in any clear order. In desperation, I asked him to set aside all his notes and tell me what he was thinking about the question; and what emerged was a fast-moving visual matrix of ideas and connections. This was creative and exciting and difficult to capture. I was struck by the speed of the connections and the fact that there was nothing slow, laboured or blocked about this (Herrington, 2001).

However, it did not lend itself to an easy transformation into a linear piece of writing; it fitted more easily with a visually composite approach. I knew that this kind of thinking was not particularly valued, because academic values focus on analysing the nature of connections and the nature of relationships between them. This student's final year dissertation resembled a living jigsaw rather than an academic treatise.

It was clear to me that the ideas of what constitutes academic quality were penalising this kind of thinking. His quality as an archaeologist who was able to draw accurately on site, and whose curiosity and passion led him to produce an interesting investigation, was not really acknowledged. I felt like an academic gatekeeper, trying to explain how it all worked and yet seeing that some kinds of academic practice were alien to him (note also the comment subsequently made by a dyslexic academic, Dr Stella ni Ghallchoir Cottrell, at the Dyslexia and Effective Learning Conference at Leicester University in 1996, that it was like being asked to deconstruct the Mona Lisa in a linear form). This was a more heightened version of the role of adult literacy tutors in relation to standard English.

During this period Ellen Morgan, a colleague at the University of North London, presented a conference paper which contrasted the kind of thinking and writing commonly valued in higher education with that preferred by dyslexic learners: linear versus holistic/global processing, for example. This encouraged me to think about cognitive difference rather than having a sole emphasis on difficulties and problems. I began to really understand how the higher education context could problematise dyslexic ways of thinking.

A further example demonstrates how devastating this could be.

A had an acute memory problem and this was interpreted as reflecting a poor intellect. He was studying maths and was having great difficulty with the work and, as a result, with members of staff. He was dyslexic but also had some additional organisational and communication difficulties. These made it more difficult to unravel the main root of his difficulty with maths. However, our conversations revealed that he always had to start at the beginning with any calculation or mathematical operation. He could never start halfway through on the basis of certain assumptions. Hence he filled many notebooks every time he did a problem – painstakingly building all the steps. To academic staff his methods looked very strange and time-consuming (indeed, time-wasting).

I decided to employ a postgraduate maths student to try and assess the extent to which this was a memory issue or an understanding issue. His judgement at the end of a term was that the more difficult the maths became conceptually, the better he was at it. He was very poor at remembering to do the most mundane task and had to always start at the beginning from first

> principles. It was not a problem of understanding at all. My understanding is that he eventually obtained his degree by working with the textbooks at home.

The students who struggled and battled most forced me to think hardest about the limiting power of some academic practices.

The clusters of dyslexic students in particular disciplines made me start to analyse academic practices in those disciplines from a dyslexic perspective. For example, with regard to archaeology, I started to ask questions about archaeological imagination, the ability to read landscapes, and the importance of feel and touch when dealing with artefacts, and I realised that some disciplines were more attractive than others to dyslexic learners.

Additional layers of complexity came with international students: both contextual and in-person. Though most learned to handle the UK higher education practices sufficiently well to get through their courses, some individuals presented with more acute difficulties. I recognised that there were new questions here. Could you be dyslexic in other languages? How could I separate out the effects of second language learning from dyslexic difficulties? What assumptions was I making about mother tongue literacy? Why did I assume for some time that the first dyslexic Saudi student I worked with was using English as his second language when in fact it was his first and he had been taught in English throughout his schooling?

The pressure to identify dyslexic students mounted during this period as the Disabled Students' Allowances became more widely available. In adult basic education there had been no additional resources for dyslexic learners in general and no undue pressure to identify. In higher education I had to arrange for students to have formal tests so that they could access their DSAs and obtain appropriate technology. The only way I could reconcile my own view about the importance of learner descriptions was to create a screening device that would allow for this but that would predict a positive outcome on the test. This would limit any unnecessary testing and yet give the students the right to describe their literacy in their own terms. We created an interview structure that allowed a past–present–future discussion time frame as well as some task completion. This worked well in that it not only allowed the dyslexic issues to emerge but also acted as a starting point into a learning support curriculum for which the students had been encouraged to identify their own priorities.

What did I make of this?

I was now seeing even more heterogeneity among dyslexic learners, although severity, intractability and unpredictability did act as guides about the highest priority students . The unexpected differentials were very, very clear for the most highly able students and I could see how the idea of discrepancy could be attractive for those students and their families. The idea of very rapid and creative thinking skills, over and above general adult competence and alongside literacy difficulties, became very real for the first time. But what did this mean? Was this something quite separate from dyslexia? Was it coexisting? Was it an intrinsic part of a dyslexic sub-type? Or was it part and parcel of dyslexia in general and we had just never explored it?

The context was also more complex than in ABE. I started to realise that not only did academic practices require some unpicking for accessibility but they actually represented dominant ideas about quality in thinking and about relationships with forms of literacy. These could exclude dyslexic students from high achievement. There were also coexisting opportunities to focus on particular disciplines that offered challenge and opportunity to dyslexic learners.

At the same time as delving deeper in this way, I was conscious that national policy in relation to dyslexic students was based on a fairly simple set of ideas about dyslexia. The debates, complexities and uncertainties about dyslexia in other educational sectors seemed to have been set on one side. I did not object too much because the new policies ensured that individual help was allocated to these students. However, I was uneasy about the personal 'gap' I experienced between my own developing sense of dyslexia and the emerging policy discourse that appeared to be the necessary driver for institutional change.

It was clear to me that, in terms of the general debates about dyslexia, HE students provided a strong indication that poor literacy skills could not in themselves be seen as a sign of low intelligence or low ability. Yet I was conscious that there was no easily available mechanism that allowed feedback of this extending vision of dyslexia to the basic skills sector.

4. The University of Nottingham Learning Support Unit/Study Support Centre, 1995–2002

In 1994, I was invited to help establish a service for dyslexia learners at the University of Nottingham. With colleagues,[1] I ensured that key elements of support were in place and we were soon attracting unexpectedly large numbers. This was a period in which the student descriptions of dyslexia became even more illuminating and in which I tried to integrate my inner vision of dyslexia with my day-to-day practice. I also made a leap forward in integrating this work within broader social and cultural frameworks about literacies (Barton, 1994; Barton and Hamilton, 1998; Ivanič and Clark, 1997; Kress, 2000), within social models of disability (Barnes and Mercer, 1996), and with my readings about dyslexia (Nicholson and Fawcett, 1990; Steffert 1996, 1999; Stein, 2001; West, 1991). Attending the two ground-breaking conferences at the University of Plymouth represented an important step forward in these respects.

For this narrative, I have selected three particular observations. First, I felt I was investigating the academic practices and literacies context more deeply and discovering that these actually helped to construct disability. I found myself regularly explaining to dyslexic students not only the implicit requirements of the HE courses (the hidden curricula) but also the epistemological context in which they were being asked to operate. Students regularly struggled with where to place themselves in relation to any piece of work, and this was only resolved by my explanations about knowledge creation in general and the contesting theories about this. They could then locate their own discipline practices accordingly. It was the ultimate example of dyslexic students requiring the bigger picture in order to operate more effectively, and yet students reported that their course tutors rarely explained or discussed these broader intellectual themes.

The cultural relativity of academic practices became ever clearer in this respect. International postgraduate law students revealed the issues involved in coming to terms with different legal systems, different methods of teaching and different assessment practices (for example, the nature and wording of questions). Dyslexic students from overseas were doubly disadvantaged whenever the academic literacy practices and their rationale were implicit rather than explicit.

The extension of academic practices into vocational contexts emerged very quickly. I rapidly had to come to terms with the ways in which particular professions and disciplines, such as pharmacy, medicine, teaching, and physiotherapy among others, were viewing dyslexia. They were locked into a deficit model of dyslexia and an autonomous model of academic literacy. The starkest example was in relation to nursing training. Although academic staff were alert to equal opportunities issues, they were also obliged to guarantee that students were 'safe to practice'. Fears of what dyslexia could mean in terms of misspelt drug names, observations and instructions dominated the discourse, and staff consistently expressed concerns about allowing dyslexic nurses to practise. These ideas were so deep-rooted that even when dyslexic students demonstrated considerable literacy ability in their assignments there was always a doubt for some staff. If students had had support with assignments, comments would be made about not being able to have such support on the wards. I saw no attempts to analyse the literacy skills actually required on the wards and to teach and assess on that basis (using a New Literacies approach). I heard about many comments made during placements and during training that suggested a deficit and pejorative view of dyslexia.

Some reasons for this were clear enough: the academicisation of nursing courses, and the particularly important place of written records in medical settings. But very little attention was paid to how dyslexic nurses were managing to handle the demands on the wards and to negotiate those situated literacies. When I interviewed three qualified dyslexic nurses about their strategies they all revealed a high degree of vigilance and far more checking of everything than non-dyslexic nurses. This kind of investigative work would have been far more productive in revealing the ability of dyslexic students to practise safely. Instead, despite the sheer numbers of dyslexic nurse trainees and the best efforts of the more thoughtful staff, the basic deficit attitude to dyslexia still lingered among many.

The presence of dyslexic staff and postgraduates had very good effects in some departments: providing role models, aiding the 'normalisation' of learning differences, and challenging the connection between literacy difficulties and lack of intelligence. However, this was piecemeal and in general there was little to be gained by academics in acknowledging their dyslexia in the academic context.

In the face of this I focused on getting more evidence about the experience directly from students. It seemed to be the only way of challenging the deficit narrative from the inside. They had after all made it to university and were coping reasonably well. It seemed ridiculous for deficit narratives to hold sway in the presence of such students. I was not in a position to mount a separate research project and so focused on research in practice within a student discussion/ action group.

Student/tutor discussion/action group

This was established in addition to the individual investigative work and the group provision in the Study Support Centre. Over four years the group varied greatly in membership and range of topics covered. Members were invited to identify their own priorities for discussion/ action and their ideas for change at Nottingham. They were offered a supportive setting in which to explore dyslexia. From my own point of view it was a research in practice exercise on two levels: exploring the possibilities of this kind of group in a higher education setting, and also recording the 'knowledge' students had about dyslexia, which was so vital for how dyslexia was viewed within this institution and within the sector as a whole. Notes were made at each meeting and circulated to members ahead of subsequent meetings.

The results were important. Students devised pointers for dyslexia-friendly text, discovered the range of dyslexic characteristics from each other (and that some individual clusters were more problematic than others in the HE context), found their assumptions challenged (some science student members did not realise that you could study English literature and be dyslexic) and discovered the variability in academic practice between departments (some were already highly sensitive to dyslexic issues). They learned about handling dyslexia in relation to certain course tasks and communicating with non-dyslexic academics about dyslexia, and worked with library staff to produce real changes in the library service. They focused on experimental work with employers – inviting representatives from major employers into the university to discuss issues of disclosure and recruitment, and undertaking awareness-raising with them rather than being passive recipients of graduate recruitment presentations.

Above all, their discussions about how they experienced text were ground-breaking in revealing important visual–spatial descriptions (Herrington, 2001). This confirmed my earlier finding about the role of learners in my own training: I was forced to consider how fast-moving, multi-dimensional visual–spatial thinking actually impinged on reading processes as well as the writing processes noted above . . .

Staff development

Alongside this work, staff development activity began to take an increasing amount of time. I attempted to introduce dyslexia to academics as an area of academic interest like any other, rather than just as awareness-raising or as a set of instructions about how they were to teach some generalised dyslexic person. I developed an organic, staff-centred model of staff development that invited staff to identify their own priorities for action in relation to disability in general and to dyslexia in particular. This 'colleagual' approach (Boud and McDonald, 1981) led to discussions with academic staff in different disciplines about the nature of academic literacies and the learning outcomes they sought, and I felt that this was a valuable way of engaging the staff intellectually in creating the changes that they would 'own'. These were small beginnings, however, and though I encountered exceptional academics who devised excellent case studies showing how the curriculum could be reshaped whilst maintaining 'standards' (Herrington and Haines, 2001; Herrington with Simpson, 2002), deficit models of dyslexia and the desire for technicist solutions often dominated. I was aware of the inherent challenge of my role: acting as a change agent and yet believing strongly in staff and students governing their own change.

What did I make of this?

By 2002 I had what appeared to me to be an integrated, social interactive view of dyslexia: one which explored in-person differences within an overall critical communication framework. My approach to students was explicitly one of creating a sense of mutuality in communication and exploration. This involved encouraging student descriptions, supplementing core methods with new insights about how literacy learning could best occur (for example, physical movement in relation to cognitive activity), deconstructing academic literacy and learning practices (Jones, Turner and Street, 1999; Costello and Mitchell, 1995) with the students to reveal the bigger epistemological conventions and their ontological underpinnings, and actively trying to connect dyslexic thinking and learning with larger cultural developments. I had to problematise myself in this. I did not have the advantage of being dyslexic myself, and yet I was having to represent to other non-dyslexic people how dyslexia was experienced (see Dale and Taylor, 2001). Hence, even by this point I did not always know how best to create a way forward with every dyslexic student.

However, I continued to pursue the questions I had noted while at Leicester. What were all the broader parameters of this syndrome? What exactly were the enhanced visual–spatial skills being claimed? If we could produce evidence about these (and I was increasingly seeing them in higher education), were they just advanced skills which coincidentally accompanied dyslexia rather than skills that were intrinsic to it? Did this explain why some students did not appear to possess enhanced skills? Or should we be thinking about sub-types of dyslexia which involved these mixes of skills? I could see why some researchers wanted to focus on minimally defining characteristics, but there seemed an enormous gap between the idea that everyone had a defining difficulty in managing sounds in words and the range of issues and strengths which dyslexic students presented.

A second related question was about boundaries. The higher education context required formal identification and a push to separate the dyslexic from the non-dyslexic. The reality was a more mixed situation in which students revealed some or all of the classic dyslexic characteristics and to varying degrees. I therefore found it much more helpful to see dyslexic characteristics as at one end of a set of continua (Herrington and Hunter-Carsch, 2001). All students could be placed on these, and the different types of profiles could point the way forward to developing effective methods.

Further, I was beginning to draw conclusions about the practice of the discrepancy principle. I could see that this was a possible way of identifying dyslexia; the unusual/unexpected mix of characteristics, at whatever level, was often a trigger for further exploration. However, it was clear that discrepancy in relation to IQ was not acceptable. I had encountered dyslexic characteristics (in particular literacy difficulties) across the ability range, and to draw a line at average IQ and to label everyone above the line as dyslexic and those below as 'other' (slow learners in general) seemed bizarre.

However, if dyslexia was present across the ability range, I felt some concern about a possible logical conclusion that those who failed to make it to higher education simply did not have the ability and so nothing more was required. Yet I felt instinctively that we were seeing the very, very able dyslexic students in these two traditional higher education institutions, and that many of those who would have been on a par with the average non-dyslexic entrant often simply did not make it.

The logical conclusion of my work in higher education was that if literacy difficulties by themselves did not signify poor ability, why did so many basic skills policy-makers continue to think that they did, and why were so many academic practices in higher education still based on the idea of measuring understanding solely via particular literacy practices, when the able people with literacy difficulties would clearly struggle to express their ability in that way? Making 'reasonable adjustments' could only make inroads into this if fairer ways of assessing were developed, and I wanted to know if the more creative academics would lead the way in devising dyslexia-friendly curricula and assessment or whether the pressure of numbers in most institutions would lead to a minimum response approach. In 2002 I felt hopeful that the general mechanisms for developing teaching quality in higher education (Learning Teaching Subject Networks, for example) might encourage staff to choose the former.

Questioning the story . . . revisiting the narrative

This has been a story of how I tried to make sense of dyslexia in relation to literacy and learning through research in practice activity. Here I shall revisit it, with three particular questions.

Am I aware of leaving out any important observations?

After reading the story, I was immediately aware of what I had not mentioned – and not only relatively unimportant matters. For example, my exploration, limited as it still is, into the enhanced skills associated with dyslexia owes much to the work of Susan Parkinson of the Arts Dyslexia Trust. The issue of designing adequate research tools to investigate whether or not visual–spatial skills were always present was cast into a completely different light by her observation that if tests of visual–spatial skills were designed by non-dyslexic people they might not necessarily be able to capture the skills involved. This reinforced my view that dyslexic people must not only be engaged in determining the research agenda but must also contribute in terms of methods.

What kind of professional development narrative is evident in this?

The narrative so far shows a practice-centred, non-linear, multi-layered research and practice dynamic. Successive phases of development projects, research in practice and investigative practice are clear, as is the coexistence and overlapping of the three. Here an investigative practitioner is seen as drawing on the research of others, building research with students into everyday practice, setting up new research and development projects and continuing to explore research questions about dyslexia throughout. It offers a model for building research into the infrastructure of practice in order to pursue long-term research questions. It can also be seen as incorporating a range of theoretical insights about learning, research and change.

Why did I work in this way? Essentially my priority was one of making research work for the development of ideas, concepts and methods to enrich practice. My primary motivation has always been to find ways of understanding what the professional challenge of this field actually is in relation to adult literacy and numeracy students, and for me the overarching idea that made most sense was that of investigative practice: tutors and students asking critical questions about literacy, numeracy and ESOL, and drawing in many different kinds of research activity (for some this would amount to a version of critical pedagogy) in order to drive progress. My view was that though dyslexia can seem just like a case in point in relation to literacy difficulties, we simply did not know enough about the case, nor about the light it could shed on all our practice, and there was an additional set of narratives that learners had a right to question from a position of knowledge. I felt it important to find some way of connecting these narratives with the broader situated literacies work and started to think about dyslexia as involving distinctive literacies.

Do the outcomes support the initial rationale for constructing the narrative?

The pattern of professional development was indeed 'organic' in the sense of a mix of 'constitutional, in the structure of something, fundamental, integral; the coordination of integral parts' (Collins Concise Dictionary) and ongoing organismal development. It has involved the dynamic evaluation of fundamental ideas, the integral, persistent, questioning underlying all processes; the processing and integration of emotions; and specific connections between the episodes. For example, the research work with ABE students led developmentally (not just cumulatively) to the research in practice group at Nottingham and to further transformation. However, the term 'organic' here also includes the idea of unevenness and uncertainty in relation to the various academic discourses about dyslexia. I can now see how my uncertainties were useful in pushing me on in the exploration.

But what about new knowledge from the dyslexic learners? I was frequently aware that what I was hearing, especially about how literacy, memory and time were being experienced, was not available in the published literature. This was both exciting and worrying. Why were these kinds of descriptions not being gathered systematically by researchers when they were so illuminating about learning preferences? Why were the implications of these descriptions for assessment and teaching processes not always considered? How, for example, could one possibly interpret a student's reading performance if one knew nothing about how the student reacted to text visually (either experiencing visual disturbance or switching modally between visual and textual processing)? The research in practice only allowed me to capture a certain amount given the speed at which the descriptions were emerging, and so these remained context bound for some time. Some of the descriptions have now been published and have reached a wider audience, but I am left with a niggling discomfort that much more sustained research on dyslexic descriptions of literacy and learning is needed if we are to really address the still live questions about parameters, boundaries and dyslexic heterogeneity. I was aware of being at the contested positivist/constructivist interface in terms of developing knowledge about dyslexia, and over time became more confident about the constructivist position.

Finally, the narrative shows me edging towards seeing dyslexia in a multi-dimensional way, that is, wanting to use the very thinking style that students had described to me. I had started to ask myself if it could be better understood by visualising a multi-dimensional moving web, with the individual at the centre and in which certain types of literacy technology could slow down the speed of communication between the dimensions. The fast-moving connections with questions and routes to answers appeared to exist independently of literacy. Could the problem be that non-visual thinkers were less able to think in this way and so could not really see dyslexia as a whole? It seemed increasingly clear to me that, if different ways of seeing dyslexia were not understood and articulated in professional and employment contexts, deficit models of dyslexia would continue to dominate.

Note

1. Key colleagues at Nottingham University were Linda Basey, Jayne Biernat, Christine Carter, Mark Dale, Carole East, Ann Hurford, Lesley Morrice and Barbara Taylor.

References

Abbs, P. (1976) *Root and Blossom: Essays on the Philosophy, Practice and Politics of English Teaching*. Heinemann, p.148; quoted in Graham, R. (1991).

Barnes, C. and Mercer, G. (1996) *Exploring the Divide: Illness and Disability*. The Disability Press.

Barton, D. (1994) *Literacy: An Introduction to the Ecology of Written Language*. Blackwell.

Barton, D. and Hamilton, M. (1998) *Local Literacies*. Routledge.

Brain, S. (1988) 'Autobiographical writing: Understated or overplayed', in B. Street and J. McCaffery (eds) *Literacy Research in the UK: Adult and School Perspectives*. RaPAL.

British Psychological Society (1988) Division of Educational and Child Psychology Occasional Papers, Vol. 7 No. 3.

Clandinin, D. J. and Connelly, F. M. (2000) *Narrative Enquiry: Experience and Story in Qualitative Research*. Jossey-Bass.

Clarke, J. (1996) *Making Sense of Experience: The Role of Narrative in Research and Practice. Lifelong Literacies*, Papers from the RaPAL Conference, Gatehouse.

Costello, P. J. M. and Mitchell, S. (1995) *Competing and Consensual Voices: The Theory and Practice of Argument*. Multilingual Matters Ltd.

Dale, M. and Taylor, B. (2001) 'How adult learners make sense of their dyslexia', *Disability & Society*, Vol. 16 No. 7, pp. 997–1008.

Erben, M. (ed.) (1998) *Biography and Education: A Reader*. Falmer Press.

Freire, P. (1979) *Pedagogy of the Oppressed*. Penguin.

Graham, R. (1991) *Reading and Writing the Self: Autobiography in Education and the Curriculum*. Teachers College Press.

Herrington, M. (1985) Report for ALBSU on the Distance Learning Special Development Project.

Herrington, M. (1993) 'Learning at home', in M. Hamilton, D. Barton and R. Ivanič (eds), *Worlds of Literacy*. Multilingual Matters, pp. 182–7.

Herrington, M. (1995) Dyslexia: Old dilemmas and new policies. *RaPAL Bulletin* No. 27, Summer.

Herrington, M. (2001) 'Adult dyslexia: Partners in learning', in M. Hunter-Carsch and M. Herrington (eds), *Dyslexia and Effective Learning in Secondary and Tertiary Education*. Whurr.

Herrington, M. (2001) 'Dyslexia: The continuing exploration. Insights for literacy educators', *RaPAL Bulletin*, Vol. 46.

Herrington, M. and Haines, C. (2001) 'Exploring academic staff interest in disability matters via an e-mail discussion group', *Skill Journal*, Spring, pp. 11–16.

Herrington, M. and Hunter-Carsch, M. (2001) 'A social interactive model of specific learning difficulties', in M. Hunter-Carsch (ed.) *Dyslexia: A Psycho-Social Perspective*. London: Whurr, pp. 107–33.

Herrington, M. with Simpson, D. (2002) 'Making reasonable adjustments with disabled students in HE'. Online at: www.nottingham.ac.uk/ssc/staff/research

Huberman, M. (1995) 'Professional careers and professional development: Some intersections', in T. R. Guskey and M. Huberman (eds), *Professional Development in Education, New Paradigms and Practices*. Teachers College Press.

Ivanič, R. (1998) *Writing and Identity. The Discoursal Construction of Identity in Academic Writing*. John Benjamins.

Ivanič, R. and Clark, R. (1997) *The Politics of Writing*. Routledge.

Ivanič, R., Aitchison, M. and Weldon, S. (1999) 'Bringing ourselves into our writing', *RaPAL Bulletin* No. 28/29, Autumn 1995/Spring 1996.

Jones, C., Turner, J. and Street , B.V. (eds) (1999) *Students Writing in the University. Cultural and Epistemological Issues*. John Benjamins.

Klein, C. (1986) ALBSU special development project report. ALBSU.

Kress, G. (2000), 'The futures of literacy', *RaPAL Journal* No. 42.

Mace, J. (1992) *Talking about Literacy: Principles and Practice of Adult Literacy Education*. Routledge.

Miles, T. R. (1983) *Dyslexia: The Pattern of Difficulties*. Collins Educational.

Nicholson, R. I. and Fawcett, A. J. (1990) 'Automaticity: A new framework for dyslexia research?', *Cognition* Vol. 33, pp. 159–82.

RaPAL Doing Research (1989) *RaPAL Bulletin* No. 9, Summer.

Schmidt, P. (1997) *Beginning in Retrospect: Writing and Reading a Teacher's Life*. Teachers College Press.

Snowling, M. (1987) *Dyslexia: A Cognitive Developmental Perspective*. Blackwell.

Steffert, B. (1996) 'Sign minds and design minds: The trade-off between visual spatial skills and linguistic skills', in *Dyslexia in Higher Education, Learning Along the Continuum*. 2nd International Conference Proceedings, University of Plymouth, pp. 53–69.

Steffert, B. (1999) 'Visual–spatial ability and dyslexia, Part 1 pp. 8–49 and Part 2 pp. 127–167', in I. Padgett (ed.), *Visual–Spatial Ability and Dyslexia*. Central St. Martins College of Art and Design, The London Institute.

Stein, J. (2001) 'The magnocellular theory of developmental dyslexia', *Dyslexia*, Vol. 7 No. 1, January–March, pp. 12–36.

Strawson, G. (2004) *The Guardian*, 10 January 2004.

Street, B.V. (1988) 'Comparative perspectives on literacy research', in J. McCaffery and B.V. Street, *Literacy Research in the UK: Adult and School Perspectives*. RaPAL.

Usher, R. (1998) 'The story of the self: Education, experience and autobiography', in M. Erben (ed.), *Biography and Education: A Reader*, SRES Series 19. Falmer Press, pp. 18–31.

West, L. (2003) 'Learning to be a doctor: An auto/biographical study', *Dolnoslaska Szkola Wyzsza Edukacji*, pp. 141–56.

West, T. (1991) *In the Mind's Eye: Visual Thinkers, Gifted People with Learning Difficulties, Computer Images and the Ironies of Creativity*. Prometheus.

Index

Page numbers in **bold** denote text written by the author named.

N

O

P

T

U

V

W